Some Integrals

$$\int_0^\infty x^{n-1}e^{-x}\,dx = \Gamma(n) \qquad \Gamma(n) = (n-1)\Gamma(n-1) = (n-1)!$$

$$(2n)!! = 2^n \cdot n! \qquad (2n+1)!! = \frac{(2n+1)!}{2^n \cdot n!}$$

$$\int_0^\infty e^{-ax^2}\,dx = \frac{1}{2}\left(\frac{\pi}{a}\right)^{1/2}$$

$$\int_0^\infty x^{2n}e^{-ax^2}\,dx = \frac{1\cdot 3\cdot 5\cdots(2n-1)}{2^{n+1}a^n}\left(\frac{\pi}{a}\right)^{1/2} \qquad (n \text{ positive integer})$$

$$\int_0^\infty x^{2n+1}e^{-ax^2}\,dx = \frac{n!}{2a^{n+1}} \qquad (n \text{ positive integer})$$

$$\int_0^\infty e^{-ax}\cos bx\,dx = \frac{a}{a^2+b^2}$$

$$\int_0^\infty e^{-ax}\sin bx\,dx = \frac{b}{a^2+b^2}$$

$$\int_0^\infty e^{-a^2x^2}\cos bx\,dx = \left(\frac{\pi}{4a^2}\right)^{1/2}e^{-b^2/4a^2}$$

$$\int_0^1 x^{n-1}(1-x)^{m-1}dx = 2\int_0^\pi \sin^{2n-1}\theta\,\cos^{2m-1}\theta\,d\theta = B(n,m) = \frac{\Gamma(n)\Gamma(m)}{\Gamma(n+m)}$$

$$\frac{2}{\pi^{1/2}}\int_0^x e^{-u^2}du = \mathrm{erf}\,(x) = 1 - \mathrm{erfc}\,(x)$$

$$\int_0^a \sin\frac{n\pi x}{a}\sin\frac{m\pi x}{a}\,dx = \int_0^a \cos\frac{n\pi x}{a}\cos\frac{m\pi x}{a}\,dx = \frac{a}{2}\delta_{nm}$$

$$\int_0^a \cos\frac{n\pi x}{a}\sin\frac{m\pi x}{a}\,dx = 0 \quad (m \text{ and } n \text{ integers})$$

$$\int_0^\pi \frac{dx}{a+b\cos x} = \frac{\pi}{(a^2-b^2)^{1/2}} \qquad a > b \geq 0$$

$$\int_0^{\pi/2}\frac{dx}{a^2\sin^2 x + b^2\cos^2 x} = \frac{\pi}{2ab}$$

$$J_n(x) = \frac{1}{i^n\pi}\int_0^\pi e^{ix\cos\theta}\cos n\theta\,d\theta$$

$$K(k) = \int_0^{\pi/2}\frac{d\theta}{(1-k^2\sin^2\theta)^{1/2}} = \frac{\pi}{2}\left[1+\left(\frac{1}{2}\right)^2 k^2 + \left(\frac{1\cdot 3}{2\cdot 4}\right)^2 k^4 + \cdots\right]$$

$$E(k) = \int_0^{\pi/2}(1-k^2\sin^2\theta)^{1/2}d\theta = \frac{\pi}{2}\left[1-\left(\frac{1}{2}\right)^2 k^2 - \left(\frac{1\cdot 3}{2\cdot 4}\right)^2 k^4 + \cdots\right]$$

SOLUTIONS

to accompany
McQuarrie's

MATHEMATICAL METHODS
For Scientists and Engineers

Carole H. McQuarrie

UNIVERSITY SCIENCE BOOKS
Sausalito, California

University Science Books
www.uscibooks.com

Compositor: Carole McQuarrie
Production Manager & Design: Side by Side Studios
Printer & Binder: Edwards Brothers

This book is printed on acid-free paper.

ISBN 1-891389-37-8

Printed in the United States of America
10 9 8 7 6 5 4 3 2

Contents

iv   Contents

CHAPTER 1

Functions of a Single Variable

1.1 Functions

1. (a) Since y must be real, the domain of x is $x^2 \leq 16$, or $-4 \leq x \leq 4$.

(b) In this case, x may take any value, so the domain of x is $-\infty < x < \infty$.

(c) The argument of the logarithm must be positive, so the domain of x is $x > 0$.

(d) The function is undefined at $x = 1$, but y is defined at all other values of x. Thus, the domain of x consists of all values except $x = 1$.

3.

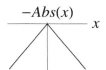
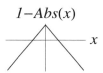

6.

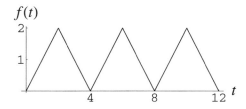

9. The condition of the problem reads

$$[(c+x)^2 + y^2]^{1/2} + [(c-x)^2 + y^2]^{1/2} = 2a$$

Squaring both sides once gives

$$2(c^2 + x^2 + y^2) + 2[(c^2 - x^2)^2 + 2y^2(c^2 + x^2) + y^4]^{1/2} = 4a^2$$

Bringing $2(c^2 + x^2 + y^2)$ onto the right side and squaring again gives

$$(a^2 - c^2)x^2 + a^2y^2 = a^2(a^2 - c^2)$$

Let $b^2 = a^2 - c^2$ and divide through by a^2b^2 to obtain

$$\frac{x^2}{a^2} + \frac{y^2}{b^2} = 1$$

which is the equation of an ellipse. Note that $x = \pm a$ when $y = 0$ and $y = \pm b$ when $x = 0$. The points $(\pm c, 0)$ are called the *foci* of the ellipse and a and b are the *semiaxes*, a being called the *semimajor* axis and b the *semiminor* axis if $a > b$. The *eccentricity* of the ellipse is $e = c/a$. Note that the ellipse reduces to a circle if $e = 0$.

12. Set up a right triangle. Then $\sin \alpha = x$ and $\cos \beta = x$, and since $\alpha + \beta = \dfrac{\pi}{2}$, we have

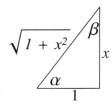

$$\alpha + \beta = \sin^{-1} x + \cos^{-1} x = \frac{\pi}{2}.$$

Similarly, using

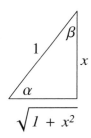

we have $\tan \alpha = x$ and $\cot \beta = x$, so we have $\alpha + \beta = \dfrac{\pi}{2} = \tan^{-1} x + \cot^{-1} x$.

15. (a) We use the addition formula for $\tan z$:

$$\tan[2(x + \tau)] = \frac{\tan 2x + \tan 2\tau}{1 - \tan 2x \tan 2\tau}$$

and the right side is equal to $\tan 2x$ if $\tan 2\tau = 0$, or when $\tau = \pi/2$.
(b) We use the addition formula for $\cos z$:

$$\cos(x + \tau) = \cos x \cos \tau - \sin x \sin \tau \qquad \text{or} \qquad |\cos(x + \tau)| = |\cos x \cos \tau - \sin x \sin \tau|$$

If $\tau = \pi$, then $|\cos(x + \tau)| = |\cos x|$.
(c) $f(x) = \sin x / x$ is not periodic.

18. Add $-|x| \le x \le |x|$ and $-|y| \le y \le |y|$ to obtain $-(|x| + |y|) \le x + y \le |x| + |y|$, and then take the absolute magnitude of both sides.

21. Solve $y = \dfrac{x + 1}{x - 1}$ for x to obtain $x = \dfrac{1 + y}{y - 1}$.

24.

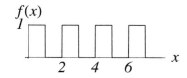

1.2   Limits

1. Consider $|x - a| < \delta$. If $x > a$, then we have $x - a < \delta$, or $x < a + \delta$. If $x < a$, then we have $a - x < \delta$, or $a - \delta < x$. Thus we have $a - \delta < x < a + \delta$.

3. We have that

$$
\begin{aligned}
\lim_{\theta \to 0} \frac{1 - \cos\theta}{\theta} &= \lim_{\theta \to 0} \frac{1 - \cos\theta}{\theta} \cdot \left(\frac{1 + \cos\theta}{1 + \cos\theta}\right) \\
&= \lim_{\theta \to 0} \frac{1 - \cos^2\theta}{\theta(1 + \cos\theta)} = \lim_{\theta \to 0} \frac{\sin^2\theta}{\theta(1 + \cos\theta)} \\
&= \lim_{\theta \to 0} \frac{\sin\theta}{\theta} \lim_{\theta \to 0} \frac{\sin\theta}{1 + \cos\theta} = (1)(0) = 0
\end{aligned}
$$

6. (a) Using $f(x) = x^2$, we have

$$
\lim_{h \to 0} \frac{(x + h)^2 - x^2}{h} = \lim_{h \to 0} \frac{2xh + h^2}{h} = \lim_{h \to 0} (2x + h) = 2x
$$

(b) Using $f(x) = 1/x$, we have

$$
\lim_{h \to 0} \frac{\dfrac{1}{x + h} - \dfrac{1}{x}}{h} = -\lim_{h \to 0} \frac{h}{h(x + h)x} = -\lim_{h \to 0} \frac{1}{x(x + h)} = -\frac{1}{x^2}
$$

(c) Using $f(x) = \sin x$, we have

$$
\begin{aligned}
\lim_{h \to 0} \frac{\sin(x + h) - \sin x}{h} &= \lim_{h \to 0} \frac{\sin x \cos h + \cos x \sin h - \sin x}{h} \\
&= \lim_{h \to 0} \frac{\cos x \sin h}{h} = \cos x \lim_{h \to 0} \frac{\sin h}{h} = \cos x
\end{aligned}
$$

(d) Using $f(x) = \cos x$, we have

$$
\begin{aligned}
\lim_{h \to 0} \frac{\cos(x + h) - \cos x}{h} &= \lim_{h \to 0} \frac{\cos x \cos h - \sin x \sin h - \cos x}{h} \\
&= \lim_{h \to 0} \frac{-\sin x \sin h}{h} = -\sin x
\end{aligned}
$$

9. Let $x < 0$, then

$$
\lim_{x \to 0-} \frac{x}{|x|} = \lim_{\epsilon \to 0} \frac{-\epsilon}{|\epsilon|} = -1
$$

For $x > 0$,

$$\lim_{x \to 0+} \frac{x}{|x|} = \lim_{\epsilon \to 0} \frac{\epsilon}{|\epsilon|} = 1$$

12. Starting with $f(x) \le g(x) \le h(x)$, we have

$$0 \le g(x) - f(x) \le h(x) - f(x)$$

and

$$\lim_{x \to a} [\, h(x) - f(x)\,] = L - L = 0$$

Because $g(x) - f(x) \le h(x) - f(x)$ and $\lim[\, h(x) - f(x)\,] = 0$, we also have

$$\lim_{x \to a} [\, g(x) - f(x)\,] = 0$$

(To prove this statement, let $\epsilon > 0$ with $|\, h(x) - f(x)\,| < \epsilon$ whenever $0 < |\, x - a\,| < \delta$. Therefore, $|\, g(x) - f(x)\,| \le |\, h(x) - f(x)\,| < \epsilon$ whenever $0 < |\, x - a\,| < \delta$, and so $\lim_{x \to a} [\, g(x) - f(x)\,] = 0$.)

To continue the proof, let $g(x) = [\, g(x) - f(x)\,] + f(x)$. Now

$$\lim_{x \to a} g(x) = \lim_{x \to a} [\, g(x) - f(x)\,] + \lim_{x \to a} f(x) = 0 + L = L$$

1.3   Continuity

1. We need to prove that $|x^2 - f(2)| = |x^2 - 4| < \epsilon$ when $|x - 2| < \delta$. Substitute $x = 2 \pm \delta$ to get $|\delta^2 \pm 4\delta| < \epsilon$, or, using the triangle inequality, $|\delta|^2 + 4|\delta| < \epsilon$. Solving for δ gives $|\delta| < 2[(1+\epsilon/4)^{1/2}-1]$. Multiply and divide by $(1+\epsilon/4)^{1/2}+1$ to get $\delta < \epsilon/[1 + (1 + \epsilon/4)^{1/2}] < \epsilon$.

3. Use the intermediate value theorem and the fact that $f(1) = 5$ and $f(-1) = -3$. [$f(x) = 0$ at $x = -0.476$.]

6. First write $x \csc x = \dfrac{x}{\sin x}$. Both the numerator and denominator are continuous for all values of x, so $f(x)$ is continuous for all values of x except when $\sin x = 0$, or for $x = 0, \ \pm\pi, \ \pm 2\pi, \dots$.

9. The following figures illustrate the difference in the behavior of the two functions at their points of discontinuity.

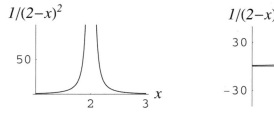

12. To be continuous at $x = \pi/2$, we require that $-\sin\dfrac{\pi}{2} = \alpha \sin\dfrac{\pi}{2} + \beta$ or that $\alpha + \beta = -1$. To be continuous at $x = 3\pi/2$, we require that $\alpha \sin\dfrac{3\pi}{2} + \beta = 0$ or that $\alpha = \beta$. Thus, we find that $\alpha = \beta = -1/2$. The following figure shows this result.

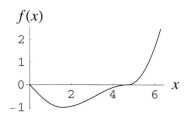

15. A function $h(x)$ is continuous if $\lim\limits_{x \to a} h(x) = h(a)$. Let $h(x) = f(x)/g(x)$ and use the fact that

$$\lim_{x \to a} \frac{f(x)}{g(x)} = \frac{\lim\limits_{x \to a} f(x)}{\lim\limits_{x \to a} g(x)} = \frac{f(a)}{g(a)}$$

provided that $g(a) \neq 0$.

1.4   Differentiation

1. (a) $f(x) = (2 + x)e^{-x^2}$
   $f'(x) = e^{-x^2} - 2x(2 + x)e^{-x^2} = (1 - 4x - 2x^2)e^{-x^2}$

(b) $f(x) = \dfrac{\sin x}{x}$
   $f'(x) = \dfrac{\cos x}{x} - \dfrac{\sin x}{x^2}$

(c) $f(x) = x^2 \tan 2x$
   $f'(x) = 2x \tan 2x + x^2(2\sec^2 2x) = 2x^2 \sec^2 2x + 2x \tan x$

3. (a) Let $y(u) = \tan^{-1} u$ and $u = e^{-x}$. Then

$$\frac{dy}{dx} = \frac{dy}{du}\frac{du}{dx} = \frac{1}{1 + u^2} \cdot (e^{-x}) = -\frac{e^{-x}}{1 + e^{-2x}}$$

(b) Let $y(u) = \ln u$ and $u = \sec x + \tan x$. Then

$$\frac{dy}{dx} = \frac{dy}{du}\frac{du}{dx} = \frac{1}{\sec x + \tan x}(\sec x \tan x + \sec^2 x) = \sec x$$

(c) For $f(x) = x^{\sin x}$, consider $\ln f(x) = \sin x \cdot \ln x$. Then $\dfrac{d \ln f(x)}{dx} = \dfrac{f'(x)}{f(x)} = \cos x \cdot \ln x + \dfrac{\sin x}{x}$

and so $f'(x) = x^{\sin x}\left(\cos x \cdot \ln x + \dfrac{\sin x}{x}\right)$.

6. Let a be any value of x in the open interval $(0, 1)$. Then

$$f'(a) = \lim_{h \to 0}\left\{\frac{(a + h)^3 - a^3}{h}\right\} = \lim_{h \to 0}(3a^2 + 3ah + h^2) = 3a^2$$

Now we must look at the end points

$$f'(0) = \lim_{h \to 0+} \left\{ \frac{f(0+h) - f(0)}{h} \right\} = \lim_{h \to 0+} h^2 = 0$$

and

$$f'(1) = \lim_{h \to 1-} \left\{ \frac{f(1+h) - f(1)}{h} \right\} = \lim_{h \to 1-} (3 + 3h + h^2) = 3$$

Therefore $f(x) = x^3$ is differentiable in $[0, 1]$.

9. $f(x) = 1 + x^{2/3}$ and $f'(x) = \dfrac{2}{3x^{1/3}}$ and so $f(x)$ is not differentiable at $x = 0$. The graph of $f(x) = 1 + x^{2/3}$ below shows that the minimum value of $f(x)$ is $x = 0$.

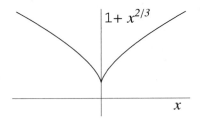

12. Consider $\ln y = v(x) \ln u(x)$. Then,

$$\frac{d \ln y}{dx} = \frac{1}{y} \frac{dy}{dx} = \frac{dv}{dx} \ln u + \frac{v}{u} \frac{du}{dx}$$

and

$$\frac{dy}{dx} = u^v (\ln u) \frac{dv}{dx} + v u^{v-1} \frac{du}{dx}$$

15. Start with

$$f'(0) = \lim_{\Delta x \to 0} \frac{(0 + \Delta x)^2 \sin \dfrac{1}{0 + \Delta x} - 0}{\Delta x} = \lim_{\Delta x \to 0} \Delta x \sin \frac{1}{\Delta x}$$

But $\left| \sin \dfrac{1}{\Delta x} \right| \leq 1$, and so $f'(0) = 0$. Now

$$f'(x) = \frac{d}{dx} \left(x^2 \sin \frac{1}{x} \right) = 2x \sin \frac{1}{x} - \cos \frac{1}{x}$$

The limit does not exist as $x \to 0$ because the limit of $\cos(1/x)$ does not exist. Therefore, $f'(0) \neq \lim_{x \to 0} f'(x)$ in this case.

18. The distance of any point of the curve to the origin is given by $D = (x^2 + y^2)^{1/2} = \left(x^2 + \dfrac{1}{x} \right)^{1/2}$. The condition $\dfrac{dD}{dx} = \dfrac{x - \dfrac{1}{2x^2}}{\left(x^2 + \dfrac{1}{x} \right)^{1/2}} = 0$ implies that $x = \left(\dfrac{1}{2} \right)^{1/3}$ and $y = \pm 2^{1/6}$ are the points on the curve $xy^2 = 1$ that are closest to the origin.

21. Differentiating $\rho(\lambda, T)$ with respect to λ,

$$\frac{d\rho}{d\lambda} = \frac{8\pi hc}{e^{hc/\lambda k_{\mathrm{B}} T} - 1} \left[-\frac{5}{\lambda^6} + \frac{\frac{hc}{\lambda^2 k_{\mathrm{B}} T} e^{hc/\lambda k_{\mathrm{B}} T}}{\lambda^5 (e^{hc/\lambda k_{\mathrm{B}} T} - 1)} \right]$$

Setting $d\rho/d\lambda = 0$ yields

$$5 = \frac{hc}{\lambda_{\max} k_{\mathrm{B}} T} \frac{e^{hc/\lambda_{\max} k_{\mathrm{B}} T}}{(e^{hc/\lambda_{\max} k_{\mathrm{B}} T} - 1)}$$

Let $u = hc/\lambda_{\max} k_{\mathrm{B}} T$ and rearrange to get $e^{-u} + \dfrac{u}{5} = 1$. Solving for u yields $u = 4.965$ or $\lambda_{\max} T = hc/4.965 k_{\mathrm{B}} = 2.90 \times 10^{-3} \, \mathrm{m \cdot K}$.

24. Start with $h(x) = f(x)g(x)$. Then

$$\frac{h(x + \Delta x) - h(x)}{\Delta x} = \frac{f(x + \Delta x)g(x + \Delta x) - f(x)g(x)}{\Delta x}$$

$$= \frac{[f(x + \Delta x)g(x + \Delta x) - g(x)f(x + \Delta x)] + [g(x)f(x + \Delta x) - f(x)g(x)]}{\Delta x}$$

$$= f(x + \Delta x)\frac{g(x + \Delta x) - g(x)}{\Delta x} + g(x)\frac{f(x + \Delta x) - f(x)}{\Delta x}$$

and so

$$\lim_{\Delta x \to 0} \frac{h(x + \Delta x) - h(x)}{\Delta x} = \frac{d}{dx}f(x)g(x) = f(x)\frac{d}{dx}g(x) + g(x)\frac{d}{dx}f(x) = f(x)\frac{dg}{dx} + g(x)\frac{df}{dx}$$

1.5 Differentials

1. Take $y = f(x) = x^{1/3}$ with $x = 125$ and $\Delta x = -5$. Equation 3 gives

$$f(x + \Delta x) = f(120) = 120^{1/3} \approx f(125) + f'(125)\Delta x = 5 + f'(125)\Delta x$$

$$= 5 + \frac{-5}{3(125)^{2/3}} = 5 - \frac{5}{3 \cdot 25} = 4.9333$$

The "correct" answer to four-decimal places is 4.9324.

3. For $y(x) = x^3 + x^2 + 6x - 3$,

$$\Delta y = y(x + \Delta x) - y(x)$$
$$= (x + \Delta x)^3 + (x + \Delta x)^2 + 6(x + \Delta x) - 3 - x^3 - x^2 - 6x + 3$$
$$= (3x^2 + 2x + 6)\Delta x + (3x + 1)(\Delta x)^2 + (\Delta x)^3$$

and

$$\frac{\Delta y}{\Delta x} = 3x^2 + 2x + 6 + (3x + 1)(\Delta x) + (\Delta x)^2$$

Thus $\epsilon = \dfrac{\Delta y}{\Delta x} - \dfrac{dy}{dx} = (3x + 1)\Delta x + (\Delta x)^2 \to 0$ as $\Delta x \to 0$.

6. For $y(x) = x^{1/2}$, $\Delta y = (x + \Delta x)^{1/2} - x^{1/2} = (4.35)^{1/2} - (4.00)^{1/2} = 0.086$ at $x = 4.00$ and $\Delta x = 0.35$.

$$dy = f'(x)\Delta x = \frac{\Delta x}{2x^{1/2}} = \frac{0.35}{4.00} = 0.0875$$

Therefore, $\Delta y - dy = -0.0015$ and $\dfrac{\Delta y - dy}{\Delta y} = -0.017$

9. (a) If $y(x) = (x^2 - 2)^{1/3}$, then $dy = \dfrac{2x\,dx}{3(x^2 - 2)^{2/3}}$.

(b) If $y(x) = \sin x^{1/2}$, then $dy = \dfrac{\cos x^{1/2}\,dx}{2x^{1/2}}$.

(c) If $y(x) = e^{\cos x}$, then $dy = -e^{\cos x} \sin x\,dx$.

1.6   Mean Value Theorems

1. Use $\sin x = x - \dfrac{x^3}{3!} + \dfrac{x^5}{5!} - \dfrac{x^7}{7!} + R_n$ where the remainder term is

$$R_n = \frac{1}{(n+1)!} \left[\frac{d^{n+1}\sin x}{dx^{n+1}}\right]_{x=\xi} x^{n+1} \leq \frac{x^{n+1}}{(n+1)!}$$

With $x = \pi/4$, $R_n = 3.6 \times 10^{-6}$ when $n = 7$. Thus we calculate

$$\sin \frac{\pi}{4} = \frac{\pi}{4} - \frac{(\pi/4)^3}{3!} + \frac{(\pi/4)^5}{5!} - \frac{(\pi/4)^7}{7!} = 0.707\ 106$$

The exact value to six significant figures is $1/\sqrt{2} = 0.707\ 107$.

3. $f(x) < 0$ if x_0 is chosen suitably negative such that $x^3 + px < -|q|$ and $f(x) > 0$ for x_1 suitably positive such that $x^3 + px > |q|$. Therefore, there is at least one root between x_0 and x_1. But $f(x)$ is always increasing because $f'(x) = 3x^2 + p > 0$ for all x. Therefore, there is only the one root.

6. (a) $\lim\limits_{x\to 0}(\csc x - \cot x) = \lim\limits_{x\to 0}\left(\dfrac{1}{\sin x} - \dfrac{\cos x}{\sin x}\right) = \lim\limits_{x\to 0}\dfrac{1 - \cos x}{\sin x} = \lim\limits_{x\to 0}\dfrac{\sin x}{\cos x} = 0$

(b) $\lim\limits_{x\to 0}\dfrac{1 - \cos x}{x^2} = \lim\limits_{x\to 0}\dfrac{\sin x}{2x} = \lim\limits_{x\to 0}\dfrac{\cos x}{2} = \dfrac{1}{2}$

(c) $\lim\limits_{x\to 0+}\dfrac{\ln \sin x}{\ln \tan x} = \lim\limits_{x\to 0+}\left(\dfrac{\cos x}{\sin x} \cdot \cos x \sin x\right) = \lim\limits_{x\to 0+}\cos^2 x = 1$

(d) $\lim\limits_{x\to 0}\dfrac{\ln(1 + x)}{x} = \lim\limits_{x\to 0}\dfrac{1}{1 + x} = 1$

9. Let $y = x^{1/x}$. Taking the logarithm of both sides yields $\ln y = \dfrac{\ln x}{x}$. Now $\lim\limits_{x\to\infty}\dfrac{\ln x}{x} = 0$, and so $\lim\limits_{x\to\infty} x^{1/x} = 1$

12. This is essentially the same as Problem 9.

15. With the definition of $F(x)$ given in the statement of the problem, $F(a) = F(b)$ and so $F'(x) = f'(x) - \dfrac{f(b) - f(a)}{g(b) - g(a)} g'(x)$. By Rolle's theorem, there is a point c in (a, b) such that $F'(c) = 0$, and so

$$\frac{f'(c)}{g'(c)} = \frac{f(b) - f(a)}{g(b) - g(a)}$$

1.7 Integration

1. Start with $d(uv) = u\,dv + v\,du$ and integrate to obtain $\displaystyle\int_a^b u\,dv = uv\Big]_a^b - \int_a^b v\,dv.$

3. (a) Let $x = a\cos\theta$, with $dx = -a\sin\theta\,d\theta$. Thus

$$\int \frac{dx}{(a^2 - x^2)^{1/2}} = -\int \frac{\sin\theta\,d\theta}{\sin\theta} = -\theta = -\cos^{-1}\frac{x}{a}$$

(b) Let $x = a\tan\theta$, with $dx = a\sec^2\theta\,d\theta$. Thus

$$\int_0^a \frac{dx}{(a^2 + x^2)^{1/2}} = \int_0^{\pi/4} \sec\theta\,d\theta = \ln\tan\left(\frac{\pi}{4} + \frac{\theta}{2}\right)\Big]_0^{\pi/4}$$

$$= \ln\tan\frac{3\pi}{8} - \ln\tan\frac{\pi}{4} = \ln\tan\frac{3\pi}{8} = 0.8814$$

(c) Let $x = \tan\theta$, with $dx = \sec^2\theta\,d\theta$. Thus

$$\int_0^1 \frac{dx}{(1 + x^2)^2} = \int_0^{\pi/4} \cos^2\theta\,d\theta = \frac{\pi}{8} + \frac{1}{4}$$

(d) Let $x = a\cos\theta$, with $dx = -a\sin\theta\,d\theta$. Thus

$$\int_0^a (a^2 - x^2)^{1/2}\,dx = -a^2 \int_{\pi/2}^0 \sin^2\theta\,d\theta = a^2\left[\frac{\theta}{2} - \frac{\sin 2\theta}{4}\right]_0^{\pi/2} = \frac{a^2\pi}{4}$$

6. The area is given by $A = \displaystyle\int_{-a}^a (a^2 - x^2)^{1/2}\,dx$. Let $x = a\sin\theta$ to obtain

$$A = \int_{-a}^a (a^2 - x^2)^{1/2}\,dx = a^2 \int_0^\pi \cos^2\theta\,d\theta = \frac{a^2\pi}{2}$$

9. Start with $G(x) = \displaystyle\int_{u(x)}^{v(x)} f(t)\,dt$ and $G'(x) = \displaystyle\lim_{\Delta x \to 0} \frac{G(x + \Delta x) - G(x)}{\Delta x}$. So

$$G(x + \Delta x) - G(x) = \int_{u(x+\Delta x)}^{v(x+\Delta x)} f(z)\,dz - \int_{u(x)}^{v(x)} f(z)\,dz$$

$$= \int_{v(x)}^{v(x+\Delta x)} f(z)\,dz - \int_{u(x)}^{u(x+\Delta x)} f(z)\,dz$$

According to Equation 10, $\displaystyle\int_{v(x)}^{v(x+\Delta x)} f(z)\,dz = f\{v(\xi)\}\{v(x+\Delta x) - v(x)\}$ where $x \le \xi \le$

$x + \Delta x$ and $\displaystyle\int_{u(x)}^{u(x+\Delta x)} f(z)\,dz = f\{u(\xi)\}\{u(x+\Delta x) - u(x)\}$. Equation 16 results when we

substitute all these equations into the equation for $G'(x)$ above.

12.

$$
\begin{array}{ccccccc}
1 & 2 & 3 & \cdots & n-1 & n \\
n & n-1 & n-2 & \cdots & 2 & 1 \\
\hline
\end{array}
$$

$$
\underbrace{n+1 \quad n+1 \quad n+1 \quad \cdots \quad n+1 \quad n+1}_{n \text{ times}}
$$

The sum of each column (made up of two rows) is $n+1$; there are n such columns and two

sets, so $\displaystyle S = \sum_{j=1}^{n} j = \frac{n(n+1)}{2}$.

15. Because $g(x) - f(x) \ge 0$ in $[a,b]$, $\displaystyle\int_a^b [\,g(x) - f(x)\,]\,dx \ge 0$, or $\displaystyle\int_a^b g(x)\,dx \ge \int_a^b f(x)\,dx$

because every term in the corresponding Riemann sum is ≥ 0.

18. Applying the triangle inequality to the Riemann sum corresponding to the integral,
we have

$$
\left| \sum_{j=1}^{n} f(\xi_j)\Delta x_j \right| \le \sum_{j=1}^{n} |\,f(\xi_j)\Delta x_j| = \sum_{j=1}^{n} |f(\xi_j)||\Delta x_j|
$$

Taking the limit preserves this result.

21. As $x \to 0$, $\sin u^2 \to u^2$, and so the limit is equal to $\displaystyle\frac{1}{x^3}\int_0^x u^2\,du = \frac{1}{3}$.

1.8   Improper Integrals

1. $\displaystyle\int_0^\infty \frac{dx}{x^2+1} = \lim_{b\to\infty}\int_o^b \frac{dx}{x^2+1} = \lim_{b\to\infty}\tan^{-1} b = \frac{\pi}{2}$

3. First note that $\text{sech}\, x = \dfrac{1}{\cosh x} = \dfrac{2}{e^x + e^{-x}}$. Now choose $c = 0$ in Equation 3:

$$
\begin{aligned}
\int_0^\infty \text{sech}\, x\,dx &= 2\lim_{b\to\infty}\int_0^b \frac{dx}{e^x + e^{-x}} = 2\lim_{b\to\infty}\int_0^b \frac{e^x\,dx}{e^{2x}+1} \\
&= 2\lim_{b\to\infty}\int_1^{e^b} \frac{du}{u^2+1} = 2\lim_{b\to\infty}\left[\tan^{-1} u\right]_1^{e^b} \\
&= 2\lim_{b\to\infty}(\tan^{-1} e^b - \tan^{-1} 1) = \pi - \frac{\pi}{2} = \frac{\pi}{2}
\end{aligned}
$$

Similarly $\displaystyle\int_{-\infty}^{0} \operatorname{sech} x\, dx = \frac{\pi}{2}$ and so $\displaystyle\int_{-\infty}^{\infty} \operatorname{sech} x\, dx = \pi$.

6. Because $\sec x = \dfrac{1}{\cos x}$, $\sec x$ is unbounded at $x = \pi/2$.

$$
\begin{aligned}
\int_{0}^{\pi/2} \sec x\, dx &= \lim_{\epsilon \to 0} \int_{0}^{\pi/2 - \epsilon} \sec x\, dx = \lim_{\epsilon \to 0} \Big[\ln(\sec x + \tan x) \Big]_{0}^{\pi/2 - \epsilon} \\
&= \lim_{\epsilon \to 0} \ln \left[\sec\left(\frac{\pi}{2} - \epsilon\right) + \tan\left(\frac{\pi}{2} - \epsilon\right) \right] = \infty
\end{aligned}
$$

The integral diverges.

9. $\dfrac{1}{1+x} < \dfrac{1}{x}$ for $x \geq 1$, and $\displaystyle\int_{1}^{\infty} \frac{dx}{x}$ diverges. Therefore, $\displaystyle\int_{1}^{\infty} \frac{dx}{1+x}$ diverges.

12. Take $g(x) = 1/x^3$, for which $\displaystyle\int_{1}^{\infty} \frac{dx}{x^3}$ converges, and so $\displaystyle\int_{1}^{\infty} \frac{dx}{(x^6 + 1)^{1/2}}$ converges.

15. In this case, $K = \displaystyle\lim_{x \to 4} \frac{(x-4)^p}{x^3(x-2)^{1/2}(x-4)^{1/2}} = 0$ for $p = 3/4$ (which is less than unity) and so the integral converges.

18. $\displaystyle\lim_{x \to \infty} x^2 e^{-x^2 + 6x} = 0$, so the integral converges (p test with $p = 2$).

21. For large values of x, $\dfrac{\sinh ax}{\sinh \pi x} \to e^{(a - \pi)x}$. We know from Example 2 that the integral converges if $a - \pi < 0$ and diverges if $a - \pi \geq 0$.

1.9 Uniform Convergence of Integrals

1. Choose $M(t) = e^{-\alpha t}$ because $e^{-xt} \leq e^{-\alpha t}$ if $x \geq \alpha > 0$. Because $\displaystyle\int_{0}^{\infty} e^{-\alpha t}\, dt = 1/\alpha$ is finite for $\alpha > 0$, $\displaystyle\int_{0}^{\infty} e^{-xt}\, dt$ converges uniformly for $x \geq \alpha > 0$.

3. Differentiate $\displaystyle\int_{0}^{\infty} e^{-\alpha t}\, dt = \frac{1}{\alpha}$ with respect to α n times.

6. Choose $M(t) = \dfrac{1}{1 + x^2}$ because $\dfrac{\cos \alpha x}{1 + x^2} \leq \dfrac{1}{1 + x^2}$ for all real values of α ($|\cos \alpha x| \leq 1$.)

9. Yes, the second integral is uniformly convergent. $\displaystyle\int_{0}^{\infty} \frac{dx}{1 + x^2} = \frac{\pi}{2}$ and $\displaystyle\int_{0}^{\infty} \frac{\cos \alpha x\, dx}{1 + x^2} = \frac{\pi}{2} e^{-\alpha}$ for $\alpha > 0$.

12.

$$
\begin{aligned}
\int_{0}^{\infty} e^{-ax} \frac{\sin x}{x}\, dx &= \int_{0}^{\infty} e^{-ax} \left(\int_{0}^{1} \cos xz\, dz \right) dx \\
&= \int_{0}^{1} \left(\int_{0}^{\infty} e^{-ax} \cos xz\, dx \right) dz \\
&= \int_{0}^{1} \frac{a}{a^2 + z^2}\, dz = \tan^{-1} \frac{1}{a} = \cot^{-1} a
\end{aligned}
$$

15. Differentiate $\displaystyle\int_0^\infty e^{-u}\cos xu\,du = \dfrac{1}{1+x^2}$ with respect to x to obtain

$$-\int_0^\infty ue^{-u}\sin xu\,du = -\frac{2x}{(1+x^2)^2}.$$

CHAPTER 2

Infinite Series

2.1   Infinite Sequences

1. $\lim\limits_{n\to\infty} \dfrac{3n^2 - 6n + 2}{n^2 + 1} = \lim\limits_{n\to\infty} \dfrac{3 - \dfrac{6}{n} + \dfrac{2}{n^2}}{1 + \dfrac{1}{n^2}} = 3$

3. Let $y = \ln e^{1/n} = \dfrac{1}{n}$, then $\lim\limits_{n\to\infty} y = 0$, and so $\lim\limits_{n\to\infty} e^{1/n} = 1$.

6. Let $y = \ln \sqrt[n]{n} = \dfrac{1}{n} \ln n$. But $\dfrac{1}{n} \ln n \to 0$ as $n \to \infty$ by l'Hôpital's rule, so $y = \sqrt[n]{n} \to 1$.

9.   (a)   Choose $f(x) = \dfrac{4x - 1}{6x + 2}$.   Then $f'(x) = \dfrac{9}{(6x + 2)^2} > 0$, and so the sequence is increasing.

(b)   Choose $f(x) = \dfrac{\ln x}{x}$.   Then $f'(x) = \dfrac{1 - \ln x}{x^2} < 0$, and so the sequence is decreasing if $x > 2$.

(c)   Choose $f(x) = \dfrac{x + 2}{x}$.   Then $f'(x) = -\dfrac{2}{x^2} < 0$, and so the sequence is decreasing.

(d)   Choose $f(x) = \ln \dfrac{1 + x}{x}$.   Then $f'(x) = -\dfrac{1}{x(x + 1)} < 0$, and so the sequence is decreasing.

These sequences are plotted against n in the following figures. (See also the next page.)

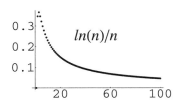

12. (a) $s_n = \dfrac{\cosh n}{\sinh n}$.   Therefore, $\lim\limits_{n\to\infty} s_n = \dfrac{e^n}{e^n} = 1$.

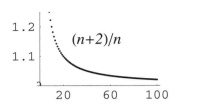

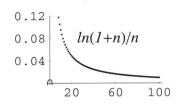

(b) $s_n = (2/n)^{1/n}$. Consider $\ln s_n = \dfrac{1}{n} \ln \dfrac{2}{n}$. Then $\lim\limits_{n\to\infty} \ln s_n = \lim\limits_{n\to\infty} \dfrac{1}{n} \ln \dfrac{2}{n} = \lim\limits_{x\to 0} x \ln 2x = 0$ by l'Hôpital's rule. Therefore, $\lim\limits_{n\to\infty} s_n = 1$.

(c) $s_n = \left(1 + \dfrac{1}{n}\right)^n$. Consider $\ln s_n = n \ln \left(1 + \dfrac{1}{n}\right)$. Then $\lim\limits_{n\to\infty} s_n = \lim\limits_{x\to 0} \dfrac{\ln(1+x)}{x} = 1$, and $\lim\limits_{n\to\infty} s_n = e$.

(d) Using the result of part (c), we write $\lim\limits_{n\to\infty} s_n = \left(\dfrac{n-1}{n+1}\right)^n = \dfrac{\lim\limits_{n\to\infty} \left(1 - \dfrac{1}{n}\right)^n}{\lim\limits_{n\to\infty} \left(1 + \dfrac{1}{n}\right)^n} = \dfrac{e^{-1}}{e} = e^{-2}$.

15. We'll prove only the first case. Since $\{s_n\}$ is convergent, it is bounded, and so $|s_n| \leq M$ for all n. If $b \neq 0$, there is an N_1 such that $|s_n - a| < \dfrac{\epsilon}{2|b|}$ if $n > N_1$ or $|b| \, |s_n - a| < \dfrac{\epsilon}{2}$.

(If $b = 0$, this last line is still satisfied.) Also, there is an N_2 such that $|t_n - b| < \dfrac{\epsilon}{2M}$ if $n > N_2$. Let $N = \max{(N_1, N_2)}$. Then

$$
\begin{aligned}
|s_n t_n - ab| &= |s_n(t_n - b) + b(s_n - a)| \leq |s_n(t_n - b)| + |b(s_n - a)| \\
&= |s_n||t_n - b| + |b||s_n - a| \leq M\left(\dfrac{\epsilon}{2M}\right) + \dfrac{\epsilon}{2} = \epsilon
\end{aligned}
$$

with $n > N = \max{(N_1, N_2)}$.

2.2   Convergence and Divergence of Infinite Series

1. This is just a geometric series, Equation 2, with $x = \dfrac{1}{3}$, so $S = \dfrac{1}{1 - \frac{1}{3}} = \dfrac{3}{2}$.

3. Write $0.142\,857\,142\,857\ldots$ as $0.142\,857(1 + 10^{-6} + 10^{-12} + \cdots) = \dfrac{0.142\,857}{0.999\,999} = \dfrac{1}{7}$.

6. If you use partial fractions, then $\dfrac{1}{n(n+1)} = \dfrac{1}{n} - \dfrac{1}{n+1}$. So the nth partial sum is

$$
\begin{aligned}
S_n &= \dfrac{1}{1 \cdot 2} + \dfrac{1}{2 \cdot 3} + \dfrac{1}{3 \cdot 4} + \dfrac{1}{4 \cdot 5} + \cdots \\
&= \left(\dfrac{1}{1} - \dfrac{1}{2}\right) + \left(\dfrac{1}{2} - \dfrac{1}{3}\right) + \left(\dfrac{1}{3} - \dfrac{1}{4}\right) + \left(\dfrac{1}{4} - \dfrac{1}{5}\right) + \left(\dfrac{1}{n} - \dfrac{1}{n+1}\right) \\
&= 1 - \dfrac{1}{n+1}
\end{aligned}
$$

and $S_n \to 1$ as $n \to \infty$. This cancellation property is called *telescoping*.

9. Using partial fractions, we write

$$\frac{1}{n(n+1)(n+2)} = \frac{1/2}{n} - \frac{1}{n+1} + \frac{1/2}{n+2}$$

$$= \frac{1}{2}\left(\frac{1}{n} - \frac{1}{n+1}\right) - \frac{1}{2}\left(\frac{1}{n+1} - \frac{1}{n+2}\right)$$

Therefore,

$$\sum_{n=1}^{\infty} \frac{1}{n(n+1)(n+2)} = \frac{1}{2}\left[\left(1 - \frac{1}{2}\right) + \left(\frac{1}{2} - \frac{1}{3}\right) + \cdots\right] - \frac{1}{2}\left[\left(\frac{1}{2} - \frac{1}{3}\right) + \left(\frac{1}{3} - \frac{1}{4}\right) + \cdots\right]$$

$$= \frac{1}{2} - \frac{1}{4} = \frac{1}{4}$$

12. Each series is a geometric series, so $|r| < 1$.

(a) $2|x| < 1 \Longrightarrow |x| < 1/2$

(b) $|x - 1| < 1 \Longrightarrow 0 < x < 2$

(c) $\left|\dfrac{2x-1}{3}\right| < 1 \Longrightarrow -1 < x < 2$

(d) $|e^x| < 1 \Longrightarrow x < 0$

2.3   Tests for Convergence

1. No. Take $f(x) = \dfrac{\ln x}{x}$. Then

$$\int_0^\infty \frac{\ln x}{x} dx = \left|\lim_{a\to\infty} \int_1^a \frac{\ln x}{x} dx\right| = \lim_{a\to\infty}\left|\frac{(\ln x)^2}{2}\right|_1^a = \infty$$

3. $R_n = \displaystyle\sum_{k=n+1}^{\infty} u_k$. Refer to Figure 2.4 to see that $\displaystyle\int_k^{k+1} f(x)dx \le u_k \le \int_{k-1}^{k} f(x)dx$. Sum from $k = n+1$ to $k = \infty$, $\displaystyle\int_{n+1}^{\infty} f(x)dx \le R_n \le \int_n^{\infty} f(x)dx$.

6. Use partial fractions:

$$\frac{1}{(2n-1)(2n+1)} = \frac{a}{2n-1} + \frac{b}{2n+1} = \frac{a(2n+1) + b(2n-1)}{(2n-1)(2n+1)}$$

where $2a + 2b = 0$ and $a - b = 1$. Thus, $a = 1/2$ and $b = -1/2$, and

$$\frac{1}{(2n-1)(2n+1)} = \frac{1/2}{2n-1} - \frac{1/2}{2n+1}$$

Summation gives

$$S = \frac{1}{2}\left[\sum_{n=1}^{\infty}\left(\frac{1}{2n-1} - \frac{1}{2n+1}\right)\right] = \frac{1}{2}\left[\left(1 - \frac{1}{3}\right) + \left(\frac{1}{3} - \frac{1}{5}\right) + \left(\frac{1}{5} - \frac{1}{7}\right) + \cdots\right] = \frac{1}{2}$$

9. $S = \left(1 - \dfrac{1}{3}\right) + \left(\dfrac{1}{2} - \dfrac{1}{4}\right) + \left(\dfrac{1}{3} - \dfrac{1}{5}\right) + \left(\dfrac{1}{4} - \dfrac{1}{6}\right) + \left(\dfrac{1}{5} - \dfrac{1}{7}\right) + \cdots = \dfrac{3}{2}.$ Thus $\displaystyle\sum_{n=1}^{\infty} \dfrac{1}{n(n+2)} =$

12. The p-test with $p = 1$ shows that the series $\displaystyle\sum_{n=1}^{\infty} \dfrac{1}{n + \sqrt{n}}$ does not converge.

The p-test with $p = 3/2$ shows that the series $\displaystyle\sum_{n=1}^{\infty} \dfrac{1}{n + n^{3/2}}$ does converge.

2.4   Alternating Series

1.

N	v_{N+1}	S_N	$R = S - S_N$	$|R_N|$
1	0.06250	1.00000	−0.05297	0.05297
2	0.01235	0.93750	+0.00953	0.00953
3	0.00391	0.94985	−0.00281	0.00281
4	0.00160	0.94594	+0.00109	0.00109
5		0.94754	−0.00051	0.00051

3. We want $\dfrac{n}{4^n} < 0.0001$, or $\dfrac{4^n}{n} > 10\,000$, or $n \geq 9$. Thus, 8 terms will suffice ($S = 0.159973$.) The sum is actually equal to $4/25 = 0.1600000$.

6. The following figure was created in *Mathematica*.

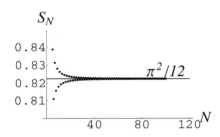

9. $\displaystyle\int_2^{\infty} \dfrac{dx}{x \ln x} = \Big[\ln(\ln x)\Big]_2^{\infty} \to \infty.$ The series does not converge.

12. (a) conditionally convergent; not absolutely convergent (p-test with $p = 1$)

(b) conditionally convergent; not absolutely convergent (p-test with $p = 1/2$)

(c) absolutely convergent (p-series with $n = 2$)

(d) absolutely convergent $\left(\displaystyle\lim_{n\to\infty} \left|\dfrac{u_{n+1}}{u_n}\right| = \dfrac{1}{2} < 1\right)$

(e) conditionally convergent; not absolutely convergent (first write $\sqrt{1+n} - \sqrt{n}$ as

$\dfrac{1}{\sqrt{n+1} + \sqrt{n}}$ by multiplying by $\dfrac{\sqrt{1+n} + \sqrt{n}}{\sqrt{n+1} + \sqrt{n}}$. Then use the ratio test,

$$\lim_{n\to\infty} \left|\dfrac{\sqrt{1+n} + \sqrt{n}}{\sqrt{2+n} + \sqrt{n+1}}\right| = 1$$

(f) diverges; $\lim_{n\to\infty} |u_n| = 1$

15. The ratio test gives us $\lim_{n\to\infty} \left| x \left(\dfrac{n+1}{n} \right) \right| = |x|$ so that the series converges absolutely with $|x| < 1$. The root test gives us the same result $\lim_{n\to\infty} \sqrt[n]{|nx^n|} = |x| \lim_{n\to\infty} \sqrt[n]{n} = |x|$.

18. Start with

$$R_{2N} = (v_{2N+1} - v_{2N+2}) + (v_{2N+3} - v_{2N+4}) + (v_{2N+5} - v_{2N+6}) + \cdots$$

Note that $R_{2N} \geq 0$ because each term in parentheses is ≥ 0. Also, note that $0 \leq R_{2N} \leq v_{2N+1}$ because of the relation

$$R_{2N} = v_{2N+1} - (v_{2N+2} - v_{2N+3}) - (v_{2N+4} - v_{2N+5}) - \cdots$$

where once again each term in parentheses is positive. Now write

$$R_{2N+1} = (-v_{2N+2} + v_{2N+3}) + (-v_{2N+4} + v_{2N+5}) + \cdots$$

and see that $R_{2N+1} \leq 0$ because each term in parentheses is negative. Finally, write R_{2N+1} as

$$R_{2N+1} = -v_{2N+2} + (v_{2N+3} - v_{2N+4}) + (v_{2N+5} - v_{2N+6}) + \cdots$$

and see that $R_{2N+1} \geq -v_{2N+2}$ because each term in parentheses is positive. Thus we have $-v_{2N+2} \leq R_{2N+1} \leq 0$ and $0 \leq R_{2N} \leq v_{2N+1}$, or $|R_N| \leq v_{N+1}$.

2.5   Uniform Convergence

1. We use the ratio test in each case.

(a) $\lim_{n\to\infty} \left| \dfrac{u_{n+1}}{u_n} \right| = \lim_{n\to\infty} \left| \dfrac{x^{2n+2}}{(2n+2)!} \cdot \dfrac{(2n)!}{x^{2n}} \right| = \lim_{n\to\infty} \left| \dfrac{x^2}{4n^2 + 6n + 2} \right| = 0$, so the series converges for all values of x.

(b) $\lim_{n\to\infty} \left| \dfrac{u_{n+1}}{u_n} \right| = \lim_{n\to\infty} \left| \dfrac{(x-2)n}{2(n+1)} \right| = \dfrac{|x-2|}{2}$. The series converges absolutely if $|x-2| < 2$, or if $0 < x < 4$. The series diverges at $x = 4$ (harmonic series) and converges conditionally at $x = 0$ (alternating harmonic series.)

(c) $\lim_{n\to\infty} \left| \dfrac{u_{n+1}}{u_n} \right| = \lim_{n\to\infty} \left| \dfrac{x(n+1)^2}{n^2} \right| = |x|$. The series converges absolutely if $|x| < 1$. It also converges at both end points (p-series with $p = 2$).

(d) $\lim_{n\to\infty} \left| \dfrac{(x+1)\sqrt{n+1}}{\sqrt{n}} \right| = |x+1|$. The series converges for $|x+1| < 1$, or $-2 < x < 0$. The series diverges at $x = 0$ (p-series with $p = 1/2$) and converges conditionally at $x = -2$.

3. The ratio test gives us $\lim_{n\to\infty} \left| \dfrac{u_{n+1}}{u_n} \right| = \left(\dfrac{x}{2} \right)^2 \left| \dfrac{1}{(n+1)^2} \right| = 0$, so the series converges for all values of x.

6. Use the Weierstrass M test with $M_n = \dfrac{1}{n^2}$ in part (a) and $M_n = \dfrac{1}{n^2}$ in part (b).

9. The series is uniformly convergent (the Weierstrass M test with $M_n = 1/n^2$) and each term in the series is continuous.

12. Start with $S(x) = 1 + x + x^2 + \cdots = \dfrac{1}{1-x}$. Then $\dfrac{dS}{dx} = 1 + 2x + 3x^2 + \cdots = \dfrac{1}{(1-x)^2}$. The term-by-term differentiation is valid because $S(x)$ is an uniformly convergent series for $|x| < 1$.

15. $\displaystyle\lim_{n\to\infty} \int_0^1 nxe^{-nx^2}\,dx = -\frac{1}{2}\int^n e^{-u}\,du = \frac{1}{2}(1 - e^{-n}) = \frac{1}{2} \ne \int_0^1 \lim_{n\to\infty} S_n(x)\,dx = \int_0^1 0\,dx = 0.$
The two limiting processes give different results because $S_n(x)$ does not converge uniformly to $S(x) = 0$. The following figure shows why this is so. The function $S_n(x)$ has a maximum value of $(n/2)^{1/2}e^{-1/2}$ at $x = 1/\sqrt{2n}$ so as $n \to \infty$, $S_n(x)$ cannot be made arbitrarily small for all x.

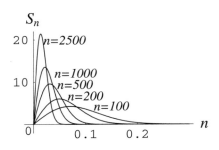

2.6   Power Series

1. Use the ratio test:

(a) $\displaystyle\lim_{n\to\infty}\left|\frac{u_{n+1}}{u_n}\right| = \lim_{n\to\infty}|(n+1)(x-1)| = \infty$ unless $x = 1$.

(b) $\displaystyle\lim_{n\to\infty}\left|\frac{u_{n+1}}{u_n}\right| = \lim_{n\to\infty}\left|\frac{x}{n+1}\right| = 0$ for all x.

3. Use the ratio test:

(a) $\displaystyle\lim_{n\to\infty}\left|\frac{u_{n+1}}{u_n}\right| = \lim_{n\to\infty}\left|\frac{x^2}{(2n+3)(2n+2)}\right| = 0$, and so the interval of convergence is $-\infty < x < \infty$.

(b) $\displaystyle\lim_{n\to\infty}\left|\frac{u_{n+1}}{u_n}\right| = \lim_{n\to\infty}\left|\frac{n^2(x-2)}{(n+1)^2}\right| = |x - 2| < 1$ or $1 < x < 3$. The series converges at both end points (p-series with $p = 2$), and so the interval of convergence is $1 \le x \le 3$.

6. Use the series $\dfrac{1}{1+x^4} = \displaystyle\sum_{n=0}^{\infty}(-1)^n x^{4n}$. Integrating term by term gives

$$\int_0^{\pi/4} \frac{dx}{1+x^4} = \sum_{n=0}^{\infty}(-1)^n \int_0^{\pi/4} x^4\,dx = \sum_{n=0}^{\infty}(-1)\frac{(\pi/4)^{4n+1}}{4n+1}$$

$$= 1 - \frac{(\pi/4)^5}{5} + \frac{(\pi/4)^9}{9} - \frac{(\pi/4)^{13}}{13} + \frac{(\pi/4)^{17}}{17} + \cdots$$

This is an alternating series, so we choose n such that $\dfrac{(\pi/4)^{4n+1}}{4n+1} < 0.0001$, or $n = 6$. The sum of 6 terms gives $0.735\ 702$, and a numerical integration yields $0.735\ 677$.

9. Start with

$$S(x) = x - \frac{x^3}{3!} + \frac{x^5}{5!} - \frac{x^7}{7!} + \cdots$$

Then

$$S'(x) = 1 - \frac{x^2}{2!} + \frac{x^4}{4!} - \frac{x^6}{6!} + \cdots = C(x)$$

Similarly,

$$C'(x) = -x + \frac{x^3}{3!} - \frac{x^5}{5!} + \cdots = -S(x)$$

Also,

$$S^2(x) = x^2 + \frac{x^6}{36} - \frac{x^4}{3} + \frac{x^6}{60} + O(x^7)$$

and

$$C^2(x) = 1 + \frac{x^4}{4} - x^2 + \frac{x^4}{12} - \frac{x^6}{360} - \frac{x^6}{24} + O(x^7)$$

Therefore,

$$
\begin{aligned}
S^2(x) + C^2(x) &= 1 + (x^2 - x^2) + x^4\left(\frac{1}{4} - \frac{1}{3} + \frac{1}{12}\right) + x^6\left(\frac{1}{36} + \frac{1}{60} - \frac{1}{360} - \frac{1}{24}\right) + \cdots \\
&= 1 + 0 + 0 + 0 + \cdots
\end{aligned}
$$

It turns out that $S(x) = \sin x$ and $C(x) = \cos x$.

12. Substituting $\varepsilon_n = \left(n + \dfrac{1}{2}\right) h\nu$ into $\varepsilon_{\mathrm{vib}} = (1 - e^{-h\nu/k_\mathrm{B}T}) \displaystyle\sum_{n=0}^{\infty} \varepsilon_n e^{-h\nu/k_\mathrm{B}T}$ gives

$$\frac{\varepsilon_{\mathrm{vib}}}{1 - e^{-h\nu/k_\mathrm{B}T}} = \frac{h\nu}{2} \sum_{n=0}^{\infty} e^{-h\nu/k_\mathrm{B}T} + \sum_{n=0}^{\infty} nh\nu e^{-h\nu/k_\mathrm{B}T}$$

The first summation is a geometric series and the second is the derivative of a geometric series, so we have

$$\frac{\varepsilon_{\mathrm{vib}}}{1 - e^{-h\nu/k_\mathrm{B}T}} = \frac{h\nu}{2(1 - e^{-h\nu/k_\mathrm{B}T})} + \frac{h\nu e^{-h\nu/k_\mathrm{B}T}}{(1 - e^{-h\nu/k_\mathrm{B}T})^2}$$

Therefore, $\varepsilon_{\mathrm{vib}} = \dfrac{h\nu}{2} + \dfrac{h\nu e^{-h\nu/k_\mathrm{B}T}}{1 - e^{-h\nu/k_\mathrm{B}T}}$.

2.7   Taylor Series

1. The ratio test shows that $\displaystyle\sum_{n=0}^{\infty} \frac{x^n}{n!}$ converges for all x, so $\dfrac{x^n}{n!}$ must go to zero as $n \to \infty$.

3. We know that $(x + y)^N = \displaystyle\sum_{N_1=0}^{N} \frac{N!}{N_1!(N-N_1)!} x^{N_1} y^{N-N_1}$. If we set $x = y = 1$, then we have

$$(1+1)^N = \sum_{N_1=0}^{N} \frac{N!}{N_1!(N-N_1)!} = \sum_{k=0}^{N} \frac{N!}{k!(N-k)!} = 2^N$$

6. Starting with $f(x) = \ln x$, we have $f'(x) = \dfrac{1}{x}$, $f''(x) = -\dfrac{1}{x^2}$, $\ldots$, $f^{(n)}(x) = \dfrac{(-1)^{n+1}(n-1)!}{x^n}$ and so $f^{(n)}(1) = (-1)^{n+1}(n-1)!$. The remainder, given by Equation 2,

$$R_n(\xi, x) = f^{(n+1)}(\xi) \frac{(x-1)^{n+1}}{(n+1)!} = \frac{(-1)^{n+2}}{n+1} \left(\frac{x-1}{\xi}\right)^{n+1}$$

where ξ lies between x and 1, goes to zero as $n \to \infty$. Therefore, Equation 3 gives

$$\ln x = \sum_{n=1}^{\infty} \frac{(-1)^{n+1}}{n}(x-1)^n$$

9. Substitute $-x^2$ for u in $e^u = \displaystyle\sum_{n=0}^{\infty} \frac{u^n}{n!}$ and then multiply by x to get

$$\sum_{n=0}^{\infty} \frac{x(-x^2)^n}{n!} = \sum_{n=0}^{\infty} (-1)^n \frac{x^{2n+1}}{n!}$$

12. Start with $(1+x)^\alpha = 1 + \displaystyle\sum_{n=1}^{\infty} \frac{\alpha(\alpha-1)(\alpha-2)\ldots(\alpha-n+1)x^n}{n!}$.
Then

$$
\begin{aligned}
\alpha(1+x)^{\alpha-1} &= \alpha + \sum_{n=2}^{\infty} \frac{\alpha(\alpha-1)(\alpha-2)\ldots(\alpha-n+1)x^{n-1}}{(n-1)!} \\
&= \alpha \left[1 + \sum_{n=1}^{\infty} \frac{(\alpha-1)(\alpha-2)\ldots(\alpha-1-n+1)}{n!} x^n \right] \\
&= \alpha(1+x)^{\alpha-1}
\end{aligned}
$$

15. (a) The Maclaurin series for $\sin x$ is $\sin x = x - \dfrac{x^3}{3!} + \dfrac{x^5}{5!} - \dfrac{x^7}{7!} + \cdots$, and so

$$1 + x\sin x = 1 + x^2 - \frac{x^4}{3!} + \frac{x^6}{5!} - \cdots$$

(b) The Maclaurin series for $\sin x$ is $\sin x = x - \dfrac{x^3}{3!} + \dfrac{x^5}{5!} - \dfrac{x^7}{7!} + \cdots$, and so

$$\frac{\sin x}{x} = 1 - \frac{x^2}{3!} + \frac{x^4}{5!} - \frac{x^6}{6!} + \cdots$$

(c) The Maclaurin series for $\cos x$ is $\cos x = 1 - \dfrac{x^2}{2!} + \dfrac{x^4}{4!} - \dfrac{x^6}{6!} + \cdots$, and so

$$\frac{1 - \cos x}{x^2} = \frac{1}{2!} - \frac{x^2}{4!} + \frac{x^4}{4!} + \cdots$$

18. Starting with $f(x) = e^{-1/x^2}$, we have $f'(x) = \dfrac{2}{x^3} e^{-1/x^2} \to 0$ as $x \to 0$ and

$f''(x) = -\dfrac{2}{x^4} e^{-1/x^2} + \dfrac{4}{x^6} e^{-1/x^2} \to 0$ as $x \to 0$. All derivatives contain the factor e^{-1/x^2},

and so all derivatives vanish as $x \to 0$ because $\dfrac{1}{x^n} e^{-1/x^2} \to 0$ as $x \to 0$ for any value of n.

21. The Taylor series is

$$x^3 \ln x = (x-1) + \frac{5}{2}(x-1)^2 + \frac{11}{6}(x-1)^3 + \frac{1}{4}(x-1)^4$$
$$+ \frac{1}{20}(x-1)^5 + \frac{1}{60}(x-1)^6 + O((x-1)^7)$$

2.8 Applications of Taylor Series

1. Yes. $\dfrac{e^{-x} - 1 + x}{x^2} = \displaystyle\sum_{n=2}^{\infty} \frac{(-1)^n x^{n-2}}{n!}$ converges uniformly and can be integrated term by term.

3. No. $\dfrac{\tan^{-1} x}{x^2} = \dfrac{1}{x} - \dfrac{x}{3} + \dfrac{2x^3}{15} + O(x^5)$, but $\displaystyle\int_0^1 \frac{dx}{x}$ diverges.

6. 0.503 098

9. Expand $\sin \sqrt{x}$ about $x = 0$ to obtain

$$\int_0^1 \left[\sum_{n=0}^{\infty} (-1)^n \frac{(\sqrt{x})^{2n+1}}{(2n+1)!} \right] dx = \sum_{n=0}^{\infty} \frac{(-1)^n}{(2n+1)!} \int_0^1 x^{n+\frac{1}{2}} dx = \sum_{n=0}^{\infty} \frac{(-1)^n}{(2n+1)!} \cdot \frac{2}{2n+3}$$

The $n = 4$ term is equal to 5×10^{-7}, so three terms will give five-decimal-place accuracy (0.60234).

12. Use the geometric series with $x = -u^4$ to get

$$f(x) = \int_0^x u \left[\sum_{n=0}^{\infty} (-u^4)^n \right] du = \sum_{n=0}^{\infty} (-1)^n \int_0^x u^{4n+1} du = \sum_{n=0}^{\infty} \frac{(-1)^n x^{4n+2}}{4n+2}$$

15. $\displaystyle\int_0^\infty e^{-a^2 x^2} \cos bx\, dx = \sum_{n=0}^{\infty} \frac{(-1)^n b^{2n}}{(2n)!} \int_0^\infty x^{2n} e^{-a^2 x^2} dx = \sum_{n=0}^{\infty} \frac{(-1)^n b^{2n}}{(2n)!} \cdot \frac{1 \cdot 3 \cdot 5 \cdots (2n-1)\sqrt{\pi}}{2^{2n+1} a^{2n+1}}$

But

$$\frac{1 \cdot 3 \cdot 5 \cdots (2n-1)}{(2n)(2n-1)(2n-2)\cdots(1)} = \frac{1}{(2n)(2n-2)(2n-4)\cdots(2)} = \frac{1}{2^n n!}$$

so

$$\int_0^\infty e^{-a^2 x^2} \cos bx\, dx = \frac{\sqrt{\pi}}{2a} \sum_{n=0}^\infty \frac{(-1)^n}{n!} \left(\frac{b^2}{4a^2}\right)^n = \frac{\sqrt{\pi}}{2a} e^{-b^2/4a^2}$$

18. The Maclaurin expansion of $\ln(1 + \kappa R)$ is

$$\ln(1 + \kappa R) = \kappa R - \frac{(\kappa R)^2}{2} + \frac{(\kappa R)^3}{3} - \frac{(\kappa R)^4}{4} + \frac{(\kappa R)^5}{5} - \cdots$$

and so

$$\ln(1 + \kappa R) - \kappa R + \frac{(\kappa R)^2}{2} = \frac{(\kappa R)^3}{3} - \frac{(\kappa R)^4}{4} + \frac{(\kappa R)^5}{5} - \cdots$$

and

$$\frac{3}{(\kappa R)^3}\left[\ln(1 + \kappa R) - \kappa R + \frac{(\kappa R)^2}{2}\right] = 1 - \frac{3\kappa R}{4} + \frac{3}{5}(\kappa R)^2 - \cdots$$

2.9   Asymptotic Expansions

1. We have that $I(x) = \int_x^\infty e^{-u^2}\, du$. Integrate by parts with "u" $= \dfrac{1}{u}$ and "dv" $= ue^{-u^2}\, du$ to obtain

$$I(x) = \left[-\frac{e^{-u^2}}{2u}\right]_x^\infty - \frac{1}{2}\int_x^\infty \frac{e^{-u^2}}{u^2}\, du = \frac{e^{-x^2}}{2x} - \frac{1}{2}\int_x^\infty \frac{e^{-u^2}}{u^2}\, du$$

Now integrate by parts again with "u" $= \dfrac{1}{u^3}$ and "dv" $= ue^{-u^2}\, du$ to obtain

$$I(x) = \frac{e^{-x^2}}{2x} - \frac{1}{2}\left[-\frac{1}{2}\frac{e^{-u^2}}{u^3}\right]_x^\infty + \frac{1}{2}\int_x^\infty \frac{3}{2}\frac{e^{-u^2}}{u^4}\, du = \frac{e^{-x^2}}{2x} - \frac{1}{4}\frac{e^{-x^2}}{x^3} + \frac{3}{4}\int_x^\infty \frac{e^{-u^2}}{u^4}\, du$$

and so on.

3. Let $xt = u$. Then

$$\int_0^\infty \frac{te^{-xt}}{t^2 + 1}\, dt = \frac{1}{x^2}\int_0^\infty \frac{ue^{-u}\, du}{1 + \dfrac{u^2}{x^2}} = \frac{1}{x^2}\sum_{n=0}^\infty \frac{(-1)^n}{x^{2n}}\int_0^\infty u^{2n+1}e^{-u}\, du$$

$$\approx \frac{1}{x^2}\sum_{n=0}^\infty (-1)^n \frac{(2n+1)!}{x^{2n}}$$

6. Equation 12 is $xe^x E_1(x) = \displaystyle\sum_{n=0}^\infty \frac{(-1)^n n!}{x^n}$

The successive terms are

1	-0.05	0.005	-0.00075
0.00015	-0.0000375	0.00001125	-3.93750×10^{-6}
1.57500×10^{-6}	-7.08750×10^{-7}	3.54375×10^{-7}	-1.94906×10^{-7}
1.16944×10^{-7}	-7.60134×10^{-8}	5.32094×10^{-8}	-3.99071×10^{-8}
3.19256×10^{-8}	-2.71368×10^{-8}	2.44231×10^{-8}	-2.32020×10^{-8}
2.32020×10^{-8}	-2.43621×10^{-8}	2.67983×10^{-8}	-3.0818×10^{-8}
3.69816×10^{-8}	-4.62270×10^{-8}		

The first 25 partial sums are

1.000000	0.950000	0.955000	0.954250	0.954400
0.954363	0.954374	0.95437	0.954371	0.954371
0.954371	0.954371	0.954371	0.954371	0.954371
0.954371	0.954371	0.954371	0.954371	0.954371
0.954371	0.954371	0.954371	0.954371	0.954371

9. Integrate $F(s) = \displaystyle\int_0^\infty f(x)e^{-sx}\, dx$ by parts letting "u" $= f(x)$ and "dv" $= e^{-sx}\, dx$ to obtain

$$
\begin{aligned}
F(s) &= \left[f(x)\left(-\frac{1}{s}e^{-sx}\right) \right]_0^\infty + \frac{1}{s}\int_0^\infty f'(x)e^{-sx}dx \\
&= \frac{f(0)}{s} + \frac{1}{s}\left\{ \left[-\frac{f'(x)}{s}e^{-sx} \right]_0^\infty + \frac{1}{s}\int_0^\infty f''(x)e^{-sx}dx \right\} \\
&= \frac{f(0)}{s} + \frac{f'(0)}{s^2} + \frac{1}{s^2}\int_0^\infty f''(x)e^{-sx}dx
\end{aligned}
$$

and so on.

If $f(x) = \cos x$, we have

$$
F(s) = \frac{1}{s} - \frac{1}{s^3} + \frac{1}{s^5} - \frac{1}{s^5}\int_0^\infty e^{-sx}\sin x\, dx
$$

Because $|\sin x| \le 1$,

$$
\frac{1}{s^5}\int_0^\infty e^{-sx}\sin x\, dx \le \frac{1}{s^5}\int_0^\infty e^{-sx}\, dx = \frac{1}{s^6}
$$

The Maclaurin expansion of $s/(1+s^2)$ is the same as $F(s)$ above. To calculate $F(10)$ to 6-place accuracy, we use four terms in the expansion of $F(s)$ to obtain 0.099 009 900. The exact value is $10/101 = 0.099\,099\,01$. These two results differ by 10^{-8}.

CHAPTER 3

Functions Defined As Integrals

3.1   The Gamma Function

1. Let $au = t$. Then

$$\int_0^\infty e^{-au} u^{3/2} du = a^{-5/2} \int_0^\infty e^{-t} t^{3/2} dt = a^{-5/2} \Gamma(5/2) = a^{-5/2} \left[\left(\frac{3}{2}\right) \left(\frac{1}{2}\right) \Gamma\left(\frac{1}{2}\right) \right] = \frac{3\sqrt{\pi}}{4a^{5/2}}$$

3. Let $z^3 = u$. Then

$$\int_0^\infty z^{1/2} e^{-z^3} dz = \frac{1}{3} \int_0^\infty u^{1/6} u^{-2/3} e^{-u} du = \frac{1}{3} \int_0^\infty u^{-1/2} e^{-u} du = \frac{1}{3} \Gamma\left(\frac{1}{2}\right) = \frac{\sqrt{\pi}}{3}$$

6. Let $\ln \dfrac{1}{x} = u$ $(x = e^{-u})$ to write

$$\int_0^1 \frac{dx}{\sqrt{x \ln(1/x)}} = \int_\infty^0 e^{u/2} u^{-1/2}(-e^{-u} du) = \int_0^\infty u^{-1/2} e^{-u/2} du$$

Now let $u/2 = z$ to write

$$\int_0^1 \frac{dx}{\sqrt{x \ln(1/x)}} = \frac{2}{2^{1/2}} \int_0^\infty z^{-1/2} e^{-z} dz = \sqrt{2} \Gamma(1/2) = (2\pi)^{1/2}$$

9. $\displaystyle\lim_{x \to 0+} \int_0^\infty z^{x-1} dz = \lim_{x \to 0+} \int_0^\alpha z^{x-1} dz + \int_\alpha^\infty z^{x-1} e^{-z} dz$ where α is small enough that

$e^{-z} \approx 1$. Then $\displaystyle\lim_{x \to 0+} \int_0^\alpha z^{x-1} dz = \lim_{x \to 0+} \left[\frac{z^x}{x}\right]_0^\alpha = -\infty$ by l'Hôpital's rule.

12. $f(z)$ at $x = z$ is $f(x) = x \ln x - x$. $f'(z) = \dfrac{x}{z} - 1$, so $f'(x) = 0$; $f''(z) = -\dfrac{x}{z^2}$, so $f''(x) = -\dfrac{1}{x}$; and $f'''(z) = \dfrac{2x}{z^3}$, so $f'''(x) = \dfrac{2}{x^2}$. Thus,

$$f(x) = x \ln x - x - \frac{1}{2x}(z - x)^2 + \frac{1}{3}\frac{(z-x)^3}{x^2} + \cdots$$

15. $10!! = 10 \times 8 \times 6 \times 4 \times 2 = 3840$ and $7!! = 7 \times 5 \times 3 \times 1 = 105$

18. Use the result of Problem 2.7.13 with $x = -u$ and $\alpha = 1/2$ to obtain

$$(1 - u)^{-1/2} = \sum_{n=0}^{\infty} \frac{(-1)^n (n - 1/2)! (-u)^n}{n! (-1/2)!}$$

Now use Equation 8 (with $x = n + 1$) to write

$$(2n + 1)! = \frac{2^{2n+1} n! (n + 1/2)!}{\pi^{1/2}}$$

or

$$(2n + 1)(2n)! = \frac{2^{2n+1} n! (n + 1/2)(n - 1/2)!}{\pi^{1/2}}$$

or

$$(n - 1/2)! = \frac{(2n)! \pi^{1/2}}{n! 2^{2n}}$$

Now $\left(-\dfrac{1}{2}\right)! = \Gamma\left(\dfrac{1}{2}\right) = \sqrt{\pi}$, so

$$\frac{(n - 1/2)!}{(-1/2)!} = \frac{(2n)!}{n!} \frac{1}{2^{2n}} = \frac{(2n)!}{n!} \frac{1}{4^n}$$

and so $(1 - u)^{-1/2} = \displaystyle\sum_{n=0}^{\infty} \frac{(2n)!}{(n!)^2} \left(\frac{u}{4}\right)^n$.

21. Using the Maclaurin expansion of $\cos 2xt$, we have

$$
\begin{aligned}
I &= \int_0^{\infty} e^{-at^2} \cos 2xt \, dt = \sum_{n=0}^{\infty} \frac{(-1)^n}{(2n)!} \int_0^{\infty} e^{-at^2} (2xt)^{2n} dt \\
&= \sum_{n=0}^{\infty} \frac{(-1)^n (4x^2/a)^n}{(2n)!} \int_0^{\infty} (at^2)^n e^{-at^2} dt
\end{aligned}
$$

Now let $at^2 = u$ to obtain

$$I = \frac{1}{2a^{1/2}} \sum_{n=0}^{\infty} \frac{(-1)^n (4x^2/a)^n}{(2n)!} \int_0^{\infty} e^{-u} u^{n-1/2} du = \frac{1}{2a^{1/2}} \sum_{n=0}^{\infty} \frac{(-1)^n (4x^2/a)^n}{(2n)!} \Gamma\left(n + \frac{1}{2}\right)$$

But, according to Equation 8, $\dfrac{\Gamma(n + \frac{1}{2})}{(2n)!} = \dfrac{(n - 1/2)!}{(2n)!} = \dfrac{\pi^{1/2}}{2^{2n} n!}$ and so

$$I = \frac{\pi^{1/2}}{2a^{1/2}} \sum_{n=0}^{\infty} \frac{(-1)^n (x^2/a)^n}{n!} = \left(\frac{\pi}{4a}\right)^{1/2} e^{-x^2/a}$$

3.2   The Beta Function

1. Let $z = 1 - u$ in Equation 1 to write

$$B(x,y) = \int_0^1 z^{x-1}(1-z)^{y-1}dz = \int_0^1 (1-u)^{x-1}u^{y-1}du = \int_0^1 z^{y-1}(1-z)^{x-1}dz = B(y,x)$$

3. First write $\Gamma(x)$ as $\Gamma(x) = \int_0^\infty z^{x-1}e^{-z}dz$. Let $z = u^2$ to obtain $\Gamma(x) = 2\int_0^\infty u^{2x-1}e^{-u^2}du$

and $\Gamma(y) = 2\int_0^\infty v^{2y-1}e^{-v^2}dv$ and $\Gamma(x)\Gamma(y) = 4\int_0^\infty\int_0^\infty u^{2x-1}v^{2y-1}e^{-(u^2+v^2)}dudv$.   Now

transform to plane polar coordinates $(u = r\cos\theta,\ v = r\sin\theta,\ dudv = rdrd\theta)$

$$\Gamma(x)\Gamma(y) = 4\int_0^\infty r^{2x-1}r^{2y-1}e^{-r^2}\left[\int_0^{\pi/2}\cos^{2x-1}\theta\sin^{2y-1}\theta\,d\theta\right]rdr = 2B(x,y)\int_0^\infty r^{2x+2y-2}e^{-r^2}rdr$$

Let $r^2 = z$ to obtain $\Gamma(x)\Gamma(y) = B(x,y)\int_0^\infty z^{x+y-1}e^{-z}dz = B(x,y)\Gamma(x+y)$ or $B(x,y) = \dfrac{\Gamma(x)\Gamma(y)}{\Gamma(x+y)}$.

6. $\displaystyle\int_0^1 \sqrt[3]{u(1-u)}\,du = \int_0^1 u^{1/3}(1-u)^{1/3}du = B(4/3,4/3) = \dfrac{\Gamma(\frac{4}{3})\Gamma(\frac{4}{3})}{\Gamma(\frac{8}{3})} + \dfrac{\frac{1}{9}\Gamma(\frac{1}{3})\Gamma(\frac{1}{3})}{\frac{5}{3}\cdot\frac{2}{3}\Gamma(\frac{2}{3})} = \dfrac{\Gamma^2(\frac{1}{3})}{10\Gamma(\frac{2}{3})} =$

0.52999

9. Let $e^{3u} = z$ to obtain

$$\int_{-\infty}^\infty \dfrac{e^{2u}du}{1+e^{3u}} = \dfrac{1}{3}\int_0^\infty \dfrac{z^{-1/3}dz}{1+z} = \dfrac{1}{3}\dfrac{\pi}{\sin(2\pi/3)} = \dfrac{\pi}{3}\dfrac{2}{\sqrt{3}} = \dfrac{2\pi}{3\sqrt{3}}$$

where we have used Equation 5.

12. The plot of $x^{2/3} + y^{2/3} = 1$ is

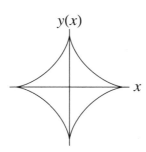

Solving the equation for y gives $y = (1 - x^{2/3})^{3/2}$.   From the plot, we see that the area bounded by the curve is given by 4 times the area in the first quadrant, or area $=$ $4\int_0^1(1-x^{2/3})^{3/2}dx$. Let $x^{2/3} = z$ and $dx = \dfrac{3}{2}z^{1/2}\,dz$, so that

$$\text{Area} = \dfrac{3}{2}\cdot 4\int_0^1 z^{1/2}(1-z)^{3/2}dz = 6B(3/2,5/2) = \dfrac{6\Gamma(3/2)\Gamma(5/2)}{\Gamma(4)} = \dfrac{3}{8}\Gamma^2(1/2) = \dfrac{3\pi}{8}$$

3.3   The Error Function

1. $\mathrm{erf}(-x) = \dfrac{2}{\sqrt{\pi}} \displaystyle\int_0^{-x} e^{-u^2}\, du.$ Let $u = -z$ to obtain $-\dfrac{2}{\sqrt{\pi}} \displaystyle\int_0^{x} e^{-z^2}\, dz = -\mathrm{erf}(x).$

3. Start with

$$\mathrm{Prob}\{-v_{x_0} \le v_x \le v_{x_0}\} = \int_{-v_{x_0}}^{v_{x_0}} f(v_x)\, dv_x = 2\int_0^{v_{x_0}} \left(\frac{m}{2\pi k_{\mathrm{B}}T}\right)^{1/2} e^{-mv_x^2/2k_{\mathrm{B}}T}\, dv_x$$

Let $u = mv_x^2/2k_{\mathrm{B}}T$ to obtain

$$\mathrm{Prob}\{-v_{x_0} \le v_x \le v_{x_0}\} = \frac{2}{\sqrt{\pi}} \int_0^{v_{x_0}(m/2k_{\mathrm{B}}T)^{1/2}} e^{-u^2}\, du = \mathrm{erf}\left\{\left(\frac{m}{2k_{\mathrm{B}}T}\right)^{1/2} v_{x_0}\right\}$$

6. The chance that an experimental observation will lie within two standard deviations from the mean is given by

$$\frac{1}{(2\pi\sigma^2)^{1/2}} \int_{-2\sigma}^{2\sigma} e^{-x^2/2\sigma^2}\, dx = \frac{2}{(2\pi\sigma^2)^{1/2}} \int_0^{2\sigma} e^{-x^2/2\sigma^2}\, dx$$

Let $x = \sigma z$ to obtain $\dfrac{2}{\sqrt{\pi}} \displaystyle\int_0^{\sqrt{2}} e^{-z^2}\, dz = \mathrm{erf}\left(\sqrt{2}\right) = 0.955.$

9. Neglect a factor of $2/\sqrt{\pi}$ for the time being and write $\dfrac{d\,\mathrm{erf}(x)}{dx} = e^{-x^2}$; $\dfrac{d^2\mathrm{erf}(x)}{dx^2} = -2xe^{-x^2}$;
$\dfrac{d^3\mathrm{erf}(x)}{dx^3} = (4x^2 - 2)e^{-x^2}$; $\dfrac{d^4\mathrm{erf}(x)}{dx^4} = (12x - 8x^3)e^{-x^2}$; and $\dfrac{d^5\mathrm{erf}(x)}{dx^5} = (16x^4 - 48x^2 + 12)e^{-x^2}$.
Thus we have $\mathrm{erf}(x) = \dfrac{2}{\sqrt{\pi}}\left(x - \dfrac{x^3}{3} + \dfrac{x^5}{10} + \cdots\right)$

12. Let $t + x^2 = u^2$ and $dt = 2u\, du$ and write

$$\int_0^\infty \frac{e^{-at}\, dt}{\sqrt{t + x^2}} = \int_x^\infty \frac{e^{-a(u^2 - x^2)}2u\, du}{u} = 2e^{ax^2}\int_x^\infty e^{-au^2}\, du$$

$$= \frac{2}{a^{1/2}}e^{ax^2}\int_{a^{1/2}x}^\infty e^{-u^2}\, du = \left(\frac{\pi}{a}\right)^{1/2} e^{ax^2}\mathrm{erfc}\left(\sqrt{a}\,x\right)$$

15. Complete the square in the exponent and write $ax + \dfrac{x^2}{4} = \left(\dfrac{x}{2} + a\right)^2 - a^2$. Now

$$\int_0^\infty e^{-ax - x^2/4}\, dx = e^{a^2}\int_0^\infty e^{-(x/2 + a)^2}\, dx = 2e^{a^2}\int_a^\infty e^{-u^2}\, du = \sqrt{\pi}e^{a^2}\mathrm{erfc}(a)$$

18. Letting $a = (1-i)/2^{1/2}$ in the integral gives $\displaystyle\int_0^\infty e^{ix^2}\, dx = \left(\frac{\pi}{2}\right)^{1/2}\frac{1}{1-i} = \left(\frac{\pi}{2}\right)^{1/2}\frac{1+i}{2}.$
Using the identity $e^{iy} = \cos y + i\sin y$, we can write the integral as

$$\int_0^\infty \cos x^2\, dx + i\int_0^\infty \sin x^2\, dx = \frac{1}{2}\left(\frac{\pi}{2}\right)^{1/2} + \frac{i}{2}\left(\frac{\pi}{2}\right)^{1/2}$$

3.4   The Exponential Integral

1. Integrate $E_n(x)$ by parts, letting "u" $= z^{-n}$ and "dv" $= e^{-xz}dz$.

$$\int_1^\infty \frac{e^{-xz}}{z^n}dz = \left|-\frac{e^{-xz}}{xz^n}\right|_1^\infty - \frac{n}{x}\int_1^\infty \frac{e^{-xz}dz}{z^{n+1}}$$

or

$$E_n(x) = \frac{e^{-x}}{x} - \frac{n}{x}E_{n+1}(x)$$

which is Equation 3.

Starting with Equation 1,

$$\frac{dE_n(x)}{dx} = -\int_1^\infty \frac{e^{-xz}}{z^{n-1}}dz = -E_{n-1}(x)$$

3. Start with $Ei(x) = \int_{-\infty}^x \frac{e^u}{u}du$ for $x > 0$. Let $u = -t$ to obtain $Ei(x) = \int_{-x}^\infty \frac{e^{-t}}{t}dt$ for $x > 0$.

6.

$$\begin{aligned}
\int_0^x \frac{1-e^{-t}}{t}dt &= \lim_{\epsilon \to 0}\left[\int_\epsilon^x \frac{dt}{t} - \int_\epsilon^z \frac{e^{-t}}{t}dt\right]\\
&= \lim_{\epsilon \to 0}\left[\ln x - \ln \epsilon - \int_\epsilon^\infty \frac{e^{-t}}{t}dt + \int_z^\infty \frac{e^{-t}}{t}dt\right]\\
&= \lim_{\epsilon \to 0}\left[\ln x - \ln \epsilon - E_1(\epsilon) + E_1(x)\right]\\
&= \lim_{\epsilon \to 0}\left[\ln x - \ln \epsilon + \gamma + \ln \epsilon + O(\epsilon) + E_1(x)\right]\\
&= \ln x + \gamma + E_1(x)
\end{aligned}$$

given that $\int_0^\infty e^{-x}\ln x\, dx = -\gamma$

9.

$$\begin{aligned}
\int_0^\infty e^{-at}\left[-\int_t^\infty \frac{\cos u}{u}du\right]dt &= -\int_0^\infty e^{-at}\left[\int_1^\infty \frac{\cos tx}{x}dx\right]dt\\
&= -\int_1^\infty \frac{1}{x}\left[\int_0^\infty e^{-at}\cos tx\, dt\right]dx = -\int_1^\infty \frac{1}{x}\frac{a}{(a^2+x^2)}dx\\
&= -\frac{1}{2a}\left|\ln \frac{x^2}{a^2+x^2}\right|_1^\infty = -\frac{1}{2a}\ln(1+a^2)
\end{aligned}$$

12. The figure on the next page shows $I(n) = \int_\epsilon^\infty e^{-x}\ln x\, dx$ calculated with *Mathematica* plotted against n, where $\epsilon = 10^{-n}$.

15. Start with $E_1(x) = \int_x^\infty \frac{e^{-u}}{u}du$, which we write as

$$E_1(x) = \lim_{\epsilon \to 0}\left[\int_\epsilon^\infty \frac{e^{-u}}{u}du - \int_\epsilon^x \frac{e^{-u}}{u}du\right]$$

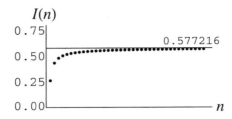

Now use the relation in Problem 14 to write

$$E_1(x) = \lim_{\epsilon \to 0} \left[\int_\epsilon^\infty e^{-x} \ln x \, dx - \ln \epsilon - \int_\epsilon^x \frac{e^{-u}}{u} du \right]$$

and the relation $\displaystyle \lim_{\epsilon \to 0} \int_\epsilon^\infty e^{-x} \ln x \, dx = -\gamma$ to write

$$E_1(x) = -\gamma - \lim_{\epsilon \to 0} \left[\ln \epsilon + \int_\epsilon^x \frac{e^{-u}}{u} du \right]$$

For small values of x, the variable u never gets large in the above integral, so we can expand the exponential and integrate term by term to obtain

$$E_1(x) = -\gamma - \lim_{\epsilon \to 0} \left[\ln \epsilon + \ln x - \ln \epsilon + \sum_{n=1}^\infty \frac{(-1)^n (x^n - \epsilon^n)}{n \, n!} \right] = -\gamma - \ln x - \sum_{n=1}^\infty \frac{(-1)^n x^n}{n \, n!}$$

3.5   Elliptic Integrals

1. The acceleration is $d^2 s/dt^2$ and the force is $-mg \sin \theta$ (see Figure 3.15). Using $s = l\theta$, we obtain $ml\dfrac{d^2\theta}{dt^2} = -mg \sin \theta$ for the equation of motion.

3. Use $\cos x = 1 - \dfrac{x^2}{2} + \cdots$ and write $\dfrac{d\theta}{(\theta_0^2 - \theta^2)^{1/2}} = \left(\dfrac{g}{l}\right)^{1/2} dt$, which gives

$$\int_{\theta_0}^\theta \frac{d\phi}{(\theta_0^2 - \phi^2)^{1/2}} = \left[\sin^{-1} \frac{\phi}{\theta_0} \right]_{\theta_0}^\theta = \sin^{-1} \frac{\theta}{\theta_0} - \frac{\pi}{2} = \left(\frac{g}{l}\right)^{1/2} t$$

or $\theta = \theta_0 \sin\left[\dfrac{\pi}{2} + \left(\dfrac{g}{l}\right)^{1/2} t \right] = \theta_0 \cos\left(\dfrac{g}{l}\right)^{1/2} t$. The period is the time it takes to go through one cycle, or when $t = 2\pi \left(\dfrac{l}{g}\right)^{1/2}$.

6. The length of the curve $y(\theta) = \cos \theta$ is given by

$$l = \int_0^{2\pi} \{1 + [\, y'(\theta)\,]^2\}^{1/2} d\theta = \int_0^{2\pi} (1 + \sin^2 \theta)^{1/2} d\theta = 4 \int_0^{\pi/2} (1 + \sin^2 \theta)^{1/2} d\theta$$

Let $\theta = \pi/2 - \phi$ to write

$$\begin{aligned} l &= 4 \int_0^{\pi/2} (1 + \cos^2 \phi)^{1/2} d\phi = 4 \int_0^{\pi/2} (2 - \sin^2 \phi)^{1/2} d\phi = 4\sqrt{2} \int_0^{\pi/2} \left(1 - \frac{1}{2} \sin^2 \phi \right)^{1/2} d\phi \\ &= 4\sqrt{2} E(1/\sqrt{2}) = 7.640 \end{aligned}$$

9. First use the identity $\cos\theta = 2\cos^2\dfrac{\theta}{2} - 1$ to write $I = \displaystyle\int_0^{\pi/2} \dfrac{d\theta}{[4 - 2\cos^2(\theta/2)]^{1/2}}$. Now

let $\dfrac{\theta}{2} = \dfrac{\pi}{2} - \phi$ to get

$$I = \int_{\pi/4}^{\pi/2} \frac{d\phi}{(1 - \frac{1}{2}\sin^2\phi)^{1/2}} = K(1/\sqrt{2}) - F(1/\sqrt{2}, \pi/4) = 1.8541 - 0.8260 = 1.0281$$

15. Let $x = \sin\theta$, so that $dx = \cos\theta\, d\theta$, or $d\theta = \dfrac{dx}{\sqrt{1 - x^2}}$,

Thus, $F(k, z) = \displaystyle\int_0^{\sin\phi=z} \dfrac{dx}{\sqrt{1 - x^2}\sqrt{1 - kx^2}}$.

Similarly, the same substitution gives $E(k, z) = \displaystyle\int_0^{\sin\phi=z} \sqrt{\dfrac{1 - k^2x^2}{1 - x^2}}\, dx$.

3.6 The Dirac Delta Function

1. Integrate by parts: $\displaystyle\int_{-\infty}^{\infty} f(x)x\delta'(x)dx = -\int_{-\infty}^{\infty} f(x)\delta(x)\, dx = -f(0)$.

Let $ax = u$:

$$\int_{-\infty}^{\infty} f(x)\delta(ax)\, dx = \frac{1}{a}\int_{-\infty}^{\infty} f(u)\delta(u)\, du$$

3. Let $ax = u$: $\displaystyle\int_{-\infty}^{\infty} f(x)\delta(ax)\, dx = \frac{1}{a}\int_{-\infty}^{\infty} f(u)\,\delta(u)\, du$

6. Let $nx = u$ and write

$$\frac{n}{\pi}\int_{-\infty}^{\infty} \frac{e^{-x^2}\, dx}{1 + n^2x^2} = \frac{1}{\pi}\int_{-\infty}^{\infty} \frac{e^{-u^2/n^2}\, du}{1 + u^2} = e^{1/n^2}\operatorname{erfc}\left(\frac{1}{n}\right) \to 1 \quad \text{as} \quad n \to \infty$$

3.7 Bernoulli Numbers and Bernoulli Polynomials

1.

$$\begin{aligned}
(1 - 2^{1-n})\zeta(n) &= \sum_{k=1}^{\infty}\frac{1}{k^n} - 2\sum_{k=1}^{\infty}\frac{1}{(2k)^n} \\[2mm]
&= \sum_{k=0}^{\infty}\frac{1}{(2k+1)^n} + \sum_{k=1}^{\infty}\frac{1}{(2k)^n} - 2\sum_{k=1}^{\infty}\frac{1}{(2k)^n} \\[2mm]
&= \sum_{k=0}^{\infty}\frac{1}{(2k+1)^n} - \sum_{k=1}^{\infty}\frac{1}{(2k)^n} = \sum_{k=1}^{\infty}\frac{(-1)^{k+1}}{k^n}
\end{aligned}$$

3. Start with the generating functions of $B_n(x + 1)$ and $B_n(x)$:

$$\begin{aligned}
\frac{te^{(x+1)t}}{e^t - 1} - \frac{te^{xt}}{e^t - 1} &= \sum_{n=0}^{\infty}[B_n(x+1) - B_n(x)]\frac{t^n}{n} = \frac{te^{xt}(e^t - 1)}{e^t - 1} = te^{xt} \\[2mm]
&= \sum_{n=0}^{\infty}\frac{x^n t^{n+1}}{n!} = \sum_{n=1}^{\infty}\frac{x^{n-1}t^n}{(n-1)!} = \sum_{n=1}^{\infty}nx^{n-1}\frac{t^n}{n!}
\end{aligned}$$

so $B_n(x+1) - B_n(x) = nx^{n-1}$.

6. The generating function of $(-1)^n B_n(-x)$ is obtained from Equation 7 by replacing t by $-t$ and x by $-x$:

$$\frac{(-t)e^{(-x)(-t)}}{e^{-t}-1} = \sum_{n=0}^{\infty} B_n(-x)\frac{(-t)^n}{n!} = \sum_{n=0}^{\infty}(-1)^n B_n(-x)\frac{t^n}{n!}$$

But

$$\frac{(-t)e^{(-x)(-t)}}{e^{-t}-1} = \frac{-te^{xt}e^t}{1-e^t} = \frac{te^{(x+1)}}{e^t-1} = \sum_{n=0}^{\infty} B_n(x+1)\frac{t^n}{n!}$$

Now we use the relation $B_n(x+1) = B_n(x) + nx^{n-1}$. (See Problem 3.)

9. Simply substitute Equation 17 into Equation 11.

12. We have that $f(x) = \dfrac{1}{x^{1/2}}$; $f'(x) = -\dfrac{1}{2x^{3/2}}$; $f''(x) = \dfrac{3}{4x^{5/2}}$; $f'''(x) = -\dfrac{15}{8x^{7/2}}$;

$f^{(4)}(x) = \dfrac{105}{16x^{9/2}}$; and $f^{(5)}(x) = -\dfrac{9\cdot 105}{32x^{11/2}}$. Thus, using Equation 15

$$\sum_{x=1}^{n}\frac{1}{x^{1/2}} \approx 2(n^{1/2}-1) + \frac{1}{2}\left(\frac{1}{n^{1/2}}+1\right) + \frac{1}{12}\left(-\frac{1}{2n^{3/2}}+\frac{1}{2}\right) - \frac{1}{720}\left(-\frac{15}{8n^{7/2}}+\frac{15}{8}\right)$$
$$+ \frac{1}{30240}\left(-\frac{9\cdot 105}{32n^{11/2}}+\frac{9\cdot 105}{32}\right) + \cdots$$

The following figure shows the exact value of the sum and its Euler-Maclaurin approximation plotted against n. The two sets of numbers are essentially indistinguishable.

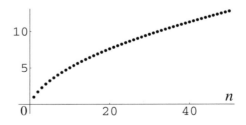

15. Using the Euler-Maclaurin summation formula (Equation 15), we have

$$\sum_{J=0}^{\infty}(2J+1)e^{-\Theta J(J+1)/T} \approx \int_{0}^{\infty} dJ(2J+1)e^{-\Theta J(J+1)/T} + \frac{1}{2}\left[f(\infty)+f(0)\right]$$
$$-\frac{1}{12}\left\{\left.\frac{df}{dJ}\right|_{J=0} - \left.\frac{df}{dJ}\right|_{J=\infty}\right\} + \frac{1}{720}\left\{\left.\frac{d^3f}{dJ^3}\right|_{J=0} - \left.\frac{d^3f}{dJ^3}\right|_{J=\infty}\right\} + \cdots$$

Taking derivatives, we have

$$\frac{df}{dJ} = -\frac{\Theta}{T}(2J+1)^2 e^{-\Theta J(J+1)/T} + 2e^{-\Theta J(J+1)/T}$$

$$\frac{d^2 f}{dJ^2} = -6\left(\frac{\Theta}{T}\right)(2J+1)e^{-\Theta J(J+1)/T} + \left(\frac{\Theta}{T}\right)^2(2J+1)^3 e^{-\Theta J(J+1)/T}$$

$$\frac{d^3 f}{dJ^3} = -12\left(\frac{\Theta}{T}\right)e^{-\Theta J(J+1)/T} + O\left[\left(\frac{\Theta}{T}\right)^2\right]$$

$$\frac{d^2 f}{dJ^4} = O\left[\left(\frac{\Theta}{T}\right)^2\right]$$

Now evaluate $f(J)$ and the derivatives at $J = 0$ and $J = \infty$ and from Equation 13, we have that

$$q \approx \frac{T}{\Theta} + \left(\frac{1}{2} - \frac{1}{6}\right) + \left(\frac{1}{12} - \frac{12}{720}\right)\frac{\Theta}{T} + O\left[\left(\frac{\Theta}{T}\right)^2\right]$$

$$= \left(\frac{T}{\Theta}\right)\left[1 + \frac{1}{3}\left(\frac{\Theta}{T}\right) + \frac{1}{15}\left(\frac{\Theta}{T}\right)^2 + \cdots\right]$$

CHAPTER 4

Complex Numbers and Complex Functions

4.1   Complex Numbers and the Complex Plane

1. (a) $z_1 z_2 = x_1 x_2 - y_1 y_2 + i(x_1 y_2 + x_2 y_2)$ and

$(z_1 z_1)^* = x_1 x_2 - y_1 y_2 - i(x_1 y_2 + x_2 y_1) = (x_1 - iy_1)(x_2 - iy_2) = z_1^* z_2^*$

(b) $\left(\dfrac{1}{z}\right)^* = \dfrac{1}{x - iy} = \dfrac{1}{z^*}$

(c) $z^n = (x + iy)(x + iy) \cdots (x + iy)$ n times.

$(z^n)^* = (x - iy)(x - iy) \cdots (x - iy)$ n times $= (z^*)^n$.

3.
$$
\begin{aligned}
|z_1 z_2| &= |(x_1 + iy_1)(x_2 + iy_2)| = |x_1 x_2 - y_1 y_2 + i(x_1 y_2 + x_2 y_1)| \\
&= [(x_1 x_2 - y_1 y_2)^2 + (x_1 y_2 + x_2 y_1)^2]^{1/2} \\
&= [x_1^2 x_2^2 - 2x_1 x_2 y_1 y_2 + y_1^2 y_2^2 + x_1^2 y_2^2 + 2x_1 x_2 y_1 y_2 + x_2^2 y_1^2]^{1/2} \\
&= [x_1^2(x_2^2 + y_2^2) + y_1^2(x_2^2 + y_2^2)]^{1/2} \\
&= [(x_1^2 + y_1^2)(x_2^2 + y_2^2)]^{1/2} = |x_1 + iy_1| |x_2 + iy_2| = |z_1| |z_2|
\end{aligned}
$$

6. $|z + i| = |x + i(y + 1)| = x^2 + (y + 1)^2$. The region $1 < |z + i| \le 3$ represents a circular annulus centered at $x = 0, y = -1$. One boundary is the circle of radius 3, and the other is the circle of radius 1. The $r = 3$ circle is included in the region, but the $r = 1$ circle is not.

9. $\operatorname{Re}(z + i) = x \ge 2$ represents the region of the complex plane to the right of and including the vertical line $x = 2$.

12. (a) We have that $\cos(\pi/4) = 1/\sqrt{2}$ and $\sin(\pi/4) = 1/\sqrt{2}$, so that $z = 2\left(\cos\dfrac{\pi}{4} + i\sin\dfrac{\pi}{4}\right) = \sqrt{2}\,(1 + i)$.

(b) We have that $\cos(2\pi/3) = -1/2$ and $\sin(2\pi/3) = \sqrt{3}/2$, so that
$z = \sqrt{3}\left(\cos\dfrac{2\pi}{3} + i\sin\dfrac{2\pi}{3}\right) = -\dfrac{\sqrt{3}}{2} + i\dfrac{3}{2}.$

(c) $\cos(3\pi/2) = 0$ and $\sin(3\pi/2) = -1$, so that $z = \cos\dfrac{3\pi}{2} + i\sin\dfrac{3\pi}{2} = -i$.

(d) $\cos(\pi/6) = \sqrt{3}/2$ and $\sin(\pi/6) = 1/2$, so that $z = 2\left(\cos\dfrac{\pi}{6} + i\sin\dfrac{\pi}{6}\right) = \sqrt{3} + i$.

15. Start by writing $z_1 = \alpha_1 + i\beta_1$ and $z_2 = \alpha_2 + i\beta_2$, so that $z_1 + z_2 = (\alpha_1 + \alpha_2) + i(\beta_1 + \beta_2)$. We have that $|z_1 + z_2| = [(\alpha_1 + \alpha_2)^2 + (\beta_1 + \beta_2)^2]^{1/2}$, $|z_1| = (\alpha_1^2 + \beta_1^2)^{1/2}$, and $|z_2| = (\alpha_2^2 + \beta_2^2)^{1/2}$. We now assert that

$$[\,(\alpha_1 + \alpha_2)^2 + (\beta_1 + \beta_2)^2\,]^{1/2} \leq (\alpha_1^2 + \beta_1^2)^{1/2} + (\alpha_2^2 + \beta_2^2)^{1/2}$$

To see if this inequality is so, square both sides (to get rid of the square root term on the left) to get

$$(\alpha_1 + \alpha_2)^2 + (\beta_1 + \beta_2)^2 \leq \alpha_1^2 + \beta_1^2 + \alpha_2^2 + \beta_2^2 + 2(\alpha_1^2 + \beta_1^2)^{1/2}(\alpha_2^2 + \beta_2^2)^{1/2}$$

or $\alpha_1\alpha_2 + \beta_1\beta_2 \leq (\alpha_1^2 + \beta_1^2)^{1/2}(\alpha_2^2 + \beta_2^2)^{1/2}$. Now square both sides again (to get rid of the square root terms) to get $2\alpha_1\alpha_2\beta_1\beta_2 \leq \beta_1^2\alpha_2^2 + \alpha_1^2\beta_2^2$ or $0 \leq (\alpha_1\beta_2 - \alpha_2\beta_1)^2$, which is indeed true.

4.2   Functions of a Complex Variable

1. (a) $z^3 = (x + iy)^3 = x^3 + 3x^2yi + 3xy^2i^2 + i^3y^3 = (x^3 - 3xy^2) + i(3x^2y - y^3)$

(b)

$$
\begin{aligned}
\frac{1}{(1-z)^2} &= \frac{1}{(1-x)^2 - y^2 - 2i(1-x)y} \\
&= \frac{1}{(1-x)^2 - y^2 - 2i(1-x)y}\,\frac{(x-1)^2 - y^2 + 2i(1-x)y}{(x-1)^2 - y^2 + 2i(1-x)y} \\
&= \frac{(1-x)^2 - y^2}{[(1-x)^2 + y^2]^2} + \frac{2i(1-x)y}{[(1-x)^2 + y^2]}
\end{aligned}
$$

(c) $\dfrac{z^*}{z} = \dfrac{x - iy}{x + iy} = \dfrac{x - iy}{x + iy} \cdot \dfrac{x - iy}{x - iy} = \dfrac{x^2 - y^2}{x^2 + y^2} - \dfrac{2xy\,i}{x^2 + y^2}$

3. $f(z) = \dfrac{x - iy}{x^2 + y^2} = \dfrac{z^*}{|z|^2} = \dfrac{1}{z}$

6. We have that $w = x^2 - y^2 + 2ixy$, thus $z = 2 + i$ ($x = 2$, $y = 1$) corresponds to $w = 3 + 4i$, and $z = 1 - i$ ($x = 1$, $y = -1$) corresponds to $w = -2i$.

9. The range of w is the reciprocal of $1/|z|$, or $|w| > 1/2$.

12. We have that $w = x^2 - y^2 - 2ixy$.

(a) For $x = 1$ and $1 < y < 3$, we have $1 - y^2 - 2iy$.

(b) For $x = 2$ and $1 < y < 3$, we have $4 - y^2 - 4iy$.

(c) For $y = 1$ and $1 < x < 2$, we have $x^2 - 1 - 2ix$.

(d) For $y = 3$ and $1 < x < 2$, we have $x^2 - 9 - 6ix$.

These curves are shown in the figure on the next page.

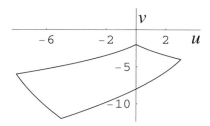

15. We have that $u(x, y) = (x^2 + y^2)^{1/2} = u_0$ and $v(x, y) = y - x = v_0$. Differentiate each with respect to x to obtain

$$\frac{x}{(x^2 + y^2)^{1/2}} + \frac{y}{(x^2 + y^2)^{1/2}}\frac{dy}{dx} = 0 \quad \text{and} \quad \frac{dy}{dx} - 1 = 0$$

The product of dy/dx does not equal -1.

4.3 Euler's Formula and the Polar Form of Complex Numbers

1. (a) $r = 6$ and $\theta_0 = \pi/2$

(b) $r = (16 + 2)^{1/2} = (18)^{1/2}$ and $\theta_0 = \tan^{-1}\left(-\frac{\sqrt{2}}{4}\right) = 340.5°$. (This angle lies in the fourth quadrant.)

(c) $r = (1+4)^{1/2} = 5^{1/2}$ and $\theta_0 = \tan^{-1} 2 = 243.4°$. (This angle lies in the third quadrant.)

(d) $r = (\pi^2 + e^2)^{1/2}$ and $\theta_0 = \tan^{-1}\frac{e}{\pi} = 40.87°$

3. Differentiate $f(\theta) = \cos\theta + i\sin\theta$ to get $f'(\theta) = -\sin\theta + i\cos\theta = if(\theta)$. Integrate to get $\ln f(\theta) = i\theta + $ constant. To determine the constant, we have that $f(0) = 1$, which leads to $\ln f(0) = 0$. Thus "constant" $= 0$ and so $f(\theta) = e^{i\theta}$.

6.
$$\int_0^{2\pi} \sin nx \sin mx\, dx = -\frac{1}{4}\left[\int_0^{2\pi}(e^{inx} - e^{-inx})(e^{imx} - e^{-imx})dx\right]$$
$$= -\frac{1}{4}\left[\int_0^{2\pi}\{e^{i(n+m)x} + e^{-i(n+m)x} - e^{i(m-n)x} - e^{i(n-m)x}\}dx\right]$$

Each of these integrals vanishes if $m \neq n$ because each integral goes over complete cycles of $\cos x$ or $\sin x$. The other integrals are similar.

9. First express $2 - i$ in polar form: $z = re^{i\theta} = 5^{1/2}e^{5.8195\,i}$. Thus,

$$z^{10} = 5^5 e^{58.1951\,i} = 5^5 e^{1.6467 i} = 5^5(-0.07584 + 0.99712\,i) = -237 + 3116\,i$$

12. We have that $i = e^{i\pi/2 + 2\pi ni}$ for $n = 0, \pm 1, \pm 2, \ldots$. The fourth root of i then is $i^{1/4} = e^{i\pi/8 + \pi ni/2}$ for $n = 0, 1, 2, 3$. Therefore, $i^{1/4} = \cos\frac{\pi}{8} + i\sin\frac{\pi}{8}$, $\cos\frac{5\pi}{8} + i\sin\frac{5\pi}{8}$, $\cos\frac{9\pi}{8} + i\sin\frac{9\pi}{8}$, $\cos\frac{13\pi}{8} + i\sin\frac{13\pi}{8}$. These roots are shown in the figure on the next page.

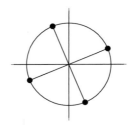

15.

$$e^{(z_1+z_2)} = 1 + (z_1 + z_2) + \frac{(z_1 + z_2)^2}{2!} + \frac{(z_1 + z_2)^3}{3!} + \cdots$$

$$= 1 + z_1 + z_2 + \frac{z_1^2}{2!} + \frac{z_2^2}{2!} + z_1 z_2 + \frac{z_1^3}{3!} + \frac{z_1^2 z_2}{2!} + \frac{z_1 z_2^2}{2!} + \frac{z_2^3}{3!} + O(z_1^4, z_2^4)$$

$$= 1 + z_1 + \frac{z_1^2}{2!} + \frac{z_1^3}{3!} + O(z_1^4) + z_2 + \frac{z_2^2}{2!} + \frac{z_2^3}{3!} + O(z_2^4)$$

$$+ z_1 z_2 + \frac{z_1^2 z_2}{2!} + \frac{z_1 z_2^2}{2!} + O(z_1^3 z_2, z_1^2 z_2^2, z_1 z_2^3)$$

$$e^{z_1} e^{z_2} = \left(1 + z_1 + \frac{z_1^2}{2!} + \frac{z_1^3}{3!} + O(z_1^4) \right) \left(1 + z_2 + \frac{z_2^2}{2!} + \frac{z_2^3}{3!} + O(z_2^4) \right)$$

$$= 1 + z_1 + \frac{z_1^2}{2!} + \frac{z_1^3}{3!} + O(z_1^4) + z_2 + \frac{z_2^2}{2!} + \frac{z_2^3}{3!} + O(z_2^4)$$

$$+ z_1 z_2 + \frac{z_1^2 z_2}{2!} + \frac{z_1 z_2^2}{2!} + O(z_1^3 z_2, z_1^2 z_2^2, z_1 z_2^3)$$

4.4   Trigonometric and Hyperbolic Functions

1. $\cos 2z = \dfrac{e^{2iz} + e^{-2iz}}{2} = \dfrac{(e^{iz} + e^{-iz})^2 - 2}{2} = 2\cos^2 z - 1$

3.

$$\cosh z = \frac{e^z + e^{-z}}{2} = \frac{e^x e^{iy} + e^{-x} e^{-iy}}{2} = \frac{e^x(\cos y + i\sin y) + e^{-x}(\cos y - i\sin y)}{2}$$

$$= \left(\frac{e^x + e^{-x}}{2} \right) \cos y + i \left(\frac{e^x - e^{-x}}{2} \right) \sin y$$

$$= \cosh x \cos y + i \sinh x \sin y$$

6. $\sinh z = \dfrac{e^z - e^{-z}}{2} = \dfrac{1}{2} \left[\displaystyle\sum_{n=0}^{\infty} \frac{z^n}{n!} - \sum_{n=0}^{\infty} \frac{(-z)^n}{n!} \right] = \dfrac{1}{2} \sum_{n=0}^{\infty} \frac{[z^n - (-z)^n]}{n!} = \sum_{n=0}^{\infty} \frac{z^{2n+1}}{(2n+1)!}$

$$\cosh z = \frac{e^z + e^{-z}}{2} = \frac{1}{2} \sum_{n=0}^{\infty} \frac{z^n + (-z)^n}{n!} = \sum_{n=0}^{\infty} \frac{z^{2n}}{(2n)!}$$

9. $\cos\left(\dfrac{\pi}{2} - 2i \right) = \cos\dfrac{\pi}{2}\cos(-2i) + \sin\dfrac{\pi}{2}\sin(2i) = \sin 2i = i\sinh(2)$

12. (a) $\operatorname{sech} iu = \dfrac{1}{\cosh iu} = \dfrac{2}{e^{iu}+e^{-iu}} = \dfrac{1}{\cos u} = \sec u$

(b) $\coth iu = \dfrac{\cosh iu}{\sinh iu} = \dfrac{e^{iu}+e^{-iu}}{e^{iu}-e^{-iu}} = -i\,\dfrac{\cos u}{\sin u} = -i\cot u$

(c) $\operatorname{sech} u = \dfrac{1}{\cosh u} = \dfrac{2}{e^{u}+e^{-u}} = \dfrac{1}{\cos iu} = \sec iu$

(d) $\sin iu = \dfrac{e^{-u}-e^{u}}{2i} = i\sinh u$

15. Start with deMoivre's formula, Equation 4.3.11,

$$e^{in\theta} = (\cos\theta + i\sin\theta)^n = \cos n\theta + i\sin n\theta$$

with $z = i\theta$ to get

$$
\begin{aligned}
e^{nz} &= \cos(-inz) + i\sin(-inz) = \cos inz - i\sin inz\\
&= \cosh nz + \sinh nz = (\cosh z + \sinh z)^n
\end{aligned}
$$

Now let $z = -i\theta$ to get

$$e^{-nz} = \cos inz + i\sin inz = \cosh nz - \sinh nz = (\cosh z - \sinh z)^n$$

18. Start with

$$\sin\frac{\pi z}{2a} = \sin\frac{\pi x}{2a}\cosh\frac{\pi y}{2a} + i\cos\frac{\pi x}{2a}\sinh\frac{\pi y}{2a}$$

The vertical line at $x=0$ gives $u(0,y)=0$ and $v(0,y)=\sinh\dfrac{\pi y}{2a}$, or the v axis from 0 to ∞. The vertical line at $x=a$ gives $u(\pi/2,y)=\cosh\dfrac{\pi y}{2a}$, or the u axis from 1 to ∞. The x axis from 0 to a gives $u(x,0)=\sin\dfrac{\pi x}{2a}$, or the u axis from 0 to 1. Other points with the region in the z-plane map into the first quadrant of the w-plane.

4.5 The Logarithms of Complex Numbers

1. Rewrite $-i$ as $-i = e^{i(3\pi/2 + 2\pi n)}$ for $n = 0, \pm 1, \pm 2, \ldots$. Then $\ln(-i) = i\left(\dfrac{3\pi}{2}+2\pi n\right)$ and $\operatorname{Ln}(-i) = \dfrac{3\pi i}{2}$.

3. First express $\operatorname{Ln}(1+i\sqrt{3})$ in polar form:

$$\operatorname{Ln}(1+i\sqrt{3}) = \ln 2 + \frac{i\pi}{3} = \left[(0.6931)^2 + \frac{\pi^2}{9}\right]^{1/2} e^{i\tan^{-1}\left(\frac{\pi/3}{0.6931}\right) + 2\pi ni} = 1.2558 e^{0.9861 i + 2\pi ni}$$

Therefore,

$$\operatorname{Ln}\{\operatorname{Ln}(1+i\sqrt{3})\} = \ln(1.2558\ldots) + 0.9861\,i = 0.22779 + 0.986\,i$$

6. Yes, because $\theta = \theta_1 + \theta_2 = \dfrac{3\pi}{4} + \pi = \dfrac{7\pi}{4}$ is less than 2π.

9. Using the result given in Problem 8, we have

$$\sin^{-1} 2 = -i \ln(2i \pm i\sqrt{3}) = -i \ln i - i \ln(2 \pm \sqrt{3}) = \frac{\pi}{2} - i \ln(2 \pm \sqrt{3})$$

12. $e^z = e^{1+\pi i} = ee^{\pi i} = -e$, so

$$\ln e^{1+i\pi} = \ln(-e) = \ln|-e| + i \arg(-e) = 1 + i(\pi + 2\pi n) \qquad n = 0,\ \pm 1,\ \pm 2, \ldots$$

4.6   Powers of Complex Numbers

1. Use Equation 5 to write $(1 - i)^{2/3} = (\sqrt{2})^{2/3} e^{i\frac{2}{3}(\theta_0 + 2\pi k)}$, where $\theta_0 = 7\pi/4$. Therefore,

$$(1 - i)^{2/3} = 2^{1/3} \left[\cos\left(\frac{7\pi}{6} + \frac{4\pi k}{3} \right) + i \sin\left(\frac{7\pi}{6} + \frac{4\pi k}{3} \right) \right]$$

where $k = 0,\ \pm 1,\ \pm 2,\ \pm 3,\ \ldots$.

3. The argument of the cosine and sine is $\theta = \dfrac{m}{n}(\theta_0 + 2\pi k)$. Note that k takes on the values $0, 1, 2, \ldots, n-1$ before θ goes through a complete cycle.

6. Use Equation 6 with $a = b = 1$ and $r = \sqrt{2}$ and $\theta_0 = \pi/4$ to write

$$(1 + i)^{i+1} = \exp\left(\frac{1}{2}\ln 2 - \frac{\pi}{4} + 2\pi k \right) \times \left[\cos\left(\frac{1}{2}\ln 2 + \frac{\pi}{4} + 2\pi k \right) + i \sin\left(\frac{1}{2}\ln 2 + \frac{\pi}{4} + 2\pi k \right) \right]$$

with $k = 0,\ \pm 1,\ \pm 2,\ \ldots$

9. $i^i = (e^{i\pi/2})^i = (e^{i[\pi/2 + 2\pi n]})^i = e^{-\pi/2} e^{-2\pi n}$ with $n = 0,\ \pm 1,\ \pm 2,\ \ldots$.

12. The region $0 \le \mathrm{Arg}\, z \le \theta_0 \le \pi/m$ maps into the region $0 \le \mathrm{Arg}\, w \le \pi$ (the upper half w-plane) under the mapping $w = z$.

CHAPTER 5

Vectors

5.1 Vectors in Two Dimensions

1. $|\mathbf{u}| = (4+1)^{1/2} = \sqrt{5}$ and $|\mathbf{v}| = (1+1)^{1/2} = \sqrt{2}$; $\mathbf{u} + \mathbf{v} = (2,1) + (-1,1) = (1,2)$; $2\mathbf{u} - 3\mathbf{v} = (4,2) - (-3,3) = (7,-1)$; and $|\mathbf{u} + \mathbf{v}| = (1+4)^{1/2} = \sqrt{5}$.

3. (a) $\dfrac{\mathbf{u}}{|\mathbf{u}|} = \left(\dfrac{3}{\sqrt{10}}, -\dfrac{1}{\sqrt{10}} \right)$; (b) $\dfrac{\mathbf{u}}{|\mathbf{u}|} = \dfrac{2}{\sqrt{13}}\mathbf{i} + \dfrac{3}{\sqrt{13}}\mathbf{j}$; (c) $\dfrac{\mathbf{u}}{|\mathbf{u}|} = \dfrac{\mathbf{i}}{\sqrt{2}} + \dfrac{\mathbf{j}}{\sqrt{2}}$

6. (a) $\cos\theta = \dfrac{\mathbf{u} \cdot \mathbf{v}}{uv} = \dfrac{-2}{2^{1/2}2^{1/2}} = -1$, or $\theta = \pi$

(b) $\cos\theta = \dfrac{\mathbf{u} \cdot \mathbf{v}}{uv} = \dfrac{0}{2^{1/2}2^{1/2}} = 0$, or $\theta = \pi/2$

9. Let $\mathbf{x}$ be the vector joining the midpoints of sides u and v. Then we have that $\dfrac{\mathbf{u}}{2} + \mathbf{x} = \dfrac{\mathbf{v}}{2}$. Thus, $\mathbf{x} = \dfrac{\mathbf{u} - \mathbf{v}}{2}$

12. Use the following figure:

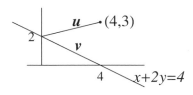

Thus we have that $\mathbf{u} + 2\mathbf{j} = 4\mathbf{i} + 3\mathbf{j}$ or $\mathbf{u} = 4\mathbf{i} + \mathbf{j}$ and $\mathbf{v} + 2\mathbf{j} = 4\mathbf{i}$ or $\mathbf{v} = 4\mathbf{i} - 2\mathbf{j}$. The angle between $\mathbf{u}$ and $\mathbf{v}$ is given by $\dfrac{\mathbf{u} \cdot \mathbf{v}}{uv} = \cos\theta = \dfrac{14}{(17 \cdot 20)^{1/2}}$, so that $\sin\theta = \left(1 - \dfrac{196}{340} \right)^{1/2} = \left(\dfrac{36}{85} \right)^{1/2}$.

The perpendicular distance from the point to the line is given by $d = u\sin\theta = (17)^{1/2} \left(\dfrac{36}{85} \right)^{1/2} = 2.683$.

15. We wish to show that $(\mathbf{u}_1 + \mathbf{u}_2) \cdot \mathbf{v} = \mathbf{u}_1 \cdot \mathbf{v} + \mathbf{u}_2 \cdot \mathbf{v}$. The figure on the following page shows that the projection of $\mathbf{u}_1$ onto $\mathbf{v}$ (in other words, $\mathbf{u}_1 \cdot \mathbf{v}$), plus the projection of $\mathbf{u}_2$ onto $\mathbf{v}$ ($\mathbf{u}_2 \cdot \mathbf{v}$) is the same as the projection of $\mathbf{u}_1 + \mathbf{u}_2$ onto $\mathbf{v}$, which is $(\mathbf{u}_1 + \mathbf{u}_2) \cdot \mathbf{v}$.

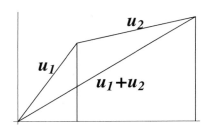

5.2   Vector Functions in Two Dimensions

1. The curve is described by $\mathbf{r} = (2t^2 + 1)\,\mathbf{i} + (t^3 + 2t)\,\mathbf{j}$. Therefore, $\mathbf{r}'(t) = 4t\,\mathbf{i} + (3t^2 + 2)\,\mathbf{j}$, which gives $\mathbf{r}'(1) = 4\,\mathbf{i} + 5\,\mathbf{j}$.

$$\mathbf{r}(1) + \mathbf{r}'(1)(t-1) = 3\,\mathbf{i} + 3\,\mathbf{j} + (4\,\mathbf{i} + 5\,\mathbf{j})(t-1) = (4t-1)\,\mathbf{i} + (5t-2)\,\mathbf{j}$$

or $x = 4t - 1$, $y = 5t - 2$. The following figure shows the curve and its tangent line at $t = 1$.

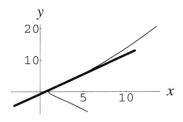

3. (a) Let $\mathbf{w}(t) = w_x(t)\,\mathbf{i} + w_y(t)\,\mathbf{j} + w_z(t)\,\mathbf{k}$ and differentiate.

(b) Write $\mathbf{u} \cdot \mathbf{w} = u_x(t)w_x(t) + u_y(t)w_y(t) + u_z(t)w_z(t)$ and differentiate.

6. Differentiate $\mathbf{r}(t) = 2\cos\,\mathbf{i} + 2\sin t\,\mathbf{j}$ to obtain $\mathbf{r}'(t) = -2\sin t\,\mathbf{i} + 2\cos t\,\mathbf{j}$. Therefore, $v(t) = (4\sin^2 t + 4\cos^2 t)^{1/2} = 2$, and $s = \displaystyle\int_0^1 v(t)\,dt = 2$. One radian of a circle of radius 2 subtends an arc of length 2.

9. In this case,

$$\frac{d\mathbf{r}}{du} = -(2\cos u \sin u)\,\mathbf{i} + (2\sin u \cos u)\,\mathbf{j} = -(\sin 2u)\,\mathbf{i} + (\sin 2u)\,\mathbf{j}$$

and so

$$\left| \frac{d\mathbf{r}}{du} \right| = \frac{ds}{du} = (2\sin^2 2u)^{1/2} = \sqrt{2}\,\sin 2u$$

Finally,

$$\mathbf{T} = \frac{d\mathbf{r}}{ds} = \frac{d\mathbf{r}}{du}\frac{du}{ds} = \frac{-(\sin 2u)\,\mathbf{i} + (\sin 2u)\,\mathbf{j}}{\sqrt{2}\,\sin 2u} = -\frac{\mathbf{i}}{\sqrt{2}} + \frac{\mathbf{j}}{\sqrt{2}}$$

(The vector $\mathbf{r}(t)$ traces out a straight line with slope -1.)

12. Let $y(x) = mx + b$ in Equation 16 to get $\kappa = 0$.

15. Equation 16 gives $\kappa = \dfrac{4}{(1+16x^2)^{3/2}}$, which equals $\dfrac{4}{(17)^{3/2}}$ at $x = 1$. Therefore,

$\rho = \dfrac{1}{\kappa} = \dfrac{(17)^{3/2}}{4} = 17.52$.

18. Differentiating $\mathbf{r}(t) = t\cos t\,\mathbf{i} + t\sin t\,\mathbf{j}$ gives $\mathbf{r}'(t) = (\cos t - t\sin t)\,\mathbf{i} + (\sin t + t\cos t)\,\mathbf{j}$. The magnitude of $\mathbf{r}'(t)$ is given by

$v = [(\cos t - t\sin t)^2 + (\sin t + t\cos t)^2]^{1/2} = [\cos^2 t + t^2\sin^2 t + \sin^2 t + t^2\cos^2 t]^{1/2} = (1+t^2)^{1/2}$

Therefore,

$$\mathbf{T}(t{=}\pi) = \frac{\mathbf{r}'(\pi)}{v} = (\cos\pi - \pi\sin\pi)\,\mathbf{i} + (\sin\pi + \pi\cos\pi)\,\mathbf{j} = \frac{-\mathbf{i} - \pi\mathbf{j}}{(1+\pi^2)^{1/2}}$$

The condition $\mathbf{N}\cdot\mathbf{T} = 0$ gives $\mathbf{N} = \pm\dfrac{\pi\mathbf{i} - \mathbf{j}}{(1+\pi^2)^{1/2}}$.

21. $\mathbf{r}(u) = a(u - \sin u)\,\mathbf{i} + a(1 - \cos u)\,\mathbf{j}$. Thus, $\mathbf{r}'(u) = \mathbf{v}(u) = a(1 - \cos u)\,\mathbf{i} + a\sin u\,\mathbf{j}$ and $v = a[(1 - \cos u)^2 + \sin^2 u]^{1/2} = a(2 - 2\cos u)^{1/2}$. Thus,

$$s = 2^{1/2}a\int_0^{2\pi}(1 - \cos u)^{1/2}du = 2^{1/2}a\cdot 4\sqrt{2} = 8a$$

5.3   Vectors in Three Dimensions

1.

$$\begin{aligned}
(u_x\,\mathbf{i} + u_y\,\mathbf{j} + u_z\,\mathbf{k})\cdot(v_x\,\mathbf{i} + v_y\,\mathbf{j} + v_k\,\mathbf{k}) &= u_xv_x\,\mathbf{i}\cdot\mathbf{i} + u_xv_y\,\mathbf{i}\cdot\mathbf{j} + u_xv_z\,\mathbf{i}\cdot\mathbf{k} \\
&\quad + u_uv_x\,\mathbf{j}\cdot\mathbf{i} + u_yv_y\,\mathbf{j}\cdot\mathbf{j} + u_yv_z\,\mathbf{j}\cdot\mathbf{k} \\
&\quad + u_zv_x\,\mathbf{k}\cdot\mathbf{i} + u_zv_y\,\mathbf{k}\cdot\mathbf{j} + u_zv_z\,\mathbf{k}\cdot\mathbf{k} \\
&= u_xv_x + u_yv_y + u_zv_z
\end{aligned}$$

3. Define the two vectors

$\mathbf{u} = (2, 0, 1) - (1, -1, 2) = (1, 1, -1)$ and $\mathbf{v} = (3, 1, -1) - (1, -1, 2) = (2, 2, -3)$

The angle between them is given by $\cos\theta = \dfrac{\mathbf{u}\cdot\mathbf{v}}{uv} = \dfrac{7}{3^{1/2}17^{1/2}} = 0.9802$, or $\theta = 0.1993$ radians = 11.42°.

6. We use Equation 18 with $\mathbf{u} = (1, 1, 1)$ and $\mathbf{v} = (2, 3, 5)$.

$$\mathbf{u}\times\mathbf{v} = \begin{vmatrix} \mathbf{i} & \mathbf{j} & \mathbf{k} \\ 1 & 1 & 1 \\ 2 & 3 & 5 \end{vmatrix} = 2\,\mathbf{i} - 3\,\mathbf{j} + \mathbf{k}, \text{ which gives } A = |\mathbf{u}\times\mathbf{v}| = (14)^{1/2}$$

9. Start with $\mathbf{u} + \mathbf{v} + \mathbf{w} = \mathbf{0}$. Now "cross" both sides with $\mathbf{u}$ to obtain

$\mathbf{0} + uv\sin(\pi - \theta_{uv})\,\mathbf{n} - uw\sin(\pi - \theta_{uw})\,\mathbf{n} = \mathbf{0} = (uv\sin\theta_{uv} - uw\sin\theta_{uw})\,\mathbf{n}$

where θ_{uv}, for example, is the "interior" angle between $\mathbf{u}$ and $\mathbf{v}$ and $\mathbf{n}$ is perpendicular to the plane of the triangle. (Whichever direction $\mathbf{n}$ points, $\mathbf{u} \times \mathbf{v}$ and $\mathbf{u} \times \mathbf{w}$ point in the opposite directions. This is why we write $-uw\sin(\pi - \theta_{uw})$ in the above line.) This equation gives $\dfrac{\sin\theta_{uv}}{w} = \dfrac{\sin\theta_{uw}}{v}$. "Crossing" both sides with $\mathbf{v}$ gives $\dfrac{\sin\theta_{uv}}{w} = \dfrac{\sin\theta_{vw}}{u}$ and so we have $\dfrac{\sin\theta_{uv}}{w} = \dfrac{\sin\theta_{vw}}{u} = \dfrac{\sin\theta_{uw}}{v}$.

12. Use (Equation 23) $\mathbf{u} \cdot (\mathbf{u} \times \mathbf{v}) = (\mathbf{u} \times \mathbf{u}) \cdot \mathbf{v} = 0$. Thus, $\mathbf{u} \times \mathbf{v}$ is perpendicular to $\mathbf{u}$ and $\mathbf{v}$.

15. $|\mathbf{u} + \mathbf{v}| = [(\mathbf{u}+\mathbf{v}) \cdot (\mathbf{u}+\mathbf{v})]^{1/2} = [u^2 + 2\mathbf{u}\cdot\mathbf{v} + v^2]^{1/2}$

$$|\mathbf{u}+\mathbf{v}|^2 = |u|^2 + |v|^2 + 2|\mathbf{u}\cdot\mathbf{v}| \le |u|^2 + |v|^2 + 2|u|\,|v| \le (|u| + |v|)^2$$

18. No. In fact, $\mathbf{v} \times \mathbf{v} = \mathbf{0}$, so that the left side is necessarily equal to the zero vector. The right side is not necessarily equal to the zero vector.

5.4   Vector Functions in Three Dimensions

1. Start with $\displaystyle\lim_{\Delta t \to 0} \dfrac{\mathbf{u}(t+\Delta t) \times \mathbf{v}(t+\Delta t) - \mathbf{u}(t) \times \mathbf{v}(t)}{\Delta t}$ and expand $\mathbf{u}(t+\Delta t)$ and $\mathbf{v}(t+\Delta t)$ in a Maclaurin expansion about Δt to write

$$\lim_{\Delta t \to 0} \dfrac{\mathbf{u}(t+\Delta t) \times \mathbf{v}(t+\Delta t) - \mathbf{u}(t) \times \mathbf{v}(t)}{\Delta t}$$

$$= \dfrac{[\mathbf{u}(t) + \Delta t\, d\mathbf{u}/dt + O(\Delta t)^2] \times [\mathbf{v}(t) + \Delta t\, d\mathbf{v}/dt + O(\Delta t)^2] - \mathbf{u}(t) \times \mathbf{v}(t)}{\Delta t}$$

$$= \dfrac{\Delta t\left[\dfrac{d\mathbf{u}}{dt} \times \mathbf{u} + \Delta t\,\mathbf{u} \times \dfrac{d\mathbf{v}}{dt} + O(\Delta t)^2\right]}{\Delta t}$$

$$= \dfrac{d\mathbf{u}}{dt} \times \mathbf{v} + \mathbf{u} \times \dfrac{d\mathbf{v}}{dt}$$

3. $\mathbf{v}(t) = \displaystyle\int \mathbf{a}(t)dt + \mathbf{c} = \dfrac{t^2}{2}\mathbf{j} + \dfrac{t^3}{3}\mathbf{k} + c_x\mathbf{i} + c_y\mathbf{j} + c_z\mathbf{k}.$
The initial condition $\mathbf{v}(0) = 3\mathbf{i} + 2\mathbf{j} - \mathbf{k}$ gives $c_x = 3$, $c_y = 2$, and $c_z = -1$. Therefore,

$$\mathbf{v}(t) = 3\mathbf{i} + \left(2 + \dfrac{t^2}{2}\right)\mathbf{j} + \left(\dfrac{t^3}{3} - 1\right)\mathbf{k}$$

Another integration gives

$$\mathbf{r}(t) = \int \mathbf{v}(t)dt + \mathbf{c} = 3t\,\mathbf{i} + \left(2t + \dfrac{t^3}{6}\right)\mathbf{j} + \left(\dfrac{t^4}{12} - t\right)\mathbf{k} + c_x\mathbf{i} + c_y\mathbf{j} + c_z\mathbf{k}$$

The initial condition $\mathbf{r}(0) = 0$ gives $\mathbf{c} = \mathbf{0}$. So,

$$\mathbf{r}(t) = 3t\,\mathbf{i} + \left(2t + \dfrac{t^3}{6}\right)\mathbf{j} + \left(\dfrac{t^4}{12} - t\right)\mathbf{k}$$

6. Use Equation 5.3.24 to write

$$\mathbf{r} \times (\boldsymbol{\omega} \times \mathbf{v}) = (\mathbf{r} \cdot \mathbf{v})\,\boldsymbol{\omega} - (\mathbf{r} \cdot \boldsymbol{\omega})\,\mathbf{v} \qquad \text{and} \qquad \boldsymbol{\omega} \times (\mathbf{r} \times \mathbf{v}) = (\boldsymbol{\omega} \cdot \mathbf{v})\,\mathbf{r} - (\mathbf{r} \cdot \boldsymbol{\omega})\,\mathbf{v}$$

But $\mathbf{r} \cdot \mathbf{v} = 0$ and $\mathbf{v} \cdot \boldsymbol{\omega} = 0$ for circular motion, so we are left with $\boldsymbol{\omega} \times (\mathbf{r} \times \mathbf{v}) = \boldsymbol{\omega} \times \mathbf{L}$.

9. For $\mathbf{r}(t) = t\,\mathbf{i} + t^2\,\mathbf{j} + t^3\,\mathbf{k}$, we have $\mathbf{v}(t) = \mathbf{i} + 2t\,\mathbf{j} + 3t^2\,\mathbf{k}$ and $v(t) = (1 + 4t^2 + 9t^4)^{1/2}$.

Therefore, $\mathbf{T}(t) = \dfrac{\mathbf{i} + 2t\,\mathbf{j} + 3t^2\,\mathbf{k}}{(1 + 4t^2 + 9t^4)^{1/2}}$

12. Start with

$$\mathbf{T}(t) = \frac{d\mathbf{r}}{dt}\frac{dt}{ds} = \frac{d\mathbf{r}}{ds} = \frac{1}{v}\frac{d\mathbf{r}}{dt}$$

Now,

$$v = \mathbf{r}'(t) = \frac{ds}{dt} = \left[\left(\frac{dx}{dt}\right)^2 + \left(\frac{dy}{dt}\right)^2 + \left(\frac{dz}{dt}\right)^2 \right]^{1/2} = (a^2 \sin^2 t + a^2 \cos^2 t + b^2)^{1/2} = (a^2 + b^2)^{1/2}$$

and so

$$\mathbf{T}(t) = \frac{1}{v}\frac{d\mathbf{r}}{dt} = \frac{-a \sin t\,\mathbf{i} + a \cos t\,\mathbf{j} + b\,\mathbf{k}}{(a^2 + b^2)^{1/2}}$$

15. Take $\mathbf{v}\times$ of Equation 26 to write $\mathbf{v} \times \mathbf{a} = \dfrac{dv}{dt}\,\mathbf{v} \times \mathbf{T} + \kappa v^2\,\mathbf{v} \times \mathbf{N}$. But $\mathbf{v} \times \mathbf{T} = \mathbf{v} \times \dfrac{\mathbf{v}}{v} = 0$ and $|\mathbf{v} \times \mathbf{N}| = v$ since $\mathbf{v}$ and $\mathbf{N}$ are perpendicular, so $\kappa = |\mathbf{v} \times \mathbf{a}|/v^3$.

18. Differentiate $\mathbf{r}(t) = t\,\mathbf{i} + t^2\,\mathbf{j} + t^3\,\mathbf{k}$ to get $\mathbf{v}(t) = \mathbf{i} + 2t\,\mathbf{j} + 3t^2\,\mathbf{k}$. The magnitude of $\mathbf{v}(t)$ is given by $v(t) = (1 + 4t^2 + 9t^4)^{1/2}$ and $\dfrac{dv}{dt} = \dfrac{4t + 18t^2}{(1 + 4t^2 + 9t^4)^{1/2}}$. Now

$$\mathbf{T}(t) = \frac{1}{v}\frac{d\mathbf{r}}{dt} = \frac{\mathbf{i} + 2t\,\mathbf{j} + 3t^2\,\mathbf{k}}{(1 + 4t^2 + 9t^4)^{1/2}}$$

and (see Equation 26)

$$\mathbf{a}_T(t) = \frac{dv}{dt}\,\mathbf{T}(t) = \frac{4t + 18t^3}{1 + 4t^2 + 9t^4}(\mathbf{i} + 2t\,\mathbf{j} + 3t^2\,\mathbf{k})$$

and $a_T(1) = 22/(14)^{1/2}$.

The normal component of $\mathbf{a}_T(t)$ is given by (see Equation 26)

$$
\begin{aligned}
\mathbf{a}_N(t) &= v^2 \kappa\,\mathbf{N} = v^2 \frac{d\mathbf{T}}{ds} = v\frac{d\mathbf{T}}{dt} \\
&= 2\,\mathbf{j} + 6t\,\mathbf{k} - \frac{4t + 18t^3}{1 + 4t^2 + 9t^4}(\mathbf{i} + 2t\,\mathbf{j} + 3t^2\,\mathbf{k})
\end{aligned}
$$

Note that $\mathbf{a}_N(t) = \mathbf{a}(t) - \mathbf{a}_T(t)$, as it must.

Finally,

$$a_N(1) = \left| 2\mathbf{j} + 6\mathbf{k} - \left(\frac{22}{14}\right)(\mathbf{i} + 2\mathbf{j} + 3\mathbf{k}) \right|$$

$$= \left[\left(\frac{11}{7}\right)^2 + \left(\frac{8}{7}\right)^2 + \left(\frac{9}{7}\right)^2 \right]^{1/2} = \left(\frac{38}{7}\right)^{1/2}$$

21. It turns out that $\mathbf{v}(t) = a_x\,\mathbf{i} + a_y\,\mathbf{j} + a_z\,\mathbf{k}$, and so $v(t) = (a_x^2 + a_y^2 + a_z^2)^{1/2}$. Therefore,

$$\mathbf{T}(t) = \frac{\mathbf{v}(t)}{v(t)} = \frac{a_x\,\mathbf{i} + a_y\,\mathbf{j} + a_z\,\mathbf{k}}{(a_x^2 + a_y^2 + a_z^2)^{1/2}}$$

Now, $\dfrac{d\mathbf{T}}{ds} = \dfrac{1}{v}\dfrac{d\mathbf{T}}{dt} = \mathbf{0}$, so $\kappa = 0$. The curve is a straight line.

24. Differentiation of $\mathbf{r}(t)$ gives $\mathbf{v}(t) = \mathbf{j} + 2t\,\mathbf{k}$ and $v(t) = (1 + 4t^2)^{1/2}$. Therefore, $\mathbf{T}(t) = \dfrac{\mathbf{v}(t)}{v(t)} = \dfrac{\mathbf{j} + 2t\,\mathbf{k}}{(1 + 4t^2)^{1/2}}$. Now,

$$\frac{d\mathbf{T}}{ds} = \frac{d\mathbf{T}}{dt}\frac{dt}{ds} = \frac{1}{v}\left[\frac{2\mathbf{k}}{(1 + 4t^2)^{1/2}} - \frac{4t\,\mathbf{j} + 8t^2\,\mathbf{k}}{(1 + 4t^2)^{3/2}} \right] = \frac{-4t\,\mathbf{j} + 2\mathbf{k}}{(1 + 4t^2)^2}$$

The curvature κ is given by $\kappa = \left|\dfrac{d\mathbf{T}}{ds}\right| = \dfrac{2}{(1 + 4t^2)^{3/2}}$, and the normal vector $\mathbf{N}(t)$ is given by $\mathbf{N}(t) = \dfrac{1}{\kappa}\dfrac{d\mathbf{T}}{ds} = \dfrac{-2t\,\mathbf{j} + \mathbf{k}}{(1 + 4t^2)^{1/2}}$. Finally,

$$\mathbf{B}(t) = \mathbf{T}(t) \times \mathbf{N}(t) = \frac{1}{1 + 4t^2}\begin{vmatrix} \mathbf{i} & \mathbf{j} & \mathbf{k} \\ 0 & 1 & 2t \\ 0 & -2t & 1 \end{vmatrix} = \mathbf{i}$$

and $\tau = 0$ because $\mathbf{B} = $ constant. (See Equation 29.)

5.5   Lines and Planes in Space

1. The vector $\mathbf{v} = (-1, 2, -1) - (2, 0, 3) = -3\,\mathbf{i} + 2\,\mathbf{j} - 4\,\mathbf{k}$ passes through the two points. Choosing the point $(2, 0, 3)$, we write

$$\mathbf{r} = 2\mathbf{i} + 3\mathbf{k} + u(-3\,\mathbf{i} + 2\,\mathbf{j} - 4\,\mathbf{k}) = (2 - 3u)\,\mathbf{i} + 2u\,\mathbf{j} + (3 - 4u)\,\mathbf{k} = x(u)\,\mathbf{i} + y(u)\,\mathbf{j} + z(u)\,\mathbf{k}$$

where $-\infty < u < \infty$. The parametric equations are

$$x(u) = 2 - 3u \qquad y(u) = 2u \qquad z(u) = 3 - 4u$$

3. $\mathbf{v} = \mathbf{k}$, and so Equation 5 gives $z - 3 = 0$, or $z = 3$.

6. The vector $\mathbf{u} \times \mathbf{v} = (5, 3, -1)$ is perpendicular to the plane. Using Equation 6, we obtain $5x + 3y - z = $ constant.

9. The vector $\mathbf{v} = (-2, -1, -1) - (0, 2, 1) = -2\,\mathbf{i} - 3\,\mathbf{j} - 2\,\mathbf{k}$ passes through the two points, and $\mathbf{r} = (3, -1, -1) - (0, 2, 1) = 3\,\mathbf{i} - 3\,\mathbf{j} - 2\,\mathbf{k}$ is a vector from the point $(3, -1, -1)$ to a point of the line. Now,

$$\mathbf{r} \times \mathbf{v} = \begin{vmatrix} \mathbf{i} & \mathbf{j} & \mathbf{k} \\ 3 & -3 & -2 \\ -2 & -3 & -2 \end{vmatrix} = 10\,\mathbf{j} - 15\,\mathbf{k}$$

and so

$$d = \left| \frac{\mathbf{r} \times \mathbf{v}}{v} \right| = \left| \frac{\sqrt{325}}{\sqrt{17}} \right| = \sqrt{\frac{325}{17}} = 4.372$$

12. Because the vector describing that point is parallel to $\mathbf{v}$.

15. The vectors normal to the two planes are $A_1\,\mathbf{i} + B_1\,\mathbf{j} + C_1\,\mathbf{k}$ and $A_2\,\mathbf{i} + B_2\,\mathbf{j} + C_2\,\mathbf{k}$, and the cosine of the angle between the two planes is given by their dot product.

CHAPTER 6

Functions of Several Variables

6.1 Functions

1. $d = \sqrt{(x_1 - x_2)^2 + (y_1 - y_2)^2 + (z_1 - z_2)^2} = \sqrt{4 + 9 + 16} = \sqrt{29}$

3. $\sin x = 0$ implies that $x = n\pi$ where $n = 0, \pm1, \pm2, \dots$. Therefore, this set of points does not constitute a connected set.

6. It bisects the second and fourth xy-quadrants and the z axis lies in the plane.

9. The level curves are

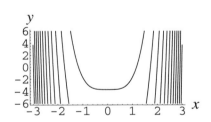

12. It is an elliptic cylinder whose axis is the z axis and its cross section parallel to the z-axis is $4x^2 + 16y^2 = 32$.

15. Let x, y, and $z = 0$ in turn.

$x = 0$ (yz-plane) $\quad \dfrac{y^2}{b^2} - \dfrac{z^2}{c^2} = 1$ is a hyperbola

$y = 0$ (xz-plane) $\quad \dfrac{x^2}{a^2} - \dfrac{z^2}{c^2} = 1$ is a hyperbola

$z = 0$ (xy-plane) $\quad \dfrac{x^2}{a^2} + \dfrac{y^2}{b^2} = 1$ is an ellipse. If we let z vary, we generate a family of ellipses. (See Figure 6.19.)

18. The surface is on the left and the level curves are on the right in the following figures.

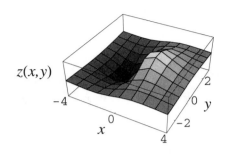
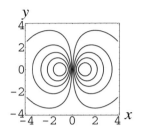

21. The surface is on the left and the level curves are on the right in the following figures.

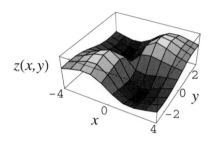
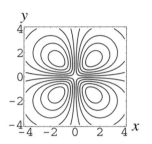

6.2 Limits and Continuity

1. We have $\left|3 \pm \dfrac{6}{\sqrt{2}}\delta + \dfrac{3}{2}\delta^2 - 1 \mp \dfrac{2}{\sqrt{2}}\delta - \dfrac{\delta^2}{2} - 2\right| < \epsilon$ or $\pm\dfrac{4}{\sqrt{2}}\delta + \delta^2 - \epsilon < 0$. Thus,

$\delta < \mp\dfrac{2}{\sqrt{2}} \pm \dfrac{1}{2}(8 + 4\epsilon)^{1/2}$ so that $\delta < \mp\dfrac{2}{\sqrt{2}} \pm \dfrac{2}{\sqrt{2}}\left(1 + \dfrac{\epsilon}{2}\right)^{1/2} < \dfrac{\epsilon}{2^{3/2}}$

3. We have that $|x+2y-5| < \epsilon$ whenever $(x-1)^2+(y-2)^2 < \delta^2$. Choose $(x-1)^2 < \dfrac{\delta^2}{2}$ and

$(y-2)^2 < \dfrac{\delta^2}{2}$, or $x < 1 \pm \dfrac{\delta}{\sqrt{2}}$ and $y < 2 \pm \dfrac{\delta}{\sqrt{2}}$. Substitute these equations into $|x + 2y - 5| < \epsilon$

and get $\left|\pm\dfrac{3\delta}{\sqrt{2}}\right| < \epsilon$, which gives $\delta < \dfrac{\sqrt{2}}{3}\epsilon$.

6. The hint says that $0 \le (x^2 - y)^2 = x^4 - 2x^2y + y^2$. Therefore, $2x^2|y| \le x^4 + y^2$, and

$$\frac{x^2y^2}{x^4 + y^2} = \frac{x^2|y|\cdot|y|}{x^4 + y^2} \le \frac{1}{2}\frac{(x^4 + y^2)|y|}{x^4 + y^2} = \frac{|y|}{2}$$

and so the limit is zero.

9. (a) $\lim\limits_{x\to 0+} \tan^{-1}\dfrac{y}{x} = \dfrac{\pi}{2}$ and $\lim\limits_{x\to 0-} \tan^{-1}\dfrac{y}{x} = -\dfrac{\pi}{2}$. The limit does not exist.

(b) Let $y = mx$, then $\displaystyle\lim_{(x,y)\to(0,0)} \frac{xy}{x^2 + y^2} = \frac{m}{1 + m^2}$. The limit does not exist.

12. Both sequential limits give 1. Letting $y = mx$ gives $\displaystyle\lim_{(x,y)\to(0,0)} \left(\frac{x^2 - y^2}{x^2 + y^2}\right)^2 = \left(\frac{1 - m^2}{1 + m^2}\right)^2$.
The two sequential limits correspond to $m = \pm 0$ and $m = \pm\infty$.

15. The limit of $f(x,y) = 1$ if x and y approach zero along the curve given by $x = t$ and $y = t$. Therefore, $f(x)$ is not continuous at $(0,0)$.

6.3 Partial Derivatives

1. (a) Starting with $f(x,y) = xe^y + y$, we have

$$\begin{aligned} f_x &= e^y & f_y &= xe^y + 1 \\ f_{xx} &= 0 & f_{yy} &= xe^y \\ f_{xy} &= e^y & f_{yx} &= e^y \end{aligned}$$

(b) Starting with $f(x,y) = y\sin x + x^2$, we have

$$\begin{aligned} f_x &= y\cos x + 2x & f_y &= \sin x \\ f_{xx} &= -y\sin x + 2 & f_{yy} &= 0 \\ f_{xy} &= \cos x & f_{yx} &= \cos x \end{aligned}$$

3. (a) Starting with $f(x,y) = x^2 e^{-y^2}$, we have

$$\begin{aligned} f_x &= 2xe^{-y^2} & f_y &= -2yx^2 e^{-y^2} \\ f_{xy} &= -4xye^{-y^2} & f_{yx} &= -4xye^{-y^2} \end{aligned}$$

(b) Starting with $f(x,y) = e^{-y}\cos xy$, we have

$$f_x = -ye^{-y}\sin xy \qquad\qquad f_{xy} = -e^{-y}\sin xy - xye^{-y}\cos xy + ye^{-y}\sin xy$$
$$f_y = -xe^{-y}\sin xy - e^{-y}\cos xy \qquad f_{yx} = -e^{-y}\sin xy - xye^{-y}\cos xy + ye^{-y}\sin xy$$

6. Starting with $f(x,y) = xy\tan(y/x)$, $x \neq 0$, we have

$$f_x = y\tan\frac{y}{x} - \frac{y^2}{x}\sec^2\frac{y}{x} \qquad\text{and}\qquad f_y = x\tan\frac{y}{x} + y\sec^2\frac{y}{x}$$

so that $xf_x + yf_y = 2xy\tan\dfrac{y}{x} = 2f$.

9. (a) Starting with $f(x,y) = \sin ax\sinh ay$, we have

$$\begin{aligned} f_x &= a\cos ax\sinh ay & f_y &= a\sin ax\cosh ay \\ f_{xx} &= -a\sin ax\sinh ay & f_{yy} &= a\sin ax\sinh ay \end{aligned}$$

and so $f_{xx} + f_{yy} = 0$.

12. Starting with $f(x, y) = x^2/(x + y)$, we have that $f_x = \dfrac{x^2 + 2xy}{(x + y)^2}$ and $f_y = -\dfrac{x^2}{(x + y)^2}$. Therefore, Equation 5 gives us $z = \dfrac{3}{4}(x - 2) - \dfrac{1}{4}(y - 2) + 1$ or $3x - y - 4z = 0$.

15. First let $(x, y) \to (0, 0)$ along the line $y = x$. Then $\displaystyle\lim_{(x,y) \to (0,0)} \frac{xy}{x^2 + y^2} = \lim_{x \to 0} \frac{x^2}{x^2 + x^2} = \frac{1}{2}$ so that the limit would equal $1/2$ if it existed. Now let $(x, y) \to (0, 0)$ along the line $y = -x$. Then $\displaystyle\lim_{(x,y) \to (0,0)} \frac{xy}{x^2 + y^2} = \lim_{x \to 0} \frac{-x^2}{x^2 + x^2} = -\frac{1}{2}$ so that the limit would equal $-1/2$. Hence the limit does not exist.

On the other hand, we have that

$$
\begin{aligned}
f_x(0, 0) &= \lim_{\Delta x \to 0} \frac{f(0 + \Delta x) - f(0, 0)}{\Delta x} = \lim_{\Delta x \to 0} \frac{\dfrac{(0 + \Delta x) \cdot 0}{\Delta x^2 + 0^2}}{\Delta x} \\
&= \lim_{\Delta x \to 0} \frac{0}{\Delta x} = \lim_{\Delta x \to 0} 0 = 0
\end{aligned}
$$

Similarly $f_y(0, 0) = 0$. Hence both f_x and f_y exist at $(0, 0)$ even though f is not continuous there.

18. For an ideal gas $P = \dfrac{RT}{V}$ and $\left(\dfrac{\partial P}{\partial T}\right)_V = \dfrac{R}{V} = \dfrac{P}{T}$. Thus, for an ideal gas $\left(\dfrac{\partial U}{\partial V}\right)_T = 0$.

For a van der Waals gas, $P = \dfrac{RT}{V - b} - \dfrac{a}{V^2}$ and $\left(\dfrac{\partial P}{\partial T}\right)_V = \dfrac{R}{V - b}$. Thus, $\left(\dfrac{\partial U}{\partial V}\right)_T = \dfrac{a}{V^2}$ for a van der Waals gas.

6.4 Chain Rule for Partial Differentiation

1. Since $x = te^{-t}$ and $y = e^{-t}$, we have that $u = x^2 y + xy^2 = t^2 e^{-3t} + te^{-3t}$ so that

$$
\frac{du}{dt} = 2te^{-3t} - 3t^2 e^{-3t} + e^{-3t} - 3te^{-3t} = (1 - t - 3t^2)e^{-3t}
$$

which is the same result as in Example 1.

3. Since $u = ye^{-x^2}$ and $x = t^2$ and $y = t^4$, we have $\dfrac{\partial u}{\partial x} = -2xye^{-x^2} = -2t^6 e^{-t^4}$ and $\dfrac{\partial u}{\partial y} = e^{-x^2} = e^{-t^4}$. Therefore,

$$
\begin{aligned}
\frac{du}{dt} &= \frac{\partial u}{\partial x}\frac{dx}{dt} + \frac{\partial u}{\partial y}\frac{dy}{dt} \\
&= (-2t^6 e^{-t^4})(2t) + (e^{-t^4})(4t^3) \\
&= 4t^3(1 - t^4)e^{-t^4}
\end{aligned}
$$

As a check, for $u(t) = t^4 e^{-t^4}$, we have that $du/dt = 4t^3 e^{-t^4} - 4t^7 e^{-t^4}$.

6. For $u(x, y) = y \cos xy$ and $y = e^{-x}$, Equation 8 says that

$$
\begin{aligned}
\frac{du}{dx} &= \frac{\partial u}{\partial x} + \frac{\partial u}{\partial y}\frac{dy}{dx} \\
&= -y^2 \sin xy + (\cos xy - xy \sin xy)(-e^{-x}) \\
&= -e^{-2x}\sin(xe^{-x}) - e^{-x}[\cos(xe^{-x}) - xe^{-x}\sin(xe^{-x})] \\
&= xe^{-2x}\sin(xe^{-x}) - e^{-x}\cos(xe^{-x}) - e^{-2x}\sin(xe^{-x})
\end{aligned}
$$

As a check, for $u(x) = e^{-x}\cos(xe^{-x})$, we have

$$
\begin{aligned}
\frac{du}{dx} &= -e^{-x}\cos(xe^{-x}) - e^{-x}\sin(xe^{-x})(e^{-x} - xe^{-x}) \\
&= xe^{-2x}\sin(xe^{-x}) - e^{-x}\cos(xe^{-x}) - e^{-2x}\sin(xe^{-x})
\end{aligned}
$$

9. We use the equation $\dfrac{\partial u}{\partial s} = \dfrac{\partial u}{\partial x}\dfrac{\partial x}{\partial s} + \dfrac{\partial u}{\partial y}\dfrac{\partial y}{\partial s}$. Now

$$
\frac{\partial u}{\partial x} = e^{x+y} = e^{ts^2 + \sin s} = \frac{\partial u}{\partial y} \qquad \frac{\partial x}{\partial s} = te^s \qquad \frac{\partial y}{\partial s} = \cos s
$$

Thus,

$$
\frac{\partial u}{\partial s} = (te^s + \cos s)e^{te^s + \sin s}
$$

As a check, for $u(s, t) = e^{te^s + \sin s}$, we have that $\dfrac{\partial u}{\partial s} = (te^s + \cos s)e^{te^s + \sin s}$.

12. Let $z_1 = x + ct$ and $z_2 = x - ct$. Therefore,

$$
\begin{aligned}
\frac{\partial u}{\partial t} &= \frac{\partial u}{\partial z_1}\frac{\partial z_1}{\partial t} + \frac{\partial u}{\partial z_2}\frac{\partial z_2}{\partial t} = c\frac{\partial u}{\partial z_1} - c\frac{\partial u}{\partial z_2} \\
\frac{\partial^2 u}{\partial t^2} &= c\frac{\partial^2 u}{\partial z_1^2}\frac{\partial z_1}{\partial t} - c\frac{\partial^2 u}{\partial z_2^2}\frac{\partial z_2}{\partial t} = c^2\left(\frac{\partial^2 u}{\partial z_1^2} + \frac{\partial^2 u}{\partial z_1^2}\right) \\
\frac{\partial u}{\partial x} &= \frac{\partial u}{\partial z_1}\frac{\partial z_1}{\partial x} + \frac{\partial u}{\partial z_2}\frac{\partial z_2}{\partial x} = \frac{\partial u}{\partial z_1} + \frac{\partial u}{\partial z_2}
\end{aligned}
$$

and

$$
\frac{\partial^2 u}{\partial x^2} = \frac{\partial^2 u}{\partial z_1^2} + \frac{\partial^2 u}{\partial z_2^2}
$$

Therefore, $\dfrac{\partial^2 u}{\partial x^2} = \dfrac{1}{c^2}\dfrac{\partial^2 u}{\partial t^2}$.

15. The n_j are extensive variables (with $\lambda = 1$), so Equation 14 gives $Y = \sum n_j \left(\dfrac{\partial Y}{\partial n_j}\right) = \sum_j n_j \overline{Y}_j$.

6.5   Differentials and the Total Differential

1. Start with $\Delta f = f(x + \Delta x, y + \Delta y) - f(x, y)$, then we have

$$
\begin{aligned}
\Delta f &= (x + \Delta x)\sin(y + \Delta y) + (x + \Delta x)^2 e^{y+\Delta y} - x\sin y - x^2 e^y \\
&= x\sin\cos\Delta y + x\cos y \sin\Delta y + \Delta x \sin y \cos\Delta y + \Delta x \cos y \sin y \\
&\quad + x^2 e^y \left[1 + \Delta y + \frac{(\Delta y)^2}{2} + O[(\Delta y)^3]\right] + 2x\Delta x e^y \left[1 + \Delta y + \frac{(\Delta y)^2}{2} + \cdots\right] \\
&\quad + (\Delta x)^2 e^y \left[1 + \Delta y + \frac{(\Delta y)^2}{2} + \cdots\right] \\
&= x\sin y \left[1 - \frac{(\Delta y)^2}{2} + \cdots\right] + x\cos y\left[\Delta y - \frac{(\Delta y)^3}{6} + \cdots\right] + \Delta x \sin y \left[1 - \frac{(\Delta y)^2}{2} + \cdots\right] \\
&\quad + \Delta x \cos y \left[\Delta y - \frac{(\Delta y)^2}{6} + \cdots\right] + x\cos y\left[\Delta y - \frac{(\Delta y)^3}{6} + \cdots\right] \\
&\quad + \Delta x \sin y \left[1 - \frac{(\Delta y)^2}{2} + \cdots\right] + \Delta x \cos y \left[\Delta y - \frac{(\Delta y)^2}{6} + \cdots\right] - x\sin y - x^2 e^y \\
&= x\sin y - \frac{x\sin y}{2}(\Delta y)^2 + x\cos y \Delta y + \sin y \Delta x + \cos y \Delta x \Delta y + x^2 e^y + x^2 e^y \Delta y \\
&\quad + \frac{x^2 e^y}{2}(\Delta y)^2 + 2xe^y \Delta x + 2xe^y \Delta x \Delta y + e^y (\Delta x)^2 + \text{higher-order terms} - x\sin y - x^2 e^y \\
&= (\sin y + 2xe^y)\Delta x + (x\cos y + x^2 e^y)\Delta y - \frac{x\sin y}{2}(\Delta y)^2 + \cos y \Delta x \Delta y \\
&\quad + \frac{x^2 e^y}{2}(\Delta y)^2 + e^y (\Delta x)^2 + 2xe^y \Delta x \Delta y + O(\Delta^3 \text{ terms}) \\
&= \frac{\partial f}{\partial x}\Delta x + \frac{\partial f}{\partial y}\Delta y + \epsilon_1 \Delta x + \epsilon_2 \Delta y + O(\Delta^3 \text{ terms})
\end{aligned}
$$

where $\epsilon_1 = \cos y\,\Delta y + e^y \Delta x + \cdots$ and $\epsilon_2 = \dfrac{x^2 e^y}{2}\Delta y + 2xe^y \Delta x - \dfrac{x\sin y}{2}\Delta y + \cdots$. Both ϵ_1 and $\epsilon_2 \to 0$ as Δx and $\Delta y \to 0$.

3. (a) For $f(x, y) = x^2 \sin y$, $df = \left(\dfrac{\partial f}{\partial x}\right)_y dx + \left(\dfrac{\partial f}{\partial y}\right)_x dy = 2x\sin y\,dx + x^2 \cos y\,dy$

(b) For $g(u, v) = (u^3 + v)e^v$, $dg = \left(\dfrac{\partial g}{\partial u}\right)_v du + \left(\dfrac{\partial g}{\partial v}\right)_u dv = 3u^2 e^v du + (u^3 + 1 + v)e^v dv$

6. The fact that $F(x, y, z)) = 0$ implies that $\dfrac{\partial F}{\partial x}dx + \dfrac{\partial F}{\partial y}dy + \dfrac{\partial F}{\partial z}dz = 0$ or that

$$
dz = -\frac{\partial F/\partial x}{\partial F/\partial z}dx - \frac{\partial F/\partial y}{\partial F/\partial z}dy
$$

for $\dfrac{\partial F}{\partial z} \neq 0$. So

$$
\frac{\partial z}{\partial x} = -\frac{\partial F/\partial x}{\partial F/\partial z} \qquad \text{and} \qquad \frac{\partial z}{\partial y} = -\frac{\partial F/\partial y}{\partial F/\partial z}
$$

for $\dfrac{\partial F}{\partial z} \neq 0$.

9. The uncertainty in the volume is given by

$$
\begin{aligned}
\Delta V &= \frac{4\pi}{3}(bc\Delta a + ac\Delta b + ab\Delta c) \\
&= \frac{4\pi}{3}\left[\,(10.0)(8.00)(0.050) + (10.0)(8.00)(0.050) + (10.0)(10.0)(0.050)\,\right] \\
&= (13.0)\frac{4\pi}{3} = 54.45
\end{aligned}
$$

The volume is 3351, and so the relative uncertainty is $\dfrac{\Delta V}{V} = 0.01625$.

12. $dx = C_V(T)dT + \dfrac{nRT}{V}dV$ is not an exact differential because $\dfrac{\partial^2 x}{\partial T \partial V} = 0 \neq \dfrac{nR}{V}$. But dx/T is an exact differential because both mixed second partial derivatives are equal to zero.

6.6   The Directional Derivative and the Gradient

1. Using the definition of $\nabla\phi$, we have $\nabla\phi = \dfrac{\partial \phi}{\partial x}\mathbf{i} + \dfrac{\partial \phi}{\partial y}\mathbf{j} = (2xy + 3y^3)\mathbf{i} + (x^2 + 9xy^2)\mathbf{j}$.

3. Using the definition of $\nabla\phi$, we have $\nabla\phi = e^x \cos y\,\mathbf{i} - e^x \sin y\,\mathbf{j} = \mathbf{i}$, or $\nabla\phi = \mathbf{i}$ at $(0,0)$. Thus, $\nabla\phi \cdot \mathbf{i} = 1$.

6. Using the definition of $\nabla\phi$, we have $\nabla\phi = 3x^2\,\mathbf{i} + 2y\,\mathbf{j} + \mathbf{k}$, or $\nabla\phi = \mathbf{k}$ at $(0,0,0)$, or $|\nabla\phi| = 1$.

9. The normals to the two curves are given by $\nabla\phi_1 = y\,\mathbf{i} + x\,\mathbf{j}$ and $\nabla\phi_2 = 2x\,\mathbf{i} - 2y\,\mathbf{j}$. The angle between the two families of curves is given by $\nabla\phi_1 \cdot \nabla\phi_2$, which equals $(2xy - 2xy) = 0$. Therefore, the two families of curves are orthogonal.

12. The gradient at the point $(5, 10)$ consists of

$$
\left(\frac{\partial z}{\partial x}\right) = -\frac{4000x}{200}e^{-(2x^2 + y^2)/200} \qquad \left(\frac{\partial z}{\partial x}\right)_0 = -100e^{-0.75}
$$

$$
\left(\frac{\partial z}{\partial y}\right) = -\frac{2000y}{200}e^{-(2x^2 + y^2)/200} \qquad \left(\frac{\partial z}{\partial y}\right)_0 = -100e^{-0.75}
$$

and so $(\nabla z)_0 = 100e^{-0.75}(-\mathbf{i} - \mathbf{j})$.

15. Let the xy-plane be perpendicular to the lines of charge, and let the lines of charge be located at $x = \pm a$. Then the equipotential surfaces are given by

$$
\frac{r_2^2}{r_1^2} = \frac{(x + a)^2 + y^2}{(x - a)^2 + y^2} = k^2 = \text{constant}
$$

(See the figure on the next page.) Expand this result and complete the square to get

$$
\left[x - a\left(\frac{k^2 + 1}{k^2 - 1}\right)\right]^2 + y^2 = \frac{4k^2 a^2}{(k^2 - 1)^2}
$$

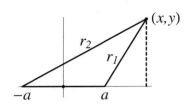

For $k > 1$ (corresponding to $\phi > 0$), the equipotential surfaces are circular right cylinders centered at $x = a\left(\dfrac{k^2+1}{k^2-1}\right) > 0$ and of radius $2ak/(k^2-1)$.

For $k < 1$ (corresponding to $\phi < 0$), they are centered at $x = a\left(\dfrac{k^2+1}{k^2-1}\right) < 0$. Cross sections of the equipotential surfaces are shown in the following figure.

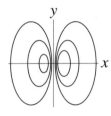

18. The normal to the surface is given by $\boldsymbol{\nabla}\phi$, which in this case is $\boldsymbol{\nabla}(xyz) = \boldsymbol{\nabla}\phi = yz\,\mathbf{i} + xz\,\mathbf{j} + xy\,\mathbf{k}$. The value of $\boldsymbol{\nabla}(xyz)$ is $= -\mathbf{i} + \mathbf{j} - \mathbf{k}$ at $(-1, 1, -1)$ and the unit vector is $\mathbf{u} = \dfrac{-\mathbf{i} + \mathbf{j} - \mathbf{k}}{\sqrt{3}}$.

21. The vector perpendicular to the surface is given by $\boldsymbol{\nabla}f = (y\cos xy)\,\mathbf{i} + (1 + x\cos xy)\,\mathbf{j}$, which is equal to $\mathbf{i} + \mathbf{j}$ at $(0, 1)$. Therefore, the tangent line is $(x - 0) + (y - 1) = 0$, or $x + y = 1$. The following figure shows the surface and the tangent line at the point $(0, 1)$.

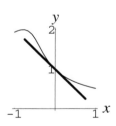

6.7   Taylor's Formula in Several Variables

1. (a) $\hat{A}2x = \dfrac{d^2}{dx^2}(2x) = 0$

(b) $\hat{A}x^2 = \left(\dfrac{d^2}{dx^2} + \dfrac{d}{dx} + 3\right)x^2 = 2 + 2x + 3x^2$

(c) $\hat{A}xy^2 = \left(\dfrac{\partial}{\partial x} + 2\dfrac{\partial}{\partial y}\right)xy^2 = y^2 + 4xy$

(d) $\hat{A}\sin xy = \left(x\dfrac{\partial}{\partial x} + y\dfrac{\partial}{\partial y}\right)\sin xy = 2xy\cos xy$

3. (a) $\hat{A}^2 f = \dfrac{d^2}{dx^2}\dfrac{d^2 f}{dx^2} = \dfrac{d^4 f}{dx^4}$

(b) $\hat{A}^2 f = \left(\dfrac{d}{dx}+x\right)\left(\dfrac{df}{dx}+xf\right) = \dfrac{d^2 f}{dx^2}+f+x\dfrac{df}{dx}+x\dfrac{df}{dx}+x^2 f = \dfrac{d^2 f}{dx^2}+2x\dfrac{df}{dx}+(1+x^2)f$

6. (a) $\left(1-\dfrac{d}{dx}\right)^2 \sin x = \left(1-\dfrac{d}{dx}\right)(\sin x - \cos x) = \sin x - \cos x - \cos x - \sin x = -2\cos x$

(b) $\left(h\dfrac{\partial}{\partial x}+k\dfrac{\partial}{\partial y}\right)^3 xy^3 = \left(h\dfrac{\partial}{\partial x}+k\dfrac{\partial}{\partial y}\right)\left(h\dfrac{\partial}{\partial x}+k\dfrac{\partial}{\partial y}\right)(hy^3+3kxy^2) =$

$\left(h\dfrac{\partial}{\partial x}+k\dfrac{\partial}{\partial y}\right)(3hky^2+3hky^2+6k^2xy) = 6hk^2y+12hk^2y+6k^3x = 18hk^2y+6k^3x$

9. We have $f(x,y) = x^2 - xy + y^2$. Thus $\dfrac{\partial f}{\partial x} = 2x - y = 1$ at $(1,1)$; $\dfrac{\partial f}{\partial y} = -x + 2y = 1$ at $(1,1)$; $f_{xx} = 2$; $f_{yy} = 2$; and $f_{xy} = -1$. All higher derivatives are equal to zero. Therefore,

$$\begin{aligned} f(x,y) &= 1 + (x-1) + (y-1) + (x-1)^2 + (y-1)^2 - (x-1)(y-1) \\ &= 1 + x - 1 + y - 1 + x^2 - 2x + 1 + y^2 - 2y + 1 - xy + x + y - 1 \\ &= x^2 + y^2 - xy \end{aligned}$$

as you would expect.

12. For $f(x,y) = \sin xy$, $f(1,1) = \sin 1$; $f_x = y\cos xy = \cos 1$ at $(1,1)$; $f_y = x\cos xy = \cos 1$ at $(1,1)$; $f_{xx} = -y^2\sin xy = -\sin 1$ at $(1,1)$; $f_{yy} = -x^2\sin xy = -\sin 1$ at $(1,1)$; and $f_{xy} = \cos xy - xy\sin xy = \cos 1 - \sin 1$ at $(1,1)$. Therefore,

$$\begin{aligned} f(x,y) &= \sin 1 + \cos 1(x-1) + \cos 1(y-1) - \dfrac{\sin 1}{2}(x-1)^2 - \dfrac{\sin 1}{2}(y-1)^2 \\ &\quad + (\cos 1 - \sin 1)(x-1)(y-1) + \cdots \end{aligned}$$

6.8   Maxima and Minima

1. $f(x) = 1 - |x|$ has a maximum value of 1 at $x = 0$. The first derivative test doesn't apply because $f(x)$ is not differentiable at $x = 0$.

3. For $f(x) = x^x$, $f'(x) = x^x(\ln x + 1) = 0$ at $x = 1/e$ and $f''(x) = x^x(\ln +1) + x^{x-1} > 0$ at $x = 1/e$. Thus, there is a minimum at $x = 1/e$. There is a maximum value of 1 at $x = 1$ in $(0,1]$, but there is no maximum value in the open interval$(0,1)$. (See the figure on the following page.)

6. The Taylor expansion of $f(x)$ about $x = a$ is

$$f(x) = f(a) + f'(a)(x-a) + \dfrac{(x-a)^{2n-1}}{(2n+1)!}f^{(2n+1)}(\xi)$$

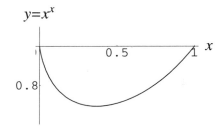

and

$$f(x) - [f(a) + f'(a)(x-a)] = \frac{(x-a)^{2n-1}}{(2n+1)!} f^{(2n+1)}(\xi)$$

The left side here is the difference between $f(x)$ and its tangent line at $x = a$. The factor $(x-a)^{2n+1}$ changes sign as x passes through a, so $f(x) > f(a) + f'(a)(x-a)$ on one side of a and $f(x) < f(a) + f'(a)(x-a)$ on the other side. Therefore, $x = a$ is an inflection point.

9. (a) $f(x,y) = xy^2(3x + 6y - 2)$. The critical points are given by the equations

$$
\begin{aligned}
f_x &= 6xy^2 + 6y^3 - 2y^2 = 0 \\
f_y &= 6x^2 y + 18xy^2 - 4xy = 0
\end{aligned}
$$

The solutions to these equations are $(0, 1/3)$, $(1/6, 1/6)$, and $(x, 0)$. The determinant D is

$$D = f_{xx}f_{yy} - f_{xy}^2 = -4y^2[27x^2 + (2-9y)^2 + 18x(3y-1)] \quad \text{and} \quad f_{xx} = 6y^2$$

Let's consider each critical point in turn.

$(0, 1/3)$: $D = -\dfrac{4}{9} < 0$ and $f_{xx} = \dfrac{2}{3} > 0$. The critical point is a saddle point.

$(1/6, 1/6)$: $D = \dfrac{1}{18} > 0$ and $f_{xx} = \dfrac{1}{6} > 0$. The critical point is a minimum.

$(x, 0)$: $D = 0$ and $f_{xx} = 0$. There is no conclusion.

(b) $f(x,y) = x^4 + y^4 - 2(x-y)^2$. The critical points are given by the equations

$$
\begin{aligned}
f_x &= 4x^3 - 4(x-y) = 0 \\
f_y &= 4(x-y) + 4y^3 = 0
\end{aligned}
$$

The solutions to these equations are $(\sqrt{2}, -\sqrt{2})$, $(-\sqrt{2}, \sqrt{2})$, and $(0,0)$. The determinant D is

$$D = f_{xx}f_{yy} - f_{xy}^2 = 48x^2(3y^2 - 1) - 48y^2 \quad \text{and} \quad f_{xx} = 12x^2 - 4$$

Let's consider each critical point in turn.

$(\sqrt{2}, -\sqrt{2})$: $D = 348 > 0$ and $f_{xx} = 20 > 0$. The critical point is a minimum.

$(-\sqrt{2}, \sqrt{2})$: $D = 348 > 0$ and $f_{xx} = 20 > 0$. The critical point is a minimum.

$(0,0)$: $D = 0$ and $f_{xx} = 4$. There is no conclusion.

12. (a) For $f(x,y) = 3x^2 + 6xy + 2y^3 + 12x - 24y$, we have that $f_x = 6x + 6y + 12$ and $f_y = 6y^2 + 6x - 24$. Both $f_x = f_y = 0$ at the points $(-5, 3)$ and $(0, -2)$. At the point

$(0, -2)$, $f_{xx} = 6 > 0$ and $D = (6)(-24) - 36 < 0$. At the point $(-5, 3)$, $f_{xx} = 6 > 0$ and $D = (6)(36) - 36 > 0$. Thus, the critical point $(0, -2)$ is a saddle point and the critical point $(-5, 3)$ is a minimum.

(b) For $f(x, y) = x^2 + y^2 + 4x - 2y + 3$, $f_x = 2x + 4$ and $f_y = 2y - 2$. Both $f_x = f_y = 0$ at the only critical point, $(-2, 1)$. At $(-2, 1)$, $f_{xx} = 2 > 0$ and $D = 4 > 0$. Thus, the critical point is a minimum.

15. We have $x + y + z = 150$ and $f(x, y, z) = xyz$. Or $f(x, y) = xy(150 - x - y)$. Thus $\dfrac{\partial f}{\partial x} = y(150 - x - y) - xy = 0$ and $\dfrac{\partial f}{\partial y} = x(150 - x - y) - xy = 0$. The solutions are $(0, 0)$, $(0, 150)$, $(50, 50)$, and $(150, 0)$. Only the $(50, 50)$ solution gives a maximum, which gives $x = y = z = 50$, or Product $= (50)^3 = 125\ 000$.

6.9   The Method of Lagrange Multipliers

1. The distance from a point of the ellipse to the origin is given by $D^2 = x^2 + y^2$. The coordinates (x, y) must satisfy the equation of the ellipse, so $g(x, y) = 3x^2 + 3y^2 + 4xy - 2 = 0$.

$$\frac{\partial D^2}{\partial x} + \lambda \frac{\partial g}{\partial x} = 2x + \lambda(6x + 4y) = 0$$

and

$$\frac{\partial D^2}{\partial y} + \lambda \frac{\partial g}{\partial y} = 2y + \lambda(6y + 4x) = 0$$

or

$$x(1 + 3\lambda) + 2\lambda y = 0 \qquad \text{and} \qquad 2\lambda x + y(1 + 3\lambda) = 0$$

Multiply the first equation by y and the second by x to get $x = \pm y$. For $x = y$, $g(x, y) = 0$ gives us $x = y = \pm \dfrac{1}{\sqrt{5}}$. For $x = -y$, $g(x, y) = 0$ gives us $x = \pm 1$, $y = \pm 1$. Now $D^2 = 2$ for $(+1, -1)$, and $(-1, 1)$, and $D^2 = 2/5$ for $x = y = \pm 1/\sqrt{5}$. Therefore, the maximum distance from the origin to the ellipse is $\sqrt{2}$ (along $x = -y$. See Figure 6.45.)

3. In this case, the volume is a rectangular parallelepiped of sides $2a$, $2b$, and $2c$, so $V = f(x, y, z) = 8xyz$. The constraint is the equation of the ellipsoid,

$$g(x, y, z) = \frac{x^2}{a^2} + \frac{y^2}{b^2} + \frac{z^2}{c^2} - 1 = 0$$

The Lagrange multiplier equations are

$$\frac{\partial f}{\partial x} + \lambda \frac{\partial g}{\partial x} = 8yz + \frac{2\lambda x}{a^2} = 0$$
$$\frac{\partial f}{\partial y} + \lambda \frac{\partial g}{\partial y} = 8xz + \frac{2\lambda y}{b^2} = 0$$
$$\frac{\partial f}{\partial z} + \lambda \frac{\partial g}{\partial z} = 8xy + \frac{2\lambda z}{c^2} = 0$$

Solving each equation for λ and equating the results yields $\dfrac{x^2}{a^2} = \dfrac{y^2}{b^2} = \dfrac{z^2}{c^2} = \dfrac{1}{3}$. Therefore,

$$V = \frac{8abc}{3^{3/2}}$$

6. Let the sides of the rectangular parallelepiped be a, b, and c. Then, the area A is $A = 2(ab + ac + bc)$ and the total edge length l is $l = 4(a + b + c)$. The three Lagrange multiplier equations are

a: $2b + 2c + 4\lambda = 0$;
b: $2a + 2c + 4\lambda = 0$;
c: $2c + 2b + 4\lambda = 0$.

Therefore,, $a = b = c$ and the body is a cube.

9. The distance from the origin to the plane is given by $f(x,y,z) = x^2 + y^2 + z^2$. The constraint is given by the equation of the plane, or by $g(x,y,z) = 3x + 2y + z - 6 = 0$. The three Lagrange multiplier equations are

$$x :\ 2x + 3\lambda = 0 \qquad y :\ 2y + 2\lambda = 0 \qquad z :\ 2z + \lambda = 0$$

The solution to these equations is $\dfrac{2x}{3} = y = 2z$, or $\dfrac{x}{3} = \dfrac{y}{2} = z$. Substituting this result into $3x + 2y + z = 6$ gives $9z + 4z + z = 6$, or $z = \dfrac{3}{7}$ and $x = \dfrac{9}{7}$ and $y = \dfrac{6}{7}$. Therefore,

$$D = \frac{(81 + 36 + 9)^{1/2}}{7} = \frac{(126)^{1/2}}{7} = \left(\frac{18}{7}\right)^{1/2}$$

12. The distance from the origin to the surface is given by $D^2 = x^2 + y^2 + z^2$. The constraint is given by the equation of the surface, or by $g(x,y,z) = x^2 + y^2 - 2xz - 4 = 0$. The three conditional minimization conditions are

$$2x + \lambda(2x - 2z) = 0 \quad \text{or} \quad (1 + \lambda) = \lambda z$$

$$2y + \lambda 2y = 0 \quad \text{or} \quad (1 + \lambda)y = 0$$

$$2z - \lambda 2x = 0 \quad \text{or} \quad z = \lambda x$$

There are two ways to satisfy these three equations, $\lambda = -1$ or $y = 0$. Let's consider the case $\lambda = -1$ first. This gives $z = 0$ and $x = 0$. The equation of the surface gives $y^2 = 4$, so that $D^2 = y^2 = 4$ or $D = 2$.

Now consider the case $y = 0$. This gives $\lambda^2 - \lambda - 1 = 0$, or $\lambda = \dfrac{1}{2} \pm \dfrac{1}{2}\sqrt{5}$. Using the equation for the surface, we have $x^2 + 0^2 - 2xz = x^2(1 - 2\lambda) = 4$, which gives us $x^2\sqrt{5} = 4$. (We reject $\lambda = \dfrac{1}{2} + \dfrac{1}{2}\sqrt{5}$ since x^2 must be positive.) Finally, then, we have

$$D^2 = x^2 + y^2 + z^2 = x^2 + \lambda^2 x^2 = \frac{4}{\sqrt{5}} + \frac{4}{\sqrt{5}}\left(\frac{1}{2} - \frac{\sqrt{5}}{2}\right)^2$$

$$= \frac{4}{\sqrt{5}}\left(1 + \frac{1}{4} - \frac{\sqrt{5}}{2} + \frac{5}{4}\right)$$

$$= \frac{2}{\sqrt{5}}(5 - \sqrt{5}) = 2.472$$

or $D = 1.572$.

15. In this case, $f(x, y, z) = z$ and the constraint is given by $g(x, y, z) = 3x^2 + 5y^2 + 2z^2 - 10xy + 2xz - 10 = 0$. The three Lagrange multiplier equations are

$$x: \ 0 - \lambda(6x - 10y + 2z) = 0 \qquad y: \ 0 - \lambda(10y - 10x) = 0 \qquad z: \ 1 - \lambda(4z + 2x) = 0$$

A solution to these three equations is given by $z = 2x = 2y = \dfrac{1}{5\lambda}$. The equation of the ellipsoid gives us $3x^2 + 5x^2 + 2(2x)^2 - 10x^2 + 2x(2x) = 10$, which implies that $x^2 = 1$, or $x = \pm 1$. Thus, $z = \pm 2$.

18. In this case, $f(x_1, x_2, \ldots, x_n) = x_1 x_2 \cdots x_n$ and the constraint is given by $x_1 + x_2 + \cdots + x_n = a$. The Lagrange multiplier equations are

$$\frac{\partial f}{\partial x_j} - \lambda \frac{\partial g}{\partial x_j} = \frac{x_1 x_2 \cdots x_n}{x_j} - \lambda = 0 \quad \text{for} \quad j = 1, \ 2, \ \ldots, \ n$$

The solution to these equations is $\lambda = \dfrac{x_1 x_2 \cdots x_n}{x_j}$, which says that the x_j are equal. The condition $x_1 + x_2 + \cdots + x_n = a$ implies that $x_j = \dfrac{a}{n}$, and so

$$f(x_1, x_2, \ldots, x_n) = x_1 x_2 \cdots x_n = \left(\frac{a}{n} \right)^n$$

6.10   Multiple Integrals

1.

$$\begin{aligned}
I &= \int_0^a dx\, x \int_0^{b(1-x/a)} dy\, y \\
&= \int_0^a dx\, x \frac{b^2}{2} \left(1 - \frac{x}{a} \right)^2 = \frac{a^2 b^2}{2} \int_0^1 du\, u(1-u)^2 = \frac{a^2 b^2}{24}
\end{aligned}$$

3.

$$\begin{aligned}
M &= \int_0^1 dx \int_0^{1-x} dy\, xy = \int_0^1 dx\, x \int_0^{1-x} dy\, y \\
&= \int_0^1 dx\, x \left[\frac{(1-x)^2}{2} \right] = \frac{1}{2} \frac{\Gamma(2)\Gamma(3)}{\Gamma(5)} = \frac{1}{24}
\end{aligned}$$

$$\bar{x} = \bar{y} = 24 \int_0^1 dx\, x^2 \int_0^{1-x} dy\, y = \frac{24}{2} \frac{\Gamma(3)\Gamma(3)}{\Gamma(6)} = \frac{2}{5}$$

6. Calculate the volume in the first octant and then multiply by 4:

$$\begin{aligned}
\frac{V}{4} &= \int_0^1 dx \int_0^{1-x} dy\, (1 - x^2 - y^2) \\
&= \int_0^1 dx \left[y(1 - x^2) - \frac{y^3}{3} \right]_0^{1-x} = \int_0^1 dx \left[(1-x)(1-x^2) - \frac{(1-x)^3}{3} \right] \\
&= \frac{2}{3} \int_0^1 dx\, (1-x)^2 (1+2x) = \frac{2}{3} \left[\int_0^1 dx\, (1-x)^2 + 2 \int_0^1 dx\, x(1-x)^2 \right] \\
&= \frac{2}{3} \left[\frac{\Gamma(3)\Gamma(1)}{\Gamma(4)} + \frac{2\Gamma(2)\Gamma(3)}{\Gamma(5)} \right] = \frac{1}{3}
\end{aligned}$$

Therefore, $V = 4/3$.

9.

$$M = \int_0^a dx \int_0^{b(1-x/a)} dy \int_0^{c(1-x/a-y/b)} dz$$

$$= \frac{bc}{2} \int_0^a \left(1 - \frac{x}{a}\right)^2 dx = \frac{abc}{2} \int_0^1 (1-u)^2 du$$

$$= \frac{abc}{2} \frac{\Gamma(3)}{\Gamma(4)} = \frac{abc}{6}$$

$$\frac{abc}{6}\overline{x} = \int_0^a dx\, x \frac{bc}{2} \left(1 - \frac{x}{a}\right)^2 = \frac{a^2 bc}{2} \int_0^1 du\, u(1-u)^2$$

$$= \frac{a^2 bc}{2} \frac{2\Gamma(2)\Gamma(3)}{\Gamma(5)} = \frac{a^2 bc}{24}$$

Thus, $\overline{x} = \dfrac{a}{4}$. By symmetry, $\overline{y} = \dfrac{b}{4}$ and $\overline{z} = \dfrac{c}{4}$.

12.

$$I = \int_0^b dy\, y \int_0^{a(1-y/b)} dx\, x$$

$$= \frac{a^2}{2} \int_0^b dy\, y \left(1 - \frac{y}{b}\right)^2 = \frac{a^2 b^2}{2} \int_0^1 du\, u(1-u)^2 = \frac{a^2 b^2}{24}$$

15. The moment of inertia is given by

$$I_z = \frac{4}{V} \int_0^h dz \int_0^R dx \int_0^{(R^2-x^2)^{1/2}} dy\, (x^2 + y^2)$$

$$= \frac{4h}{3V} \int_0^R dx\, (2x^2 + R^2)(R^2 - x^2)^{1/2} = \frac{\pi h R^4}{2V}$$

The volume of a right circular cylinder of radius R and height h is $V = h\pi r^2$, and so $I_z = R^2/2$.

18. Start with $\displaystyle\int_0^{\pi/2} dx \int_x^{\pi/2} du\, \frac{\sin u}{u}$. Changing orders of integration gives

$$\int_0^{\pi/2} du \int_0^u dx\, \frac{\sin u}{u} = \int_0^{\pi/2} \sin u\, du = 1$$

CHAPTER 7

Vector Calculus

7.1 Vector Fields

1. For $f(x, y, z) = z - 2x^2 - y^2 = 0$, $\nabla f = -4x\,\mathbf{i} - 2y\,\mathbf{j} + \mathbf{k}$, or $\nabla f = -4\,\mathbf{i} - 2\,\mathbf{j} + \mathbf{k}$ at $(1, 1, 3)$. The unit vector is $\mathbf{u} = \dfrac{-4\,\mathbf{i} - 2\,\mathbf{j} + \mathbf{k}}{\sqrt{21}}$.

3. For $\mathbf{A}(x, y, z) = xy^2\,\mathbf{i} + 2xyz\,\mathbf{j} - x^2 z\,\mathbf{k}$,

$$\operatorname{div}\mathbf{A} = \frac{\partial A_x}{\partial x} + \frac{\partial A_y}{\partial y} + \frac{\partial A_z}{\partial z} = y^2 + 2xz - x^2$$

and

$$\operatorname{curl}\mathbf{A} = \begin{vmatrix} \mathbf{i} & \mathbf{j} & \mathbf{k} \\ \dfrac{\partial}{\partial x} & \dfrac{\partial}{\partial y} & \dfrac{\partial}{\partial z} \\ xy^2 & 2xyz & -x^2 z \end{vmatrix} = -2xy\,\mathbf{i} + 2xz\,\mathbf{j} + (2yz - 2xy)\,\mathbf{k}$$

6.

$$\begin{aligned}
\operatorname{div}\frac{\mathbf{r}}{r^3} &= \frac{\partial}{\partial x}\frac{x}{(x^2 + y^2 + z^2)^{3/2}} + \frac{\partial}{\partial y}\frac{y}{(x^2 + y^2 + z^2)^{3/2}} + \frac{\partial}{\partial z}\frac{z}{(x^2 + y^2 + z^2)^{3/2}} \\
&= \left(\frac{1}{r^3} - \frac{3x^2}{r^5}\right) + \left(\frac{1}{r^3} - \frac{3y^2}{r^5}\right) + \left(\frac{1}{r^3} - \frac{3z^2}{r^5}\right) \\
&= \frac{3}{r^3} - \frac{3(x^2 + y^2 + z^2)}{r^5} = 0
\end{aligned}$$

9. Use Equation 30 to write $\nabla \times [f(r)\,\mathbf{r}] = \nabla f(r) \times \mathbf{r} + f(r)\nabla \times \mathbf{r}$. Problem 5 shows that $\nabla \times \mathbf{r} = \mathbf{0}$ $(\mathbf{r} \neq \mathbf{0}.)$ Now,

$$\begin{aligned}
\nabla f(r) &= \frac{\partial f}{\partial x}\,\mathbf{i} + \frac{\partial f}{\partial y}\,\mathbf{j} + \frac{\partial f}{\partial z}\,\mathbf{k} \\
&= \frac{\partial f}{\partial r}\frac{\partial r}{\partial x}\,\mathbf{i} + \frac{\partial f}{\partial r}\frac{\partial r}{\partial y}\,\mathbf{j} + \frac{\partial f}{\partial r}\frac{\partial r}{\partial z}\,\mathbf{k} \\
&= \left(\frac{\partial f}{\partial r}\right)\left(\frac{x}{r}\,\mathbf{i} + \frac{y}{r}\,\mathbf{j} + \frac{z}{r}\,\mathbf{k}\right) = \frac{1}{r}\left(\frac{\partial f}{\partial r}\right)\mathbf{r}
\end{aligned}$$

is colinear with $\mathbf{r}$ and so $\nabla f(r) \times \mathbf{r} = \mathbf{0}$.

12. $\operatorname{grad} \psi = \dfrac{\partial \psi}{\partial x}\mathbf{i} + \dfrac{\partial \psi}{\partial y}\mathbf{j} + \dfrac{\partial \psi}{\partial z}\mathbf{k}$ and

$$\operatorname{curl\,grad} \psi = \begin{vmatrix} \mathbf{i} & \mathbf{j} & \mathbf{k} \\ \dfrac{\partial}{\partial x} & \dfrac{\partial}{\partial y} & \dfrac{\partial}{\partial z} \\ \dfrac{\partial \psi}{\partial x} & \dfrac{\partial \psi}{\partial y} & \dfrac{\partial \psi}{\partial z} \end{vmatrix} = \mathbf{i}\left(\dfrac{\partial^2 \psi}{\partial y \partial z} - \dfrac{\partial^2 \psi}{\partial z \partial y} \right) + \text{etc.} = 0$$

15. Start with $\mathbf{u} \times \mathbf{v} = \begin{vmatrix} \mathbf{i} & \mathbf{j} & \mathbf{k} \\ u_x & u_y & u_z \\ v_x & v_y & v_z \end{vmatrix} = (u_y v_z - u_z v_y)\mathbf{i} + (u_z v_x - u_x v_z)\mathbf{j} + (u_x v_y - u_y v_x)\mathbf{k}.$

Then

$$
\begin{aligned}
\operatorname{div}(\mathbf{u} \times \mathbf{v}) &= \frac{\partial}{\partial x}(u_y v_z - u_z v_y) + \frac{\partial}{\partial y}(u_z v_x - u_x v_z) + \frac{\partial}{\partial z}(u_x v_y - u_y v_x) \\
&= v_z \frac{\partial u_y}{\partial x} + u_y \frac{\partial v_z}{\partial x} - u_z \frac{\partial v_y}{\partial x} - v_y \frac{\partial u_z}{\partial x} \\
&\quad + v_x \frac{\partial u_z}{\partial y} + u_z \frac{\partial v_x}{\partial y} - u_x \frac{\partial v_z}{\partial y} - v_z \frac{\partial u_x}{\partial y} \\
&\quad + v_y \frac{\partial u_x}{\partial z} + u_x \frac{\partial v_y}{\partial z} - u_y \frac{\partial v_x}{\partial z} - v_x \frac{\partial u_y}{\partial z} \\
&= v_x\left(\frac{\partial u_z}{\partial y} - \frac{\partial u_y}{\partial z}\right) + v_y\left(\frac{\partial u_x}{\partial z} - \frac{\partial u_z}{\partial x}\right) + v_z\left(\frac{\partial u_y}{\partial x} - \frac{\partial u_x}{\partial y}\right) \\
&\quad - u_x\left(\frac{\partial v_z}{\partial y} - \frac{\partial v_y}{\partial z}\right) - u_y\left(\frac{\partial v_x}{\partial z} - \frac{\partial v_z}{\partial x}\right) - u_z\left(\frac{\partial v_y}{\partial x} - \frac{\partial v_x}{\partial y}\right) \\
&= \mathbf{v} \cdot (\nabla \times \mathbf{u}) - \mathbf{u} \cdot (\nabla \times \mathbf{v})
\end{aligned}
$$

18. Referring to Figure 7.4, the integral over the front face $(\mathbf{n} = \mathbf{i})$ is

$$
\begin{aligned}
I_\mathrm{f} &= \int \mathbf{n} \times \mathbf{v}\, dydz = \iint (-v_z \mathbf{j} + v_y \mathbf{k})\, dydz \\
&= -\mathbf{j} \iint v_z(x + \Delta x, y, z)\, dydz + \mathbf{k} \iint v_y(x + \Delta x, y, z)\, dydz
\end{aligned}
$$

Using the mean value theorem, we can write this integral as

$$I_\mathrm{f} = -\mathbf{j}\, v_z(x + \Delta x, y_1, z_1)\Delta y \Delta z + \mathbf{k}\, v_y(x + \Delta x, y_2, z_2)\Delta y \Delta z$$

where y_1, z_1, y_2, and z_2 are coordinates lying in the front face. Similarly, the integral over the back face $(\mathbf{n} = -\mathbf{i})$ is

$$I_\mathrm{b} = \iint \mathbf{n} \times \mathbf{v}\, dydz = \mathbf{j}\, v_z(x, y_3, z_3)\Delta y \Delta z - \mathbf{k}\, v_y(x, y_4, z_4)\Delta y \Delta z$$

The sum of the integrals is

$$
\begin{aligned}
\iint \mathbf{n} \times \mathbf{v}\, dydz &= -\mathbf{j}\,[v_z(x + \Delta x, y_1, z_1)\Delta y \Delta z - v_z(x, y_3, z_3)] \\
&\quad + \mathbf{k}\,[v_y(x + \Delta x, y_2, z_2) - v_y(x, y_4, z_4)]\Delta y \Delta z
\end{aligned}
$$

Divide by $V = \Delta x \Delta y \Delta z$ and let $\Delta x \to 0$ to get

$$\lim_{V \to 0} \frac{\displaystyle\iint \mathbf{n} \times \mathbf{v}\, dydz}{V} = -\frac{\partial v_z}{\partial x}\mathbf{j} + \frac{\partial v_y}{\partial x}\mathbf{k}$$

Integrals over the other faces are evaluated in the same way to give Equation 17.

7.2   Line Integrals

1. Start with $\mathbf{F} \cdot d\mathbf{r} = -y\, dx + x^2\, dy$. Now let $x = u^2$ and $y = u$ to get $\mathbf{F} \cdot d\mathbf{r} = -u(2u\, du) + u^4\, du = (u^4 - 2u^2)\, du$. Thus $\displaystyle\int_0^1 du\,(u^4 - 2u^2) = \frac{1}{5} - \frac{2}{3} = -\frac{7}{15}$.

3. Let $x = \cos u$ and $y = \sin u$. Thus $\mathbf{A}(x,y) = y\,\mathbf{i} - x\,\mathbf{j} = \sin u\,\mathbf{i} - \cos u\,\mathbf{j}$ and

$$\oint \mathbf{A} \cdot d\mathbf{r} = \oint (A_x\, dx + A_y\, dy) = \int_0^{2\pi} (-\sin^2 u - \cos^2 u)\, du = -2\pi$$

(Note that Equation 16 gives the same result.)

6. Let $x = \cos u$ and $y = \sin u$. Thus $\mathbf{F}(x,y) = \dfrac{-y\,\mathbf{i} + x\,\mathbf{j}}{x^2 + y^2} = -\sin u\,\mathbf{i} + \cos u\,\mathbf{j}$ and

$$\oint \mathbf{F} \cdot d\mathbf{r} = \int_0^{2\pi} (\sin^2 u - \cos^2 u)\, du = 2\pi$$

Note that Equation 16 gives the same result.

9. If $\mathbf{F}(x,y) = \dfrac{x\,\mathbf{i} + y\,\mathbf{j}}{(x^2 + y^2)^{3/2}}$, then $\dfrac{\partial \phi}{\partial x} = \dfrac{x}{(x^2 + y^2)^{3/2}}$. Integrating with respect to x gives $\phi = -\dfrac{1}{(x^2 + y^2)^{1/2}} + f(y)$. Now use

$$\frac{\partial \phi}{\partial y} = \frac{y}{(x^2 + y^2)^{3/2}} + f'(y) = \frac{y}{(x^2 + y^2)^{3/2}}$$

to see that $f(y) = $ constant. Therefore, $\phi(x,y) = $ constant $- \dfrac{1}{(x^2 + y^2)^{1/2}}$, as you can readily check by taking the gradient of $\phi(x,y)$.

12. (a) Start with $\kappa = \displaystyle\oint \mathbf{v} \cdot d\mathbf{r} = \oint v_0\, dx$. Now let $x = \cos u$ to write

$$\kappa = \oint \mathbf{v} \cdot d\mathbf{r} = -\int_0^{2\pi} v_0 \sin u\, du = 0$$

(b) Letting $x = \cos u$ and $y = \sin u$, we have

$$\begin{aligned}
\kappa &= \oint \mathbf{v} \cdot d\mathbf{r} = \oint [\, 2xy\, dx + (x^2 - y^2)\, dy \,] \\
&= \int_0^{2\pi} [-2\cos u \sin^2 u + (\cos^2 u - \sin^2 u)\cos u]\, du \\
&= \int_0^{2\pi} \cos^3 u\, du - 3\int_0^{2\pi} \cos u \sin^2 u\, du = 0
\end{aligned}$$

15. Start with

$$\iint\limits_{R} \frac{\partial P}{\partial y} dx dy = \int_a^c dx \int_{y_1(x)}^{y_2(x)} dy \frac{\partial P}{\partial y} = \int_a^c dx \left[P(x, y_2(x)) - P(x, y_1(x)) \right]$$

But,

$$\int_a^c dx\, P(x, y_2(x)) - \int_a^c dx\, P(x, y_1(x)) = -\int_c^a dx\, P(x, y_2(x)) - \int_a^c dx\, P(x, y_1(x))$$
$$= -\oint P\, dx$$

Similarly,

$$\iint\limits_{R} \frac{\partial P}{\partial x} dx dy = \int_b^d dy\, Q(y, x_2(y)) - \int_b^d dy\, Q(y, x_1(y))$$
$$= \int_b^d dy\, Q(y, x_2(y)) + \int_d^b dy\, Q(y, x_1(y)) = \oint Q\, dy$$

Therefore,

$$\iint\limits_{R} \left(\frac{\partial \psi}{\partial x} - \frac{\partial P}{\partial y} \right) dx dy = \oint (P\, dx + Q\, dy)$$

18. Letting $x = a \cos \theta$ and $y = b \sin \theta$ in $A = \frac{1}{2} \oint (x\, dy - y\, dx)$ gives

$$A = \frac{1}{2} \oint (ab \cos^2 \theta + ab \sin^2 \theta)\, d\theta = \frac{ab}{2} \int_0^{2\pi} d\theta = \pi ab$$

7.3   Surface Integrals

1. (a) $x = u \quad y = v^2 \quad z = v$

(b) $x = u \quad y = v \quad z = u + 2v - 2$

(c) $x = v \cos u \quad y = v \sin u \quad z = v$

(d) $x = v \cos u \quad y = v \sin u \quad z = v^2$

3. No. $x = e^u \geq 0$ for all u. So $\mathbf{r}(u, v)$ represents only the half with $x \geq 0$.

6. We use Equation 7 with $z = h(x, y) = 6 - 2x - 3y$, $\dfrac{\partial h}{\partial x} = -2$, and $\dfrac{\partial h}{\partial y} = -3$. Thus,

$$A = \iint\limits_{R} (1 + h_x^2 + h_y^2)^{1/2}\, dx dy = \iint\limits_{R} \sqrt{14}\, dx dy$$

To determine the limits on x and y, realize that the projection of the plane onto the xy-plane is $2x + 3y = 6$. Therefore,

$$A = \sqrt{14} \int_0^3 dx \int_0^{2 - \frac{2}{3}x} dy = \sqrt{14} \int_0^3 dx \left(2 - \frac{2}{3}x \right) = \sqrt{14}\,(6 - 3) = 3\sqrt{14}$$

9. Use Equation 7 with $z = h(x,y) = 4 - 2x - \frac{4}{3}y$, $\frac{\partial h}{\partial x} = -2$, and $\frac{\partial h}{\partial y} = -\frac{4}{3}$. Thus

$$(1 + h_x^2 + h_y^2)^{1/2} = \left(1 + 4 + \frac{16}{9}\right)^{1/2} = \frac{\sqrt{61}}{3}$$

and

$$
\begin{aligned}
I &= \frac{\sqrt{61}}{3} \iint_R \left(z + 2x + \frac{4}{3}y\right) dx\, dy \\
&= \frac{\sqrt{61}}{3} \iint_R \left(\frac{12 - 6x - 4y}{3} + 2x + \frac{4}{3}y\right) dx\, dy
\end{aligned}
$$

To determine the limits on x and y, realize that the projection of the plane onto the xy-plane is $y = 3 - \frac{3}{2}x$. Therefore,

$$I = \frac{4\sqrt{61}}{3} \iint_R dx\, dy = \frac{4\sqrt{61}}{3} \int_0^2 \int_0^{3 - \frac{3}{2}x} dy = 4\sqrt{61}$$

12. Let the faces of the cube be at $x = 0$, $x = 1$; $y = 0$, $y = 1$; and $z = 0$, $z = 1$, and consider each of the six faces of the cube separately.

1. $\mathbf{n} = \mathbf{i}$; $\mathbf{F} \cdot \mathbf{n} = x = 1$
2. $\mathbf{n} = -\mathbf{i}$ $x = 0$; $\mathbf{F} \cdot \mathbf{n} = 0$
3. $\mathbf{n} = \mathbf{j}$; $\mathbf{F} \cdot \mathbf{n} = y = 1$
4. $\mathbf{n} = -\mathbf{j}$; $\mathbf{F} \cdot \mathbf{n} = -y = 0$
5. $\mathbf{n} = \mathbf{k}$; $\mathbf{F} \cdot \mathbf{n} = 2$
6. $\mathbf{n} = -\mathbf{k}$; $\mathbf{F} \cdot \mathbf{n} = -2$

The sum over the six faces gives a total of 2.

15.

$$
\begin{aligned}
I &= \sqrt{3} \int_0^1 du\, u \int_0^{1-u} (v - uv - v^2)\, dv \\
&= \sqrt{3} \int_0^1 du\, u \left[\frac{v^2}{2}(1 - u) - \frac{v^3}{3}\right]_0^{1-u} \\
&= \frac{\sqrt{3}}{6} \int_0^1 du\, u(1 - u)^3 = \frac{1}{6}\frac{\Gamma(2)\Gamma(4)}{\Gamma(6)} = \frac{3 \cdot 2}{6 \cdot 5 \cdot 4 \cdot 3 \cdot 2} = \frac{\sqrt{3}}{120}
\end{aligned}
$$

7.4 The Divergence Theorem

1. See Figure 7.39 with $(1,0,0) \to (a,0,0)$, $(0,1,0) \to (0,a,0)$, and $(0,0,1) \to (0,0,a)$ for the limits:

$$
\begin{aligned}
\iiint_V dx\, dy\, dz &= \int_0^a dx \int_0^{a-x} dy \int_0^{a-x-y} dz = \int_0^a dx \int_0^{a-x} dy\, (a - x - y) \\
&= \int_0^a dx \left[\frac{(a-x)^2}{2}\right] = \frac{a^3}{6}
\end{aligned}
$$

3. (See Figure 7.39 for the limits.) $I = \int_0^1 dz\, z \int_0^{1-z} dy\, y \int_0^{1-y-z} dx\, x = \dfrac{1}{720}$

6. For $\mathbf{F}(x,y) = x\,\mathbf{i} + y\,\mathbf{j} + z\,\mathbf{k}$, $\boldsymbol{\nabla} \cdot \mathbf{F} = 3$, and $\iiint_V \operatorname{div} \mathbf{F}\, dV = 3 \iiint_{\text{cube}} dx\,dy\,dz = 3$.

For the six surfaces:

$x = 1$, $\mathbf{n} = \mathbf{i}$, and $\iint_S \mathbf{F} \cdot \mathbf{n}\, dS = \iint_S x\, dy\,dz = 1$;

$x = 0$, $\mathbf{n} = -\mathbf{i}$, and $\iint_S \mathbf{F} \cdot \mathbf{n}\, dS = 0$;

with similar results over the y and z faces.

9. For $\mathbf{F}(x,y) = x^3\,\mathbf{i} + y^3\,\mathbf{j} + z^3\,\mathbf{k}$, $\operatorname{div} \mathbf{F} = 3x^2 + 3y^2 + 3z^2$. The integral of div $\mathbf{F}$ over the cylinder is given by

$$
\begin{aligned}
I &= \iiint_V \operatorname{div} \mathbf{F}\, dV = 3 \iiint_V x^2 dV + 3 \iiint_V y^2 dV + 3 \iiint_V z^2 dV \\
&= 6 \iiint_V x^2\, dV + 3 \iiint_V z^2\, dV \\
&= 6 \int_{-1}^1 dz \int_{-2}^2 dx\, x^2 \int_{-(4-x^2)^{1/2}}^{(4-x^2)^{1/2}} dy + 3 \int_{-1}^1 dz\, z^2 \int_{-2}^2 dx \int_{-(4-x^2)^{1/2}}^{(4-x^2)^{1/2}} dy \\
&= 48 \int_0^2 dx\, x^2 (4 - x^2)^{1/2} + 8 \int_0^2 dx \int_0^{(4-x^2)^{1/2}} dy \\
&= 48\pi + 8\pi = 56\pi
\end{aligned}
$$

We can verify this result by calculating $\iint_S \mathbf{F} \cdot \mathbf{n}\, dS$. We use the formula (see Equation 7.3.3)

$$
I_c = \iint_S (\mathbf{F} \cdot \mathbf{n}) \left| \frac{\partial \mathbf{r}}{\partial u} \times \frac{\partial \mathbf{r}}{\partial v} \right| du\,dv
$$

Parametize the cylindrical surface by $\mathbf{r}(u,v) = u\,\mathbf{i} + (4-u^2)^{1/2}\,\mathbf{j} + v\,\mathbf{k}$, and so

$$
\frac{\partial \mathbf{r}}{\partial u} \times \frac{\partial \mathbf{r}}{\partial v} = -\frac{u}{(4-u^2)^{1/2}}\,\mathbf{i} - \mathbf{j} \quad \text{or} \quad \left| \frac{\partial \mathbf{r}}{\partial u} \times \frac{\partial \mathbf{r}}{\partial v} \right| = \frac{2}{(4-u^2)^{1/2}}
$$

The outward normal vector on the curved surface of the cylinder is given by

$$
\mathbf{n} = \frac{\boldsymbol{\nabla}\phi}{|\boldsymbol{\nabla}\phi|} = \frac{\boldsymbol{\nabla}(x^2 + y^2)}{|\boldsymbol{\nabla}\phi|} = \frac{x\,\mathbf{i} + y\,\mathbf{j}}{(x^2 + y^2)^{1/2}} = \frac{x\,\mathbf{i} + y\,\mathbf{j}}{2}
$$

Therefore, $\mathbf{F} \cdot \mathbf{n}$ on the curved surface is given by $(x^4 + y^4)/2$. Referring to Example 7.3.2, we see that I on one half of the curved cylindrical surface is given by

$$
\begin{aligned}
I_c &= \iint_S \mathbf{F} \cdot \mathbf{n}\, dS = \iint_S \frac{x^4 + y^4}{2} \frac{2}{(4-u^2)^{1/2}}\, du\,dv \\
&= \int_{-1}^1 dv \int_{-2}^2 \frac{u^4 + (1-u^2)^2}{(4-u^2)^{1/2}}\, du = 24\pi
\end{aligned}
$$

The value of the integral over the entire curved surface is 48π, which is equal to the first term of the value of I given above. The integral over each end of the cylinder is simply $\mathbf{F} \cdot \mathbf{n} = \mathbf{k}$ times $4\pi\,\mathbf{k}$ (the area), and so the contribution from the two ends is 8π, which is equal to the second term above.

12. Interchange f and g in Green's first identity and subtract.

15. The solution to Problem 14 is the following:
We want to be able to replace $\boldsymbol{\nabla} f$ by div $\mathbf{F}$. We can do this by writing $\mathbf{F} = f\,\mathbf{c}$, where $\mathbf{c}$ is a constant vector. In this case, div $(f\,\mathbf{c}) = f\mathrm{div}\,\mathbf{c} + \mathbf{c} \cdot \boldsymbol{\nabla} f = \mathbf{c} \cdot \boldsymbol{\nabla} f$. Equation 2 gives us

$$\iiint_V \mathrm{div}\,(f\,\mathbf{c})\,dV = \iiint_V \mathbf{c} \cdot \boldsymbol{\nabla} f\,dV = \iint_S f\,\mathbf{c} \cdot \mathbf{n}\,dS$$

or

$$\mathbf{c} \cdot \left[\iiint_V \boldsymbol{\nabla} f\,dV - \iint_S f\,\mathbf{n}\,dS \right] = 0$$

Because $\mathbf{c}$ is an arbitrary constant vector, the terms in brackets must equal zero. To prove the assertion of Problem 15, simply let $f = 1$.

18. Apply Equation 2 to the geometry suggested in the statement of the problem to write

$$\iiint_V \mathrm{div}\,\mathbf{v}\,dV = \int_{\text{top}} \mathbf{k} \cdot \mathbf{v}\,dS + \int_{\text{bottom}} -\mathbf{k} \cdot \mathbf{v}\,dS + \int_{\text{side}} \mathbf{n} \cdot \mathbf{v}\,dS$$

The first two terms on the right are equal to zero because $\mathbf{k} \cdot \mathbf{v} = 0$. Thus, we have

$$\iiint_V \mathrm{div}\,\mathbf{v}\,dV = \iint_R \mathbf{n} \cdot \mathbf{v}\,dS$$

Since div $\mathbf{v}$ and $\mathbf{n} \cdot \mathbf{v}$ do not vary with z, we have $dV = h\,dS$ and $dS = h\,ds$, and so

$$h \iint_R \mathrm{div}\,\mathbf{v}\,dS = h \int_C \mathbf{n} \cdot \mathbf{v}\,ds$$

For another proof, start with Green's theorem in a plane (Equation 7.2.15):

$$\oint_C (P\,dx + Q\,dy) = \iint_R \left(\frac{\partial Q}{\partial x} - \frac{\partial P}{\partial y} \right) dx dy$$

Now let $\mathbf{w} = P\,\mathbf{i} + Q\,\mathbf{j}$, so that $P\,dx + Q\,dy = \mathbf{w} \cdot d\mathbf{r} = \mathbf{w} \cdot \dfrac{d\mathbf{r}}{ds}\,ds = \mathbf{w} \cdot \mathbf{T}\,ds$, where $\mathbf{T}$ is the tangent vector to C. (See the figure on the next page.) If $\mathbf{n}$ is the outward normal vector to C, then $\mathbf{T} = \mathbf{k} \times \mathbf{n}$, where $\mathbf{k}$ is the normal perpendicular vector to R. So far then, we have

$$P\,dx + Q\,dy = \mathbf{w} \cdot \mathbf{T}\,ds = \mathbf{w} \cdot (\mathbf{k} \times \mathbf{n})\,ds = (\mathbf{w} \times \mathbf{k}) \cdot \mathbf{n}\,ds$$

Furthermore,

$$\mathbf{v} = \mathbf{w} \times \mathbf{k} = (P\mathbf{i} + Q\mathbf{j}) \times \mathbf{k} = Q\mathbf{i} - P\mathbf{j}$$

and

$$\frac{\partial Q}{\partial x} - \frac{\partial P}{\partial y} = \operatorname{div} \mathbf{v}$$

Therefore, Green's theorem in a plane says that

$$\iint_{R} \operatorname{div} \mathbf{v}\, dS = \oint_{C} \mathbf{n} \cdot \mathbf{v}\, ds$$

7.5   Stokes's Theorem

1. For $\mathbf{v} = z\,\mathbf{i} + x\,\mathbf{j} + y\,\mathbf{k}$, $\boldsymbol{\nabla} \times \mathbf{v} = \mathbf{i} + \mathbf{j} + \mathbf{k}$. The outward unit normal vector to S is given in terms of $\phi(x, y, z) = z^2 + x^2 + y^2 - 1$, or $\mathbf{n} = \dfrac{\boldsymbol{\nabla} \phi}{|\boldsymbol{\nabla} \phi|} = x\,\mathbf{i} + y\,\mathbf{j} + z\,\mathbf{k}$. The dot product of $\boldsymbol{\nabla} \times \mathbf{v}$ and $\mathbf{n}$ is $(\boldsymbol{\nabla} \times \mathbf{v}) \cdot \mathbf{n} = x + y + z$ and so

$$\begin{aligned}
\iint_{S} (\boldsymbol{\nabla} \times \mathbf{v}) \cdot \mathbf{n}\, dS &= \iint_{S} x\, dS + \iint_{S} y\, dS + \iint_{S} z\, dS \\
&= 0 + 0 + \iint_{S} \frac{dx\,dy}{\mathbf{n} \cdot \mathbf{k}} = \pi
\end{aligned}$$

where the first two integrals equal zero by symmetry.

Now, $\oint \mathbf{v} \cdot d\mathbf{r} = \displaystyle\int_{0}^{2\pi} (z\, dx + x\, dy + y\, dz)$ where the integral is taken around the unit circle in the xy-plane (the capping surface). Therefore,

$$\oint \mathbf{v} \cdot d\mathbf{r} = 0 + \int_{0}^{2\pi} x\, dy + 0 = \int_{0}^{2\pi} \cos^2 \theta\, d\theta = \pi$$

where the first and third integrals equal zero because z and $dz = 0$.

3. The outward unit normal vector is given by $\boldsymbol{\nabla}\phi/|\boldsymbol{\nabla}\phi|$ where $\phi(x, y, z) = x^2 + y^2 + z^2 - 1$, or $\mathbf{n} = x\,\mathbf{i} + y\,\mathbf{j} + z\,\mathbf{k}$. Therefore, $\mathbf{r} \cdot \mathbf{n} = x^2 + y^2 + z^2 = 1$ and $\displaystyle\iint_{S} \mathbf{r} \cdot \mathbf{n}\, dS = \iint_{S} dS = 4\pi$,

where 4π is the surface area of a unit sphere.

6. Let's consider $\oint \mathbf{F} \cdot d\mathbf{r}$ first.

$$\mathbf{F} \cdot d\mathbf{r} = (2y + z)\, dx + (x - z)\, dy + (y - x)\, dz$$

Along path 1, z and $dz = 0$, so $\mathbf{F} \cdot d\mathbf{r} = 2y\, dx + x\, dy$. Also, $x + y = 1$ along this path, so $\mathbf{F} \cdot \mathbf{r} = 2(1 - x)\, dx - x\, dx = (2 - 3x)\, dx$. Thus,

$$\int_1 \mathbf{F} \cdot d\mathbf{r} = \int_1^0 (2 - 3x)\, dx = -\frac{1}{2}$$

Along path 2, x and $dx = 0$ and $y + z = 1$. So

$$\int_2 \mathbf{F} \cdot d\mathbf{r} = \int (-z\, dy + y\, dz) = \int_1^0 (y - 1)\, dy - \int_1^0 y\, dy = -\int_1^0 dy = 1$$

Along path 3, x and $dx = 0$ and $x + z = 1$. So $\mathbf{F} \cdot d\mathbf{r} = z\, dx - x\, dz = (1 - x)\, dx + x\, dx = dx$ and $\int_3 \mathbf{F} \cdot d\mathbf{r} = 1$. The sum along the three paths is $3/2$.

Now consider $\iint_S \mathrm{curl}\, \mathbf{F} \cdot d\mathbf{S}$. The unit normal vector to the surface $x + y + z = 1$ is $\mathbf{n} = \dfrac{\mathbf{i} + \mathbf{j} + \mathbf{k}}{\sqrt{3}}$ and $\mathrm{curl}\, \mathbf{F} = 2\,\mathbf{i} + 2\,\mathbf{j} - \mathbf{k}$. Therefore, $\mathrm{curl}\, \mathbf{F} \cdot \mathbf{n} = \sqrt{3}$. The plane is an equilateral triangle with sides equal to $\sqrt{2}$, so its area is $\sqrt{3}/2$. Therefore, $\iint_S \mathrm{curl}\, \mathbf{F} \cdot d\mathbf{S} = \sqrt{3} \cdot \dfrac{\sqrt{3}}{2} = \dfrac{3}{2}$.

9. The quantities z and dz are equal to zero along all three paths, so $\mathbf{F} \cdot d\mathbf{r} = y^2\, dx - x^2\, dy$.

Along the path from $(0,0)$ to $(1,0)$, $\displaystyle\int_1 \mathbf{F} \cdot d\mathbf{r} = \int_0^1 0\, dx = 0$

Along the path $(1,0)$ to $(1,1)$, $\displaystyle\int_2 \mathbf{F} \cdot d\mathbf{r} = -\int_0^1 dy = -1$

Along the path from $(1,1)$ to $(0,0)$, $\displaystyle\int_3 \mathbf{F} \cdot d\mathbf{r} = \int (x^2\, dx - x^2\, dx) = 0$

for a total of -1.

We can verify this result using Equation 1. $\mathrm{curl}\, \mathbf{F} = -\mathbf{i} - 2(x + y)\,\mathbf{k}$, so $\mathrm{curl}\, \mathbf{F} \cdot \mathbf{k} = -2(x + y)$. Therefore, according to Equation 1,

$$\oint \mathbf{F} \cdot d\mathbf{r} = -2 \int_0^1 dx \int_0^x dy\, (x + y) = -3 \int_0^1 x^2\, dx = -1$$

12. First we write

$$\mathrm{curl}\, \mathbf{F} = \mathrm{curl}\, [(z - 2y)\,\mathbf{i} + (3x - 4y)\,\mathbf{j} + (z + 3y)\,\mathbf{k}] = 3\,\mathbf{i} + \mathbf{j} + 5\,\mathbf{k}$$

The unit normal vector to the triangle formed by the points $(0,0,0)$, $(1,0,0)$, and $(1,1,0)$ is given by $\dfrac{\nabla(x+y+z-1)}{|\nabla\phi|} = \mathbf{n} = \dfrac{\mathbf{i}+\mathbf{j}+\mathbf{k}}{\sqrt{3}}$. Therefore, curl $\mathbf{F}\cdot\mathbf{n} = \dfrac{9}{\sqrt{3}}$. As in Problem 6, the area of the surface is $\sqrt{3}/2$, and so

$$\iint\limits_{S} \text{curl } \mathbf{F}\cdot\mathbf{n}\, dS = \frac{9}{\sqrt{3}}\iint\limits_{S} dS = \frac{9}{\sqrt{3}}\frac{\sqrt{3}}{2} = \frac{9}{2}$$

As a check, we evaluate $\oint \mathbf{F}\cdot d\mathbf{r}$.

$$\mathbf{F}\cdot d\mathbf{r} = (z-2y)\,dx + (3x-4y)\,dy + (z+3y)\,dz$$

Along path 1, z and dz equal zero and $x+y=1$. Therefore,

$$\begin{aligned}
\int \mathbf{F}\cdot d\mathbf{r} &= \int [-2y\,dx + (3x-4y)\,dy\,] \\
&= \int_{1}^{0}[-2(1-x)\,dx - (3x-4+4x)\,dx\,] = \int_{1}^{0}[\,2\,dx - 5x\,dx\,] = \frac{1}{2}
\end{aligned}$$

Along path 2, x and dx equal zero, and $y+z=1$. Therefore,

$$\begin{aligned}
\int \mathbf{F}\cdot d\mathbf{r} &= \int [-4y\,dy + (z+3y)\,dz\,] \\
&= \int_{1}^{0}[-4y\,dy - (1+2y)\,dy\,] = \int_{1}^{0}(-1-6y)\,dy = 4
\end{aligned}$$

Along path 3, y and dy equal zero, and $x+z=1$. Therefore,

$$\int \mathbf{F}\cdot d\mathbf{r} = \int (z\,dx + z\,dz) = \int_{1}^{0}[\,(1-x)\,dx - (1-x)\,dx\,] = 0$$

The sum of the three path integrals gives $\oint \mathbf{F}\cdot d\mathbf{r} = \dfrac{9}{2}$.

15. First write curl $\mathbf{F} = \text{curl}\,[\,(y-x)\,\mathbf{i} + (x-z)\,\mathbf{j} + (x-y)\,\mathbf{k}\,] = -\mathbf{j}$. The unit normal vector to the surface $x+2y+z=2$ is given by $\dfrac{\nabla(x+2y+z-2)}{|\nabla\phi|} = \mathbf{n} = \dfrac{\mathbf{i}+2\mathbf{j}+\mathbf{k}}{\sqrt{6}}$. Therefore, curl $\mathbf{F}\cdot\mathbf{n} = -2/\sqrt{6}$, and

$$\iint\limits_{S} \text{curl } \mathbf{F}\cdot\mathbf{n}\, dS = -\frac{2}{\sqrt{6}}(\text{area}) = -\frac{2}{\sqrt{6}}(\sqrt{6}) = -2$$

We'll now check this result by evaluating $\oint \mathbf{F}\cdot d\mathbf{r}$

$$\mathbf{F}\cdot d\mathbf{r} = (y-x)\,dx + (x-z)\,dy + (x-y)\,dz$$

Along path 1, z and dz equal zero, and $x+2y=2$. Therefore,

$$\int \mathbf{F}\cdot d\mathbf{r} = \int [(y-x)\,dx + x\,dy] = \int_{0}^{1}[(y-2+2y)(-2\,dy) + (2-2y)\,dy] = \int_{0}^{1}(6-8y)\,dy = 2$$

Along path 2, x and dx equal zero, and $2y + z = 2$. Therefore,

$$\int \mathbf{F} \cdot d\mathbf{r} = \int (-z\, dy - y\, dz) = \int \left[(2y-2)\, dy - y(-2\, dy) \right] = \int_1^0 (4y - 2)\, dy = 0$$

Along path 3, y and dy equal zero, and $x + z = 2$. Therefore,

$$\int \mathbf{F} \cdot d\mathbf{r} = \int (-x\, dx + x\, dz) = -2 \int_0^2 x\, dx = -4$$

The sum of the three path integrals is -2.

18. First we write $\boldsymbol{\nabla} \times \mathbf{v} = \boldsymbol{\nabla} \times (-y\,\mathbf{i} + x\,\mathbf{j}) = 2\,\mathbf{k}$. The unit normal vector to the xy-plane is $\mathbf{k}$, so we have $(\boldsymbol{\nabla} \times \mathbf{v}) \cdot \mathbf{n} = 2$. Also $\mathbf{v} \cdot d\mathbf{r} = -y\, dx + x\, dy$, and so

$$\oint \mathbf{v} \cdot d\mathbf{r} = \oint (-y\, dx + x\, dy) = \iint_S 2\, dx dy = 2A$$

21. Stokes's theorem says that $\displaystyle\int_C \mathbf{E} \cdot d\mathbf{l} = \iint_S (\boldsymbol{\nabla} \times \mathbf{E}) \cdot \mathbf{n}\, dS$. Now use Maxwell's equation to write $\displaystyle\iint_S (\boldsymbol{\nabla} \times \mathbf{E}) \cdot \mathbf{n}\, dS = -\iint_S \frac{\partial \mathbf{B}}{\partial t} \cdot \mathbf{n}\, dS$. Because S is arbitrary, we have $\boldsymbol{\nabla} \times \mathbf{E} = -\dfrac{\partial \mathbf{B}}{\partial t}$.

24. Start with curl $\mathbf{F} = $ curl $\left[-v_y(x,y)\,\mathbf{i} + v_x(x,y)\,\mathbf{j} \right] = \left(\dfrac{\partial v_x}{\partial x} + \dfrac{\partial v_y}{\partial y} \right) \mathbf{k}$. Then

$$\iint_R \text{curl } \mathbf{F} \cdot \mathbf{k}\, dx dy = \iint_R \left(\frac{\partial v_x}{\partial x} + \frac{\partial v_y}{\partial y} \right) dx dy = \iint_R \boldsymbol{\nabla} \cdot \mathbf{v}\, dx dy$$

Now use Green's theorem in a plane to write

$$\iint_R \left(\frac{\partial v_x}{\partial x} + \frac{\partial v_y}{\partial y} \right) dx dy = \oint_C (-v_y\, dx + v_x\, dy)$$

Now write the right side as

$$\oint_C (-v_y\, dx + v_x\, dy) = \oint_C \mathbf{v} \cdot \left(\mathbf{i}\frac{dy}{ds} - \mathbf{j}\frac{dx}{ds} \right) ds$$

The tangent line to a parametric curve $\mathbf{r}(s)$ is given by (Equation 5.2.3) $\dfrac{d\mathbf{r}}{ds} = \dfrac{dx}{ds}\,\mathbf{i} + \dfrac{dy}{ds}\,\mathbf{j}$ and so we see that $\mathbf{i}\dfrac{dy}{ds} - \mathbf{j}\dfrac{dx}{ds}$ is normal to the tangent line and hence normal to the curve. Thus we write the last integral as $\displaystyle\oint_C \mathbf{v} \cdot \mathbf{n}\, ds$, and so we see that $\displaystyle\iint_S \boldsymbol{\nabla} \cdot \mathbf{v}\, dx dy = \oint_C \mathbf{v} \cdot \mathbf{n}\, ds$ in two dimensions.

27. We use the result of Problem 25 with $x_0 = y_0 = 0$.

$$\int_0^x v_z(x', y, z)\, dx' = \int_0^x (y - x')\, dx' = xy - \frac{x^2}{2} = w_y$$

$$\int_0^y v_x(x_0, y', z)\, dy' = \int_0^y (z - y')\, dy' = yz - \frac{y^2}{2}$$

$$\int_0^x v_y(x', y, z)\, dx' = \int_0^x (x' - z)\, dx' = \frac{x^2}{2} - xz$$

and so

$$\mathbf{w} = \left(xy - \frac{x^2}{2} \right) \mathbf{j} + \left[(x + y)z - \frac{x^2}{2} - \frac{y^2}{2} \right] \mathbf{k}$$

As a check, we see that

$$v_x = \frac{\partial w_z}{\partial y} - \frac{\partial w_y}{\partial z} = z - y$$

and

$$v_y = \frac{\partial w_x}{\partial z} - \frac{\partial w_z}{\partial x} = x - z$$

and that

$$v_z = \frac{\partial w_y}{\partial x} - \frac{\partial w_x}{\partial y} = y - x$$

CHAPTER 8

Curvilinear Coordinates

8.1 Plane Polar Coordinates

1.

2 sinθ

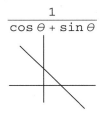

2 cosθ + sinθ

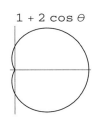

sinθ cos²θ

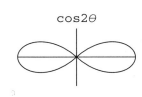

$\dfrac{1}{\cos\theta + \sin\theta}$

1 + 2 cos θ

cos2θ

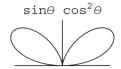

3. We use Equation 3 with $r(\theta) = \sin 2\theta$ to write

$$\frac{dy}{dx} = \frac{2\cos 2\theta \sin \theta + \sin 2\theta \cos \theta}{2\cos 2\theta \cos \theta - \sin 2\theta \sin \theta} = \frac{\sin\theta}{\cos\theta}\left(\frac{4 - 6\sin^2\theta}{6\cos^2\theta - 4}\right)$$

$$= \frac{\sin\theta}{\cos\theta}\left(\frac{6\cos^2\theta - 2}{6\cos^2\theta - 4}\right) = -1 \quad \text{at} \quad \theta = \pi/4 \quad \text{and} \quad \theta = 5\pi/4$$

The accompanying figure shows this result.

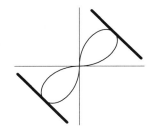

6. We use Equation 4 with

(a) $r = e^{\theta/2}$: $\cot \alpha = \dfrac{1}{r}\dfrac{dr}{d\theta} = \dfrac{1}{2}$, which gives $\alpha = 63.4°$

(b) $r = 1 + \sin \theta$: $\cot \alpha = \dfrac{\cos \theta}{1 + \sin \theta} = 0$, which gives $\alpha = 90°$

(c) $r = \sin 3\theta$: $\cot \alpha = \dfrac{3\cos 3\theta}{\sin 3\theta} = 3$, which gives $\alpha = 18.4°$

The accompanying figures show these results.

9. The area is given by $A = \dfrac{1}{2} \displaystyle\int r^2(\theta)\, d\theta = 4 \displaystyle\int_0^{\pi/4} \cos 2\theta\, d\theta = \dfrac{4}{2} = 2$

12. The curves intersect when $2\sin\theta = 1$, or at $\theta = \pi/6$ and $5\pi/6$. The total area between the two curves is given by

$$
\begin{aligned}
A &= \frac{1}{2}\int_0^{\pi/6} 4\sin^2\theta\, d\theta + \frac{1}{2}\int_{\pi/6}^{\pi/2} d\theta + \frac{1}{2}\int_{\pi/2}^{5\pi/6} d\theta + \frac{1}{2}\int_{5\pi/6}^{\pi} \sin^2\theta\, d\theta \\
&= \left(\frac{\pi}{6} - \frac{\sqrt{3}}{4}\right) + \frac{\pi}{6} + \frac{\pi}{6} + \left(\frac{\pi}{6} - \frac{\sqrt{3}}{4}\right) = \frac{2\pi}{3} - \frac{\sqrt{3}}{2}
\end{aligned}
$$

15. We have that $x^2 + y^2 + z^2 = a^2$ and $z = (a^2 - r^2)^{1/2}$. The volume is given by

$$
\begin{aligned}
V &= \int_0^{2\pi} d\theta \int_0^a zr\, dr = 2\pi \int_0^a (a^2 - r^2)^{1/2} r\, dr \\
&= \pi \int_0^{a^2} (a^2 - u)^{1/2} du = \frac{2\pi a^3}{3}
\end{aligned}
$$

8.2   Vectors in Plane Polar Coordinates

1. Start with $\mathbf{e}_\theta = -\sin\theta\, \mathbf{i} + \cos\theta\, \mathbf{j}$ (Equation 6.) Then $\dfrac{d\mathbf{e}_\theta}{d\theta} = -\cos\theta\, \mathbf{i} - \sin\theta\, \mathbf{j} = -\mathbf{e}_r$.

3. $h_r = \left[\left(\dfrac{\partial x}{\partial r}\right)^2 + \left(\dfrac{\partial y}{\partial r}\right)^2\right]^{1/2} = (\cos^2\theta + \sin^2\theta)^{1/2} = 1$

$h_\theta = \left[\left(\dfrac{\partial x}{\partial \theta}\right)^2 + \left(\dfrac{\partial y}{\partial \theta}\right)^2\right]^{1/2} = (r^2\sin^2\theta + r^2\cos^2\theta)^{1/2} = r$

6. $\mathbf{v} = \dfrac{d\mathbf{r}}{dt} = \dfrac{dx}{dt}\mathbf{i} + \dfrac{dy}{dt}\mathbf{j} = \dfrac{d}{dt}(t\,\mathbf{i} + t^2\,\mathbf{j}) = \mathbf{i} + 2t\,\mathbf{j}.$ Now, the equation $\mathbf{e}_r = \cos\theta\,\mathbf{i} + \sin\theta\,\mathbf{j}$

and $\mathbf{e}_\theta = -\sin\theta\,\mathbf{i} + \cos\theta\,\mathbf{j}$ give $\mathbf{i} = \cos\theta\,\mathbf{e}_r - \sin\theta\,\mathbf{e}_\theta$ and $\mathbf{j} = \sin\theta\,\mathbf{e}_r + \cos\theta\,\mathbf{e}_\theta$, so

$$
\begin{aligned}
\mathbf{v} &= (\cos\theta\,\mathbf{e}_r - \sin\theta\,\mathbf{e}_\theta) + 2t(\sin\theta\,\mathbf{e}_r + \cos\theta\,\mathbf{e}_\theta)\\
&= (\cos\theta + 2t\sin\theta)\,\mathbf{e}_r + (2t\cos\theta - \sin\theta)\,\mathbf{e}_\theta\\
&= \left[\frac{t}{(t^2+t^4)^{1/2}} + \frac{2t^3}{(t^2+t^4)^{1/2}}\right]\mathbf{e}_r + \left[\frac{2t^2}{(t^2+t^4)^{1/2}} - \frac{t^2}{(t^2+t^4)^{1/2}}\right]\mathbf{e}_\theta\\
&= \left[\frac{1+2t^2}{(1+t^2)^{1/2}}\right]\mathbf{e}_r + \left[\frac{t}{(1+t^2)^{1/2}}\right]\mathbf{e}_\theta
\end{aligned}
$$

9. $\nabla^2 f = \dfrac{\partial^2 f}{\partial r^2} + \dfrac{1}{r}\dfrac{\partial f}{\partial r} + \dfrac{1}{r^2}\dfrac{\partial^2 f}{\partial\theta^2} = -\dfrac{A}{r^2} + \dfrac{A}{r^2} = 0$

12. Use $r = (x^2 + y^2)^{1/2}$ and $\theta = \tan^{-1}\dfrac{y}{x}$ to write

$$
\begin{aligned}
\frac{\partial r}{\partial x} &= \frac{x}{(x^2+y^2)^{1/2}} = \frac{r\cos\theta}{r} = \cos\theta\\[2mm]
\frac{\partial r}{\partial y} &= \frac{y}{(x^2+y^2)^{1/2}} = \frac{r\sin\theta}{r} = \sin\theta\\[2mm]
\frac{\partial\theta}{\partial x} &= \frac{1}{1+\dfrac{y^2}{x^2}}\left(-\frac{y}{x}\right) = -\frac{y}{r^2} = -\frac{r\sin\theta}{r^2} = -\frac{\sin\theta}{r}\\[2mm]
\frac{\partial\theta}{\partial y} &= \frac{1}{1+\dfrac{y^2}{x^2}}\left(\frac{1}{x}\right) = \frac{x}{r^2} = \frac{\cos\theta}{r}
\end{aligned}
$$

and so

$$
\frac{\partial f}{\partial x} = \frac{\partial f}{\partial r}\frac{\partial r}{\partial x} + \frac{\partial f}{\partial\theta}\frac{\partial\theta}{\partial x} = \cos\theta\frac{\partial f}{\partial r} - \frac{\sin\theta}{r}\frac{\partial f}{\partial\theta}
$$

and

$$
\frac{\partial f}{\partial y} = \frac{\partial f}{\partial r}\frac{\partial r}{\partial y} + \frac{\partial f}{\partial\theta}\frac{\partial\theta}{\partial y} = \sin\theta\frac{\partial f}{\partial r} + \frac{\cos\theta}{r}\frac{\partial f}{\partial\theta}
$$

Now substitute $\mathbf{i} = \cos\theta\,\mathbf{e}_r - \sin\theta\,\mathbf{e}_\theta$ and $\mathbf{j} = \sin\theta\,\mathbf{e}_r + \cos\theta\,\mathbf{e}_\theta$ into $\nabla f = \dfrac{\partial f}{\partial x}\mathbf{i} + \dfrac{\partial f}{\partial y}\mathbf{j}$ to write

$$
\begin{aligned}
\nabla f &= \left(\cos\theta\frac{\partial f}{\partial r} - \frac{\sin\theta}{r}\frac{\partial f}{\partial\theta}\right)(\cos\theta\,\mathbf{e}_r - \sin\theta\,\mathbf{e}_\theta) + \left(\sin\theta\frac{\partial f}{\partial r} + \frac{\cos\theta}{r}\frac{\partial f}{\partial\theta}\right)(\sin\theta\,\mathbf{e}_r + \cos\theta\,\mathbf{e}_\theta)\\
&= \frac{\partial f}{\partial r}\mathbf{e}_r + \frac{1}{r}\frac{\partial f}{\partial\theta}\mathbf{e}_\theta
\end{aligned}
$$

8.3 Cylindrical Coordinates

1. (a) A right cylinder of radius a centered on the z axis.

(b) The yz-plane

(c) A plane perpendicular to the z axis

(d) A circular paraboloid centered on the z axis

3. Use Equation 3 to write

$$
\begin{aligned}
l &= \int_0^1 (a^2 + a^2b^2t^2 + c^2)^{1/2}dt \\
&= \frac{ab}{2}\left\{ \frac{(a^2 + c^2 + a^2b^2)^{1/2}}{ab} + \frac{a^2 + c^2}{a^2b^2}\left[\ln\left[1 + \left(\frac{a^2 + c^2 + a^2b^2}{a^2b^2}\right)^{1/2}\right] - \ln\left(\frac{a^2 + c^2}{a^2b^2}\right)^{1/2}\right]\right\} \\
&= \frac{1}{2ab}\left[ab(a^2 + c^2 + a^2b^2)^{1/2} + (a^2 + c^2)\ln\frac{ab + (a^2 + c^2 + a^2b^2)^{1/2}}{(a^2 + c^2)^{1/2}}\right]
\end{aligned}
$$

6. Substituting Equation 5 into Equation 6 gives

$$\frac{\partial \mathbf{r}}{\partial r} = \cos\theta\,\mathbf{i} + \sin\theta\,\mathbf{j} \qquad \text{and} \qquad \left|\frac{\partial \mathbf{r}}{\partial r}\right| = 1$$

$$\frac{\partial \mathbf{r}}{\partial \theta} = -r\sin\theta\,\mathbf{i} + r\cos\theta\,\mathbf{j} \qquad \text{and} \qquad \left|\frac{\partial \mathbf{r}}{\partial \theta}\right| = r$$

$$\frac{\partial \mathbf{r}}{dz} = \mathbf{k} \qquad \text{and} \qquad \left|\frac{\partial \mathbf{r}}{\partial z}\right| = 1$$

Therefore, $\mathbf{e}_r = \cos\theta\,\mathbf{i} + \sin\theta\,\mathbf{j}$, $\mathbf{e}_\theta = -\sin\theta\,\mathbf{i} + \cos\theta\,\mathbf{j}$, and $\mathbf{e}_z = \mathbf{k}$. Note that $\mathbf{e}_r \cdot \mathbf{e}_r = \mathbf{e}_\theta \cdot \mathbf{e}_\theta = \mathbf{e}_z \cdot \mathbf{e}_z = 1$ and $\mathbf{e}_r \cdot \mathbf{e}_\theta = \mathbf{e}_r \cdot \mathbf{e}_z = \mathbf{e}_\theta \cdot \mathbf{e}_z = 0$.

9. Using Equation 7, we write

$$
\begin{aligned}
\boldsymbol{\nabla} \cdot \mathbf{u} &= \left(\mathbf{e}_r \frac{\partial}{\partial r} + \mathbf{e}_\theta \frac{1}{r}\frac{\partial}{\partial \theta} + \mathbf{e}_z \frac{\partial}{\partial z}\right) \cdot (u_r\,\mathbf{e}_r + u_\theta\,\mathbf{e}_\theta + u_z\,\mathbf{e}_z) \\
&= \mathbf{e}_r \cdot \left(\frac{\partial u_r}{\partial r}\mathbf{e}_r + u_r\frac{\partial \mathbf{e}_r}{\partial r}\right) + \mathbf{e}_r \cdot \left(\frac{\partial u_\theta}{\partial r}\mathbf{e}_\theta + u_\theta\frac{\partial \mathbf{e}_\theta}{\partial r}\right) + \mathbf{e}_r \cdot \left(\frac{\partial u_z}{\partial r}\mathbf{e}_z + u_z\frac{\partial \mathbf{e}_z}{dr}\right) \\
&= \mathbf{e}_\theta \cdot \left(\frac{1}{r}\frac{\partial u_r}{\partial \theta}\mathbf{e}_r + \frac{u_r}{r}\frac{\partial \mathbf{e}_r}{\partial \theta}\right) + \mathbf{e}_\theta \cdot \left(\frac{1}{r}\frac{\partial u_\theta}{\partial \theta}\mathbf{e}_\theta + \frac{u_\theta}{r}\frac{\partial \mathbf{e}_\theta}{\partial \theta}\right) + \mathbf{e}_\theta \cdot \left(\frac{1}{r}\frac{\partial u_z}{\partial \theta}\mathbf{e}_z + \frac{u_z}{r}\frac{\partial \mathbf{e}_z}{\partial \theta}\right) \\
&= \mathbf{e}_z \cdot \left(\frac{\partial u_r}{\partial z}\mathbf{e}_r + u_r\frac{\partial \mathbf{e}_r}{\partial z}\right) + \mathbf{e}_z \cdot \left(\frac{\partial u_\theta}{\partial z}\mathbf{e}_\theta + u_\theta\frac{\partial \mathbf{e}_\theta}{\partial z}\right) + \mathbf{e}_z \cdot \left(\frac{\partial u_z}{\partial z}\mathbf{e}_z + u_z\frac{\partial \mathbf{e}_z}{\partial z}\right)
\end{aligned}
$$

Using the orthogonality of the unit vectors and the derivative relations given in Table 8.2, we see that

$$\boldsymbol{\nabla} \cdot \mathbf{u} = \frac{\partial u_r}{\partial r} + \frac{u_r}{r} + \frac{1}{r}\frac{\partial u_\theta}{\partial \theta} + \frac{\partial u_z}{\partial z}$$

12. Using $\mathbf{e}_r = \cos\theta\,\mathbf{i} + \sin\theta\,\mathbf{j}$, $\mathbf{e}_\theta = -\sin\theta\,\mathbf{i} + \cos\theta\,\mathbf{j}$, and $\mathbf{e}_z = \mathbf{k}$, we have

$$\mathbf{e}_r \times \mathbf{e}_\theta = \begin{vmatrix} \mathbf{i} & \mathbf{j} & \mathbf{k} \\ \cos\theta & \sin\theta & 0 \\ -\sin\theta & \cos\theta & 0 \end{vmatrix} = \mathbf{k} = \mathbf{e}_z$$

$$\mathbf{e}_z \times \mathbf{e}_r = \begin{vmatrix} \mathbf{i} & \mathbf{j} & \mathbf{k} \\ 0 & 0 & 1 \\ \cos\theta & \sin\theta & 0 \end{vmatrix} = -\sin\theta\,\mathbf{i} + \cos\,\mathbf{j} = \mathbf{e}_\theta$$

$$\mathbf{e}_\theta \times \mathbf{e}_z = \begin{vmatrix} \mathbf{i} & \mathbf{j} & \mathbf{k} \\ -\sin\theta & \cos\theta & 0 \\ 0 & 0 & 1 \end{vmatrix} = \cos\,\mathbf{i} + \sin\theta\,\mathbf{j} = \mathbf{e}_r$$

15. Start with $\mathbf{r}(t) = r(t)\,\mathbf{e}_r + z(t)\,\mathbf{e}_z$ to write

$$\begin{aligned} \frac{d\mathbf{r}}{dt} &= \frac{dr}{dt}\mathbf{e}_r + r\frac{d\mathbf{e}_r}{dt} + \frac{dz}{dt}\mathbf{e}_z \\ &= \frac{dr}{dt}\mathbf{e}_r + r\frac{d\theta}{dt}\frac{d\mathbf{e}_r}{d\theta} + \frac{dz}{dt}\mathbf{e}_z \\ &= \frac{dr}{dt}\mathbf{e}_r + r\frac{d\theta}{dt}\mathbf{e}_\theta + \frac{dz}{dt}\mathbf{e}_z \end{aligned}$$

18. Solve $\mathbf{e}_r = \cos\theta\,\mathbf{i} + \sin\theta\,\mathbf{j}$ and $\mathbf{e}_\theta = -\sin\theta\,\mathbf{i} + \cos\theta\,\mathbf{j}$ for $\mathbf{i}$ and $\mathbf{j}$ to get $\mathbf{i} = \cos\theta\,\mathbf{e}_r - \sin\theta\,\mathbf{e}_\theta$ and $\mathbf{j} = \sin\theta\,\mathbf{e}_r + \cos\theta\,\mathbf{e}_\theta$. Therefore,

$$\begin{aligned} \mathbf{u} &= 2z\cos\theta\,\mathbf{e}_r - 2z\sin\theta\,\mathbf{e}_\theta - r\cos\theta\sin\theta\,\mathbf{e}_r - r\cos^2\theta\,\mathbf{e}_\theta + r\sin\theta\,\mathbf{e}_z \\ &= \cos\theta(2z - r\sin\theta)\,\mathbf{e}_r - (2z\sin\theta + r\cos^2\theta)\,\mathbf{e}_\theta + r\sin\theta\,\mathbf{e}_z \end{aligned}$$

21. $\nabla\cdot\mathbf{u} = e^z r + 3r^2\cos\theta - 2rz^2\cos\theta\sin\theta$ and $\nabla\times\mathbf{u} = 2rz\sin^2\theta\,\mathbf{e}_r + (r\sin\theta - 2z^2\sin^2\theta)\,\mathbf{e}_z$.

8.4 Spherical Coordinates

1. Latitude varies from $90°$ at the North Pole to $-90°$ at the South Pole. So $\alpha = 90° - 0$. Longitude begins at the prime meridian at Greenwich, England, so let this be the "zero" for both ϕ and β, However, β increases in a *clockwise* direction looking down from the North Pole. Because ϕ increases in a *counterclockwise* direction, the relation between ϕ and β is $\beta = 360° - \phi$.

3. Start with $ds^2 = dx^2 + dy^2 + dz^2$. Then

$$\begin{aligned} dx &= \sin\theta\cos\phi\,dr + r\cos\theta\cos\phi\,d\theta - r\sin\theta\sin\phi\,d\phi \\ dy &= \sin\theta\sin\phi\,dr + r\cos\theta\sin\phi\,d\theta + r\sin\theta\cos\phi\,d\phi \\ dz &= \cos\theta\,dr - r\sin\theta\,d\theta \end{aligned}$$

Squaring and adding gives the desired result.

6. $I = \iiint x^2\,dxdydz = \int_0^{2\pi} d\phi \int_0^{\pi} d\theta \sin\theta \int_0^a dr\, r^4 = 4\pi \int_0^a r^4 dr = \dfrac{4\pi a^5}{5}.$ This result is 3 times the result in Example 2 and Problem 5 because $r^2 = x^2 + y^2 + z^2$ and

$$\iiint x^2 dxdydz = \iiint y^2 dxdydz = \iiint z^2 dxdydz$$

over the surface of a sphere centered at the origin.

9. We have that

$$\frac{\partial \mathbf{r}}{\partial r} = \sin\theta \cos\phi\, \mathbf{i} + \sin\theta \sin\phi\, \mathbf{j} + \cos\theta\, \mathbf{k}$$

$$\frac{\partial \mathbf{r}}{\partial \theta} = r\cos\theta \cos\phi\, \mathbf{i} + r\cos\theta \sin\phi\, \mathbf{j} - r\sin\theta\, \mathbf{k}$$

$$\frac{\partial \mathbf{r}}{\partial \phi} = -r\sin\theta \sin\phi\, \mathbf{i} + r\sin\theta \cos\phi\, \mathbf{j}$$

and $\dfrac{d\mathbf{r}}{dr} \cdot \dfrac{d\mathbf{r}}{d\theta} = \dfrac{d\mathbf{r}}{dr} \cdot \dfrac{d\mathbf{r}}{d\phi} = \dfrac{d\mathbf{r}}{d\theta} \cdot \dfrac{d\mathbf{r}}{d\phi} = 0$

12. Use Equation 12 to write grad $f = \boldsymbol{\nabla} f = \cos\phi\, \mathbf{e}_r - \dfrac{\sin\phi}{\sin\theta}\, \mathbf{e}_\phi = \cos\phi\, \mathbf{e}_r - \csc\theta\sin\phi\, \mathbf{e}_\phi.$

15. Using Equation 13, we write

$$\text{curl } \mathbf{u} \stackrel{?}{=} \boldsymbol{\nabla} \times \mathbf{u} = \left(\mathbf{e}_r \frac{\partial}{\partial r} + \frac{\mathbf{e}_\theta}{r}\frac{\partial}{\partial \theta} + \frac{\mathbf{e}_\phi}{r\sin\theta}\frac{\partial}{\partial \phi}\right) \times (u_r\,\mathbf{e}_r + u_\theta\,\mathbf{e}_\theta + u_\phi\,\mathbf{e}_\phi)$$

$$= \mathbf{e}_r \times \left(\frac{\partial u_r}{\partial r}\mathbf{e}_r + u_r\frac{\partial \mathbf{e}_r}{\partial r} + \frac{\partial u_\theta}{\partial r}\mathbf{e}_\theta + u_\theta\frac{\partial \mathbf{e}_\theta}{\partial r} + \frac{\partial u_\phi}{\partial r}\mathbf{e}_\phi + u_\phi\frac{\partial \mathbf{e}_\phi}{\partial r}\right)$$

$$+ \frac{\mathbf{e}_\theta}{r} \times \left(\frac{\partial u_r}{\partial \theta}\mathbf{e}_r + u_r\frac{\partial \mathbf{e}_r}{\partial \theta} + \frac{\partial u_\theta}{\partial \theta}\mathbf{e}_\theta + u_\theta\frac{\partial \mathbf{e}_\theta}{\partial \theta} + \frac{\partial u_\phi}{\partial \theta}\mathbf{e}_\phi + u_\phi\frac{\partial \mathbf{e}_\phi}{\partial \theta}\right)$$

$$+ \frac{\mathbf{e}_\phi}{r\sin\theta} \times \left(\frac{\partial u_r}{\partial \phi}\mathbf{e}_r + u_r\frac{\partial \mathbf{e}_r}{\partial \phi} + \frac{\partial u_\theta}{\partial \phi}\mathbf{e}_\theta + u_\theta\frac{\partial \mathbf{e}_\theta}{\partial \phi} + \frac{\partial u_\phi}{\partial \phi}\mathbf{e}_\phi + u_\phi\frac{\partial \mathbf{e}_\phi}{\partial \phi}\right)$$

Now use the derivative formulas given in Table 8.3 to write

$$\text{curl } \mathbf{u} = \left(0 + 0 + \frac{\partial u_\theta}{\partial r}\mathbf{e}_\phi + 0 - \frac{\partial u_\phi}{\partial r}\mathbf{e}_\theta + 0 \right) + \frac{1}{r}\left(-\frac{\partial u_r}{\partial \theta}\mathbf{e}_\phi + 0 + 0 + u_\theta\,\mathbf{e}_\phi + \frac{\partial u_\phi}{\partial \theta}\mathbf{e}_r + 0 \right)$$

$$+ \frac{1}{r\sin\theta}\left(\frac{\partial u_r}{\partial \phi}\mathbf{e}_\theta + 0 - \frac{\partial u_\theta}{\partial \phi}\mathbf{e}_r + 0 + 0 - u_\phi\sin\theta\,\mathbf{e}_\theta + u_\phi\cos\theta\,\mathbf{e}_r \right)$$

$$= \frac{1}{r\sin\theta}\left(u_\phi\cos\theta + \sin\theta\frac{\partial u_\phi}{\partial \theta} - \frac{\partial u_\theta}{\partial \phi}\right)\mathbf{e}_r + \frac{1}{r}\left(\frac{1}{\sin\theta}\frac{\partial u_r}{\partial \phi} - u_\phi - r\frac{\partial u_\phi}{\partial r}\right)\mathbf{e}_\theta$$

$$+ \frac{1}{r}\left(r\frac{\partial u_\theta}{\partial r} + u_\theta - \frac{\partial u_r}{\partial \theta}\right)\mathbf{e}_\phi = \text{Equation 21}$$

18. Let $h_r = 1$, $h_\theta = r$, $h_\phi = r\sin\theta$, $u_1 = r$, $u_2 = \theta$, and $u_3 = \phi$ in the determinant given in the problem to obtain

$$\text{curl } \mathbf{u} = \frac{1}{r^2 \sin\theta} \begin{vmatrix} \mathbf{e}_r & r\,\mathbf{e}_\theta & r\sin\theta\,\mathbf{e}_\phi \\ \dfrac{\partial}{\partial r} & \dfrac{\partial}{\partial\theta} & \dfrac{\partial}{\partial\phi} \\ u_r & ru_\theta & r\sin\theta u_\phi \end{vmatrix}$$

$$= \frac{1}{r\sin\theta}\left[\frac{\partial}{\partial\theta}(\sin\theta\,u_\phi) - \frac{\partial u_\theta}{\partial\phi}\right]\mathbf{e}_r + \frac{1}{r}\left[\frac{1}{\sin\theta}\frac{\partial u_r}{\partial\phi} - \frac{\partial(ru_\phi)}{\partial r}\right]\mathbf{e}_\theta + \frac{1}{r}\left[\frac{\partial(ru_\theta)}{\partial r} - \frac{\partial u_r}{\partial\theta}\right]\mathbf{e}_\phi$$

21. First write $\mathbf{r}\cdot\mathbf{v} = rv\cos\theta = (a^2+b^2+c^2)^{1/2}r\cos\theta$. Now,

$$\begin{aligned} I &= (a^2+b^2+c^2)^n \int_0^{2\pi} d\phi \int_0^\pi d\theta \sin\theta \int_0^R dr\, r^2 r^{2n}\cos^{2n}\theta \\ &= 2\pi(a^2+b^2+c^2)^n \frac{R^{2n+3}}{2n+3}\int_0^\pi d\theta \sin\theta\cos^{2n}\theta \\ &= 2\pi(a^2+b^2+c^2)^n \frac{R^{2n+3}}{2n+3}\int_{-1}^1 dx\, x^{2n}\cdot \\ &= 2\pi(a^2+b^2+c^2)^n\cdot\frac{2}{2n+1}\cdot\frac{R^{2n+3}}{2n+3} = \frac{4\pi(a^2+b^2+c^2)^n R^{2n+3}}{(2n+1)(2n+3)} \end{aligned}$$

8.5 Curvilinear Coordinates

1. $r = c_1$ are right cylinders centered on the z axis; $\theta = c_2$ are planes containing the z axis; and $z = c_3$ are planes perpendicular to the z axis. The r curve, the intersection of the $\theta = c_2$ and $z = c_3$ surfaces, is a straight line lying in the $z = c_3$ plane making an angle $\theta = c_2$ with the x axis.

The θ curve is a circle in the $z = c_3$ plane centered at the origin.

The z curve is a vertical line on the surface of the cylinder $r = c_1$.

3. Starting with $\mathbf{r} = r\cos\theta\,\mathbf{i} + r\sin\theta\,\mathbf{j} + z\,\mathbf{k}$, we have

$$\begin{aligned} \frac{\partial\mathbf{r}}{\partial r} &= \cos\theta\,\mathbf{i} + \sin\theta\,\mathbf{j} = \mathbf{e}_r \\ \frac{\partial\mathbf{r}}{\partial\theta} &= -r\sin\theta\,\mathbf{i} + r\cos\theta\,\mathbf{j} = r\,\mathbf{e}_\theta \\ \frac{\partial\mathbf{r}}{\partial z} &= \mathbf{k} = \mathbf{e}_z \end{aligned}$$

These three vectors are mutually orthogonal.

6. Simply use the fact that $\mathbf{j}\times\mathbf{k} = \mathbf{i}$.

9. Using $x = r\sin\theta\sin\phi$, $y = r\sin\theta\cos\phi$, and $z = r\cos\theta$, we write

$$\begin{aligned} \frac{\partial(x,y,z)}{\partial(r,\theta,\phi)} &= \begin{vmatrix} \sin\theta\sin\phi & \sin\theta\cos\phi & \cos\theta \\ r\cos\theta\sin\phi & r\cos\theta\cos\phi & -r\sin\theta \\ r\sin\theta\cos\phi & -r\sin\theta\sin\phi & 0 \end{vmatrix} \\ &= \sin\theta\sin\phi\,(-r^2\sin^2\theta\sin\phi) - \sin\theta\cos\phi\,(r^2\sin^2\theta\cos\phi) + \cos\theta\,(-r^2\sin\theta\cos\theta) \\ &= -r^2\sin^3\theta - r^2\cos^2\theta\sin\theta = -r^2\sin\theta \end{aligned}$$

and so $\left|\dfrac{\partial(x,y,z)}{\partial(r,\theta,\phi)}\right| = r^2 \sin\theta$.

12.   Let $i=1$ be r, $i=2$ be θ, and $i=3$ be ϕ in Equations 20 and 21 and use $h_r=1$, $h_\theta = r$, and $h_\phi = r\sin\theta$ to get

$$\frac{\partial \mathbf{e}_r}{\partial \theta} = \frac{\mathbf{e}_\theta}{h_r}\frac{\partial h_\theta}{\partial r} = \mathbf{e}_\theta \qquad \frac{\partial \mathbf{e}_r}{\partial \phi} = \frac{\mathbf{e}_\phi}{h_r}\frac{\partial h_\phi}{\partial r} = \sin\theta\,\mathbf{e}_\phi \qquad \frac{\partial \mathbf{e}_r}{\partial r} = -\frac{1}{h_\theta}\frac{\partial h_r}{\partial \theta}\mathbf{e}_\theta - \frac{1}{h_\phi}\frac{\partial h_r}{\partial \phi}\mathbf{e}_\phi = 0$$

$$\frac{\partial \mathbf{e}_\theta}{\partial r} = \frac{\mathbf{e}_r}{h_\theta}\frac{\partial h_r}{\partial \theta} = 0 \qquad \frac{\partial \mathbf{e}_\theta}{\partial \phi} = \frac{\mathbf{e}_\phi}{h_\theta}\frac{\partial h_\phi}{\partial \theta} = \cos\theta\,\mathbf{e}_\theta \qquad \frac{\partial \mathbf{e}_\theta}{\partial \theta} = -\frac{\mathbf{e}_r}{h_\theta}\frac{\partial h_\theta}{\partial r} - \frac{\mathbf{e}_\phi}{h_\phi}\frac{\partial h_\theta}{\partial \phi} = -\mathbf{e}_r$$

$$\frac{\partial \mathbf{e}_\phi}{\partial r} = \frac{\mathbf{e}_r}{h_\phi}\frac{\partial h_r}{\partial \phi} = 0 \qquad \frac{\partial \mathbf{e}_\phi}{\partial \theta} = \frac{\mathbf{e}_\theta}{h_\phi}\frac{\partial h_\theta}{\partial \phi} = 0 \qquad \frac{\partial \mathbf{e}_\phi}{\partial \phi} = -\frac{\mathbf{e}_r}{h_r}\frac{\partial h_\phi}{\partial r} - \frac{\mathbf{e}_\theta}{h_\theta}\frac{\partial h_\phi}{\partial \theta}$$

$$= -\sin\theta\,\mathbf{e}_r - \cos\theta\,\mathbf{e}_\theta$$

15. $dS_1 = h_\theta h_\phi\, d\theta d\phi = \sin\theta\, d\theta d\phi$, so $S_1 = \displaystyle\int_0^{2\pi} d\phi \int_0^\pi d\theta\,\sin\theta = 4\pi$.

8.6   Some Other Coordinate Systems

1. Starting with Equations 1, we have

$$
\begin{aligned}
h_\theta &= \left[\left(\frac{\partial x}{\partial \theta}\right)^2 + \left(\frac{\partial y}{\partial \theta}\right)^2 + \left(\frac{\partial z}{\partial \theta}\right)^2\right]^{1/2} \\
&= [a^2 \sinh^2\eta \cos^2\theta \cos^2\phi + a^2 \sinh^2\eta \cos^2\theta \sin^2\phi + a^2 \cosh^2\eta \sin^2\theta]^{1/2} \\
&= a[\sinh^2\eta \cos^2\theta + \cosh^2\eta \sin^2\theta]^{1/2} \\
&= a[\sinh^2\eta(1-\sin^2\theta) + (1+\sinh^2\eta)\sin^2\theta]^{1/2} \\
&= a(\sinh^2\eta + \sin^2\theta)^{1/2} \\
h_\phi &= \left[\left(\frac{\partial x}{\partial \phi}\right)^2 + \left(\frac{\partial y}{\partial \phi}\right)^2 + \left(\frac{\partial y}{\partial \phi}\right)^2\right]^{1/2} \\
&= [a^2 \sinh^2\eta \sin^2\theta \sin^2\phi + a^2 \sinh^2\eta \sin^2\theta \cos^2\phi]^{1/2} \\
&= a\sinh\eta \sin\theta
\end{aligned}
$$

3. Use Equation 8.5.25 with "1" $=\eta$, "2" $=\theta$, and "3" $=\phi$ and the result of Problem 1 to write

$$
\begin{aligned}
\nabla^2 f &= \frac{1}{a^3(\sinh^2\eta + \sin^2\theta)\sinh\eta \sin\theta}\left\{\frac{\partial}{\partial \eta}\left(a\sinh\eta \sin\theta\frac{\partial f}{\partial \eta}\right) + \frac{\partial}{\partial \theta}\left(a\sinh\eta \sin\theta\frac{\partial f}{\partial \theta}\right)\right. \\
&\quad \left. + \frac{\partial}{\partial \phi}\left[\frac{a(\sinh^2\eta + \sin^2\theta)}{\sinh\eta \sin\theta}\frac{\partial f}{\partial \phi}\right]\right\} \\
&= \frac{1}{a^2(\sinh^2\eta + \sin^2\theta)}\left(\frac{\partial^2 f}{\partial \eta^2} + \frac{\cosh\eta}{\sinh\eta}\frac{\partial f}{\partial \eta} + \frac{\partial^2 f}{\partial \theta^2} + \frac{\cos\theta}{\sin\theta}\frac{\partial f}{\partial \theta}\right) + \frac{1}{a^2\sinh^2\eta \sin^2\theta}\frac{\partial^2 f}{\partial \phi^2} = 0
\end{aligned}
$$

6. Use Equation 1 to write

$$
\begin{aligned}
r_1 &= (a^2 + 2a^2 \cosh \eta \cos \theta + a^2 \cosh^2 \eta \cos^2 \theta + a^2 \sinh^2 \eta \sin^2 \theta \cos^2 \phi \\
&\qquad + a^2 \sinh^2 \eta \sin^2 \theta \sin^2 \phi)^{1/2} \\
&= a(1 + 2 \cosh \eta \cos \theta + \cosh^2 \eta \cos^2 \theta + \sinh^2 \eta \sin^2 \theta)^{1/2} \\
&= a(1 + 2 \cosh \eta \cos \theta + \cos^2 \theta + \sinh^2 \eta)^{1/2} \\
&= a(\cosh^2 \eta + 2 \cosh \eta \cos \theta + \cos^2 \theta)^{1/2} \\
&= a(\cosh \eta + \cos \theta)
\end{aligned}
$$

$$
\begin{aligned}
r_2 &= (a^2 - 2a^2 \cosh \eta \cos \theta + a^2 \cosh^2 \eta \cos^2 \theta + a^2 \sinh^2 \eta \sin^2 \theta \cos^2 \phi \\
&\qquad + a^2 \sinh^2 \eta \sin^2 \theta \sin^2 \phi)^{1/2} \\
&= a(1 - 2 \cosh \eta \cos \theta + \cosh^2 \eta \cos^2 \theta + \sinh^2 \eta \sin^2 \theta)^{1/2} \\
&= a(1 - 2 \cosh \eta \cos \theta + \cos^2 \theta + \sinh^2 \eta)^{1/2} \\
&= a(\cosh^2 \eta - 2 \cosh \eta \cos \theta + \cos^2 \theta)^{1/2} \\
&= a(\cosh \eta - \cos \theta)
\end{aligned}
$$

So $\dfrac{r_1 + r_2}{2a} = \cosh \eta = \lambda$ and $\dfrac{r_1 - r_2}{2a} = \cos \theta = \mu.$

9. The surface is described by $\lambda = \lambda_0$, so Equation 8.5.11 with "1" $= \lambda$, "2" $= \mu$, and "3" $= \phi$ and the h's given by Equation 10 give

$$
\begin{aligned}
S &= \int_0^{2\pi} d\phi \int_{-1}^{1} d\mu \, h_\mu h_\phi \\
&= 2\pi a^2 (\lambda_0^2 - 1)^{1/2} \int_{-1}^{1} d\mu \, (\lambda_0^2 - \mu^2)^{1/2} \\
&= \pi a^2 \sinh \eta_0 \left| \mu (\lambda_0^2 - \mu^2)^{1/2} + \lambda_0 \sin^{-1} \frac{\mu}{\lambda_0} \right|_{-1}^{1} \\
&= 2\pi a^2 \sinh \eta_0 \left[(\lambda_0^2 - 1)^{1/2} + \lambda_0^2 \sin^{-1} \frac{1}{\lambda_0} \right]
\end{aligned}
$$

Now use the relations $\lambda_0 = \cosh \eta_0 = c/a$ and $(\lambda_0^2 - 1)^{1/2} = \sinh \eta_0 = b/a$ to write

$$
S = 2\pi ab \left(\frac{b}{a} + \frac{c^2}{a^2} \sin^{-1} \frac{a}{c} \right) = 2\pi \left(b^2 + \frac{bc^2}{a} \sin^{-1} \frac{a}{c} \right)
$$

Now, because $\cosh \eta_0 = c/a$ and $\sinh \eta_0 = b/a$, we have that $a = (c^2 - b^2)^{1/2}$, and so S becomes

$$
S = 2\pi \left(b^2 + \frac{bc}{e} \sin^{-1} e \right)
$$

where $e = (c^2 - b^2)^{1/2}/c$ is the eccentricity.

12. Start with Equations 13 and write

$$
\begin{aligned}
\frac{\partial \mathbf{r}}{\partial \eta} &= a \sinh \eta \sin \theta \cos \phi \, \mathbf{i} + a \sinh \eta \sin \theta \sin \phi \, \mathbf{j} + a \cosh \eta \cos \theta \, \mathbf{k} \\
\frac{\partial \mathbf{r}}{\partial \theta} &= a \cosh \eta \cos \theta \cos \phi \, \mathbf{i} + a \cosh \eta \cos \theta \sin \phi \, \mathbf{j} - a \sinh \eta \sin \theta \, \mathbf{k} \\
\frac{\partial \mathbf{r}}{\partial \phi} &= -a \cosh \eta \sin \theta \sin \phi \, \mathbf{i} + a \cosh \eta \sin \theta \cos \phi \, \mathbf{j}
\end{aligned}
$$

The three vectors are orthogonal.

15. The scale factors are $h_\eta = a(\cosh^2 \eta - \sin^2 \theta)^{1/2} = h_\theta$ and $h_\phi = a \cosh \eta \sin \theta$. Substituting these into Equation 8.5.11 with "2" $= \theta$ and "3" $= \phi$ (and $\eta = \eta_0$ to specify the surface) gives

$$S = a^2 \int_0^{2\pi} d\phi \int_0^{\pi} d\theta \, (\cosh^2 \eta_0 - \sin^2 \theta)^{1/2} \cosh \eta_0 \sin \theta$$

Let $x = \cos \theta$ to write

$$
\begin{aligned}
S &= 2\pi a^2 \cosh \eta_0 \int_{-1}^{1} (\sinh^2 \eta_0 + x^2)^{1/2} dx \\
&= \pi a^2 \cosh \eta_0 \left| x(x^2 + \sinh^2 \eta_0)^{1/2} + \sinh^2 \eta_0 \ln[\, x + (x^2 + \sinh^2 \eta_0)^{1/2}\,] \right|_{-1}^{1} \\
&= \pi a^2 \cosh \eta_0 \left[2(1 + \sinh^2 \eta_0)^{1/2} + \sinh^2 \eta_0 \ln \frac{(1 + \sinh^2 \eta_0)^{1/2} + 1}{(1 + \sinh^2 \eta_0)^{1/2} - 1} \right] \\
&= \pi a^2 \left[2 \cosh^2 \eta_0 + \cosh \eta_0 \sinh^2 \eta_0 \ln \frac{1 + \cosh \eta_0}{\cosh \eta_0 - 1} \right]
\end{aligned}
$$

But $b = a \cosh \eta_0$ and $c = a \sinh \eta_0$ and the eccentricity is equal to $\dfrac{(b^2 - c^2)^{1/2}}{b} = \dfrac{1}{\cosh \eta_0}$, so

$$S = 2\pi b^2 + \frac{\pi b^2}{e} \ln \frac{1 + e}{1 - e}$$

For the case, $b = c$, $e = 0$, or

$$\lim_{e \to 0} \frac{1}{e} \ln \frac{1 + e}{1 - e} = \lim_{e \to 0} \frac{1}{e} \left(2e + \frac{e^3}{3} + \cdots \right) = 2$$

so that $S = 4\pi b^2$.

CHAPTER 9

Linear Algebra and Vector Spaces

9.1 Determinants

1.

$$\begin{vmatrix} 2 & 1 & 1 \\ -1 & 3 & 2 \\ 2 & 0 & 1 \end{vmatrix} = 2(3) - 1(-5) + 1(-6) = 5$$

$$\begin{vmatrix} 3 & 1 & 1 \\ 2 & 3 & 2 \\ 2 & 0 & 1 \end{vmatrix} = 3(3) - 1(-2) + 1(-6) = 5$$

$$\begin{vmatrix} 1 & 4 & 3 \\ -1 & 3 & 2 \\ 2 & 0 & 1 \end{vmatrix} = 1(3) - 4(-5) + 3(-6) = 5$$

3. $|A| = 1(-2) - 6(0) + 1(2) = 0$. You can see immediately that $|A| = 0$ because the first and third columns are equal.

The second determinant is equal to zero because the first column is twice the third column.

6. $|A| = 1(-\sin^2 x - \cos^2 x) = -1$.

9. $\begin{vmatrix} \cos\theta & -\sin\theta & 0 \\ \sin\theta & \cos\theta & 0 \\ 0 & 0 & 1 \end{vmatrix} = \cos\theta(\cos\theta) + \sin\theta(\sin\theta) + 0(0) = 1$

12. To interchange the rth row with the $(r+k)$th row, it takes k successive inversions to bring the rth row to the $(r+k)$th row, and then $k-1$ inversions to bring $(r+k)$th row (which is now adjacent to its original position because it has been replaced by rth row) to the original position of the rth row. Therefore, interchanging the rth row and the $(r+k)$th row requires $2k-1$ inversions, so than $|A_{r\leftrightarrow r+1}| = (-1)^{2k-1}|A| = -|A|$ since $2k-1$ must be an odd integer. Since the two rows that we have just interchanged are arbitrary, we see that $|A|$ changes sign when *any* two rows are exhanged.

15. We have that $x + y = 2$ and $3x - 2y = 5$. Therefore, Cramer's rule gives

$$x = \frac{\begin{vmatrix} 2 & 1 \\ 5 & -2 \end{vmatrix}}{\begin{vmatrix} 1 & 1 \\ 3 & -2 \end{vmatrix}} = \frac{-9}{-5} = \frac{9}{5} \text{ and } y = \frac{\begin{vmatrix} 1 & 2 \\ 3 & 5 \end{vmatrix}}{\begin{vmatrix} 1 & 1 \\ 3 & -2 \end{vmatrix}} = \frac{-1}{-5} = \frac{1}{5}$$

18.

$$\sum_{k=1}^{3} a_{k1} A_{k2} = a_{11} A_{12} + a_{21} A_{22} + a_{31} A_{32}$$

$$= -a_{11}(a_{21}a_{33} - a_{23}a_{31}) + a_{21}(a_{11}a_{33} - a_{13}a_{31}) - a_{31}(a_{11}a_{23} - a_{21}a_{13}) = 0$$

9.2   Gaussian Elimination

1. $\begin{pmatrix} 1 & 2 & -3 & 4 \\ 2 & -1 & 1 & 1 \\ 3 & 2 & -1 & 5 \end{pmatrix} \sim \begin{pmatrix} 1 & 2 & -3 & 4 \\ 0 & -5 & 7 & -7 \\ 0 & -4 & 8 & -7 \end{pmatrix} \sim \begin{pmatrix} 1 & 2 & -3 & 4 \\ 0 & -5 & 7 & -7 \\ 0 & 0 & 12/5 & -7/5 \end{pmatrix}$

and so $x_3 = -7/12$, $x_2 = 7/12$; $x_1 = 13/12$

3. $\begin{pmatrix} 1 & 1 & 0 & 1 \\ 1 & 0 & 1 & 1 \\ 2 & 1 & 1 & 0 \end{pmatrix} \sim \begin{pmatrix} 1 & 1 & 0 & 1 \\ 0 & -1 & 1 & 0 \\ 0 & -1 & 1 & -2 \end{pmatrix} \sim \begin{pmatrix} 1 & 1 & 0 & 1 \\ 0 & -1 & 1 & 0 \\ 0 & 0 & 0 & -2 \end{pmatrix}$

$0 = -2$ implies that there is no solution.

6.

$$\begin{pmatrix} 1 & 0 & 2 & -1 & 3 \\ 0 & 1 & 1 & 0 & 5 \\ 3 & 2 & 0 & -2 & -1 \\ 0 & 0 & -1 & 4 & 13 \\ 2 & 0 & -1 & 3 & 11 \end{pmatrix} \sim \begin{pmatrix} 1 & 0 & 2 & -1 & 3 \\ 0 & 1 & 1 & 0 & 5 \\ 0 & 2 & -6 & 1 & -10 \\ 0 & 0 & -1 & 4 & 13 \\ 0 & 0 & -5 & 5 & 5 \end{pmatrix} \sim \begin{pmatrix} 1 & 0 & 2 & -1 & 3 \\ 0 & 1 & 1 & 0 & 5 \\ 0 & 0 & -8 & 1 & -20 \\ 0 & 0 & -1 & 4 & 13 \\ 0 & 0 & -5 & 5 & 5 \end{pmatrix}$$

$$\sim \begin{pmatrix} 1 & 0 & 2 & -1 & 3 \\ 0 & 1 & 1 & 0 & 5 \\ 0 & 0 & -1 & 4 & 13 \\ 0 & 0 & 0 & -31 & -124 \\ 0 & 0 & 0 & -15 & -60 \end{pmatrix}$$

The last two lines give $x_4 = 4$, so working upwards, $x_3 = 3$, $x_2 = 2$, $x_1 = 1$.

9.

$$\begin{pmatrix} 1 & -2 & 1 & -1 & 2 & -7 \\ 0 & 1 & 1 & 2 & -1 & 5 \\ 1 & -1 & 2 & 2 & 2 & -1 \end{pmatrix} \sim \begin{pmatrix} 1 & -2 & 1 & -1 & 2 & -7 \\ 0 & 1 & 1 & 2 & -1 & 5 \\ 0 & 1 & 1 & 3 & 0 & 6 \end{pmatrix}$$

$$\sim \begin{pmatrix} 1 & -2 & 1 & -1 & 2 & -7 \\ 0 & 1 & 1 & 2 & -1 & 5 \\ 0 & 0 & 0 & 1 & 1 & 1 \end{pmatrix}$$

$x_4 = 1 - x_5$; $x_2 = -x_3 - 2 + 2x_5 + 5 = 3 - x_3 + 3x_5$; $x_1 = -3x_3 + 3x_5$

12. $\lambda \neq 1$. If $\lambda = 1$, the lines are parallel and there is no solution.

9.3 Matrices

1.

$$
\begin{aligned}
C &= 2A - 3B \\
&= \begin{pmatrix} 2 & 0 & -2 \\ -2 & 4 & 0 \\ 0 & 2 & 2 \end{pmatrix} - \begin{pmatrix} -3 & 3 & 0 \\ 9 & 0 & 6 \\ 3 & 3 & 3 \end{pmatrix} \\
&= \begin{pmatrix} 5 & -3 & -2 \\ -11 & 4 & -6 \\ -3 & -1 & -1 \end{pmatrix}
\end{aligned}
$$

$$
\begin{aligned}
D &= 6B - A \\
&= \begin{pmatrix} -6 & 6 & 0 \\ 18 & 0 & 12 \\ 6 & 6 & 6 \end{pmatrix} - \begin{pmatrix} 1 & 0 & -1 \\ -1 & 2 & 0 \\ 0 & 1 & 1 \end{pmatrix} \\
&= \begin{pmatrix} -7 & 6 & 1 \\ 19 & -2 & 12 \\ 6 & 5 & 5 \end{pmatrix}
\end{aligned}
$$

6. (a) $A^T = (a_{ji})$; $(A^T)^T = (a_{ij})$

(b) $(A+B)^T = (a_{ij}+b_{ij})^T = (a_{ji}+b_{ji}) = A^T + B^T$

(c) $(\alpha A)^T = (\alpha a_{ij})^T = \alpha a_{ji} = \alpha A^T$

(d) $(AB)^T = \left(\sum_k a_{ik}b_{kj}\right)^T = \left(\sum_k a_{jk}b_{ki}\right) = \left(\sum_k b_{ki}a_{jk}\right) = \left(\sum_k (b_{ik})^T(a_{kj})^T\right) = B^T A^T$

9. (a) $A = \begin{pmatrix} 1 & 0 & 2 \\ 2 & 1 & 1 \\ -1 & 0 & 1 \end{pmatrix}$; $A_{\text{cof}} = \begin{pmatrix} 1 & -3 & 1 \\ 0 & 3 & 0 \\ -2 & 3 & 1 \end{pmatrix}$; $\text{adj}(A) = \begin{pmatrix} 1 & 0 & -2 \\ -3 & 3 & 3 \\ 1 & 0 & 1 \end{pmatrix}$

(b) $A = \begin{pmatrix} 3 & 1 & 2 \\ 1 & 2 & 0 \\ 1 & 0 & 1 \end{pmatrix}$; $A_{\text{cof}} = \begin{pmatrix} 2 & -1 & -2 \\ -1 & 1 & 1 \\ -4 & 2 & 5 \end{pmatrix}$; $\text{adj}(A) = \begin{pmatrix} 2 & -1 & -4 \\ -1 & 1 & 2 \\ -2 & 1 & 5 \end{pmatrix}$

12. (a) $\begin{pmatrix} 1 & 2 \\ 3 & 6 \end{pmatrix}\begin{pmatrix} b_{11} & b_{12} \\ b_{21} & b_{22} \end{pmatrix} = \begin{pmatrix} 0 & 0 \\ 0 & 0 \end{pmatrix}$ implies that $b_{11} + 2b_{21} = 0$, $b_{12} + 2b_{22} = 0$,

$3b_{11} + 6b_{21} = 0$, and $3b_{12} + 6b_{22} = 0$. Thus, $B = \begin{pmatrix} -2 & -2 \\ 1 & 1 \end{pmatrix}$ and $AB = \begin{pmatrix} 0 & 0 \\ 0 & 0 \end{pmatrix}$

(b) $b_{11} + b_{21} = 0$ and $b_{11} + 2b_{21} = 0$. No. These equations are inconsistent and have no solution other than $b_{ij} = 0$.

15. One way to prove that $(AB)^{-1} = B^{-1}A^{-1}$ is to write $(AB)^{-1} = XY$ and multiply by AB to get $I = ABXY$. You can see by inspection that a solution to this equation is $X = B^{-1}$ and $Y = A^{-1}$, or you can multiply from the left first by A^{-1} and then by B^{-1}.

18. (a)

$$A = \begin{pmatrix} 1 & 1 & 1 \\ 0 & 1 & 1 \\ 0 & 0 & 1 \end{pmatrix}$$

$$A_{\text{cof}} = \begin{pmatrix} 1 & 0 & 0 \\ -1 & 1 & 0 \\ 0 & -1 & 1 \end{pmatrix}$$

$$A_{\text{adj}} = \begin{pmatrix} 1 & -1 & 0 \\ 0 & 1 & -1 \\ 0 & 0 & 1 \end{pmatrix}$$

$$A^{-1} = \frac{A_{\text{adj}}}{|A|} = \begin{pmatrix} 1 & -1 & 0 \\ 0 & 1 & -1 \\ 0 & 0 & 1 \end{pmatrix}$$

(b)

$$A = \begin{pmatrix} 1 & -2 & 0 \\ 0 & 1 & 2 \\ 2 & 2 & 0 \end{pmatrix}$$

$$A_{\text{cof}} = \begin{pmatrix} -4 & 4 & -2 \\ 0 & 0 & -6 \\ -4 & -2 & 1 \end{pmatrix}$$

$$A_{\text{adj}} = \begin{pmatrix} -4 & 0 & -4 \\ 4 & 0 & -2 \\ -2 & -6 & 1 \end{pmatrix}$$

$$A^{-1} = \frac{A_{\text{adj}}}{|A|} = \begin{pmatrix} \frac{1}{3} & 0 & \frac{1}{3} \\ -\frac{1}{3} & 0 & \frac{1}{6} \\ \frac{1}{6} & \frac{1}{2} & -\frac{1}{12} \end{pmatrix}$$

21.

$$AB = \begin{pmatrix} a_{11} & a_{12} \\ a_{21} & a_{22} \end{pmatrix} \begin{pmatrix} b_{11} & b_{12} \\ b_{21} & b_{22} \end{pmatrix} = \begin{pmatrix} a_{11}b_{11} + a_{12}b_{21} & a_{11}b_{12} + a_{12}b_{22} \\ a_{21}b_{11} + a_{22}b_{21} & a_{21}b_{12} + a_{22}b_{22} \end{pmatrix}$$

$$|AB| = a_{11}b_{11}a_{21}b_{12} + a_{11}b_{11}a_{22}b_{22} + a_{12}b_{21}a_{21}b_{12} + a_{12}b_{21}a_{22}b_{22}$$
$$- a_{11}b_{12}a_{21}b_{11} - a_{11}b_{12}a_{22}b_{21} - a_{12}b_{22}a_{21}b_{11} - a_{12}b_{22}a_{22}b_{21}$$

$$|A||B| = (a_{11}a_{22} - a_{21}a_{12})(b_{11}b_{22} - b_{21}b_{12})$$
$$= a_{11}a_{22}b_{11}b_{22} - a_{11}a_{22}b_{21}b_{12} - a_{21}a_{12}b_{11}b_{22} + a_{21}a_{12}b_{21}b_{12}$$

9.4   Rank of a Matrix

1. $A = \begin{pmatrix} 3 & -2 \\ -6 & 4 \\ -3 & 2 \end{pmatrix}$ and $(A \,|\, H) = \begin{pmatrix} 3 & -2 & 2 \\ -6 & 4 & -4 \\ -3 & 2 & 2 \end{pmatrix}$

$$\begin{pmatrix} 3 & -2 \\ -6 & 4 \\ -3 & 2 \end{pmatrix} \sim \begin{pmatrix} 3 & -2 \\ 0 & 0 \\ 0 & 0 \end{pmatrix} \text{ and so } r(\mathsf{A}) = 1.$$

$$\begin{pmatrix} 3 & -2 & 2 \\ -6 & 4 & 4 \\ -3 & 2 & 2 \end{pmatrix} \sim \begin{pmatrix} 3 & -2 & 2 \\ 0 & 0 & 8 \\ 0 & 0 & 0 \end{pmatrix} \text{ and so } r(\mathsf{A}\,|\,\mathsf{H}) = 2.$$

Because $r(\mathsf{A}) < r(\mathsf{A}\,|\,\mathsf{H})$, there is no solution.

3. $\mathsf{A} = \begin{pmatrix} 1 & -2 \\ 2 & -4 \\ -3 & 6 \end{pmatrix}$ and $(\mathsf{A}\,|\,\mathsf{H}) = \begin{pmatrix} 1 & -2 & 3 \\ 2 & -4 & 6 \\ -3 & 6 & -9 \end{pmatrix}$

$$\begin{pmatrix} 1 & -2 \\ 2 & -4 \\ -3 & 6 \end{pmatrix} \sim \begin{pmatrix} 1 & -2 \\ 0 & 0 \\ 0 & 0 \end{pmatrix} \text{ and so } r(\mathsf{A}) = 1.$$

$$\begin{pmatrix} 1 & -2 & 3 \\ 2 & -4 & 6 \\ -3 & 6 & -9 \end{pmatrix} \sim \begin{pmatrix} 1 & -2 & 3 \\ 0 & 0 & 0 \\ 0 & 0 & 0 \end{pmatrix} \text{ and so } r(\mathsf{A}\,|\,\mathsf{H}) = 1.$$

Because $r(\mathsf{A}) = r(\mathsf{A}\,|\,\mathsf{H}) = 1 < 2$, there is an infinite number of solutions expressible in terms of one $(2 - 1)$ parameter.

6. $\mathsf{A} = \begin{pmatrix} 1 & 1 & 0 \\ 1 & 0 & 1 \\ 2 & 1 & 1 \end{pmatrix}$ and $(\mathsf{A}\,|\,\mathsf{H}) = \begin{pmatrix} 1 & 1 & 0 & 1 \\ 1 & 0 & 1 & 1 \\ 2 & 1 & 1 & 0 \end{pmatrix}$

$$\begin{pmatrix} 1 & 1 & 0 \\ 1 & 0 & 1 \\ 2 & 1 & 1 \end{pmatrix} \sim \begin{pmatrix} 1 & 1 & 0 \\ 0 & -1 & 1 \\ 0 & -1 & 1 \end{pmatrix} \sim \begin{pmatrix} 1 & 1 & 0 \\ 0 & -1 & 1 \\ 0 & 0 & 0 \end{pmatrix} \text{ and so } r(\mathsf{A}) = 2.$$

$$\begin{pmatrix} 1 & 1 & 0 & 1 \\ 1 & 0 & 1 & 1 \\ 2 & 1 & 1 & 0 \end{pmatrix} \sim \begin{pmatrix} 1 & 1 & 0 & 1 \\ 0 & -1 & 1 & 0 \\ 0 & -1 & 1 & -2 \end{pmatrix} \sim \begin{pmatrix} 1 & 1 & 0 & 1 \\ 0 & -1 & 1 & 0 \\ 0 & 0 & 0 & -2 \end{pmatrix} \text{ and so } r(\mathsf{A}\,|\,\mathsf{H}) = 3.$$

Because $r(\mathsf{A}) = 2 < r(\mathsf{A}\,|\,\mathsf{H}) = 3$, there is no solution.

9. $\mathsf{A} = \begin{pmatrix} 1 & 0 & 2 & -1 \\ 0 & 1 & 1 & 0 \\ 3 & 2 & 0 & -2 \\ 0 & 0 & -1 & 4 \\ 2 & 0 & -1 & 3 \end{pmatrix}$ and $(\mathsf{A}\,|\,\mathsf{H}) = \begin{pmatrix} 1 & 0 & 2 & -1 & 3 \\ 0 & 1 & 1 & 0 & 5 \\ 3 & 2 & 0 & -2 & -1 \\ 0 & 0 & -1 & 4 & 13 \\ 2 & 0 & -1 & 3 & 11 \end{pmatrix}$

$$\begin{pmatrix} 1 & 0 & 2 & -1 \\ 0 & 1 & 1 & 0 \\ 3 & 2 & 0 & -2 \\ 0 & 0 & -1 & 4 \\ 2 & 0 & -1 & 3 \end{pmatrix} \sim \begin{pmatrix} 1 & 0 & 2 & -1 \\ 0 & 1 & 1 & 0 \\ 0 & 2 & -6 & 1 \\ 0 & 0 & -1 & 4 \\ 0 & 0 & -5 & 5 \end{pmatrix} \sim \begin{pmatrix} 1 & 0 & 2 & -1 \\ 0 & 1 & 1 & 0 \\ 0 & 0 & -8 & 1 \\ 0 & 0 & -1 & 4 \\ 0 & 0 & -5 & 5 \end{pmatrix} \sim$$

$$\begin{pmatrix} 1 & 0 & 2 & -1 \\ 0 & 1 & 1 & 0 \\ 0 & 0 & -1 & 4 \\ 0 & 0 & -5 & 5 \\ 0 & 0 & -8 & 1 \end{pmatrix} \sim \begin{pmatrix} 1 & 0 & 2 & -1 \\ 0 & 1 & 1 & 0 \\ 0 & 0 & -1 & 4 \\ 0 & 0 & 0 & -15 \\ 0 & 0 & 0 & -31 \end{pmatrix} \sim \begin{pmatrix} 1 & 0 & 2 & -1 \\ 0 & 1 & 1 & 0 \\ 0 & 0 & -1 & 4 \\ 0 & 0 & 0 & -15 \\ 0 & 0 & 0 & 0 \end{pmatrix} \text{ and so } r(\mathsf{A}) = 4.$$

Similarly, $(A \mid H) = \begin{pmatrix} 1 & 0 & 2 & -1 & 3 \\ 0 & 1 & 1 & 0 & 5 \\ 3 & 2 & 0 & -2 & -1 \\ 0 & 0 & -1 & 4 & 13 \\ 2 & 0 & -1 & 3 & 11 \end{pmatrix} \sim \begin{pmatrix} 1 & 0 & 2 & -1 & 3 \\ 0 & 1 & 1 & 0 & 5 \\ 0 & 2 & -6 & 1 & -10 \\ 0 & 0 & -1 & 4 & 13 \\ 0 & 0 & -5 & 5 & 5 \end{pmatrix} \sim$

$\begin{pmatrix} 1 & 0 & 2 & -1 & 3 \\ 0 & 1 & 1 & 0 & 5 \\ 0 & 0 & -1 & 4 & 13 \\ 0 & 0 & -5 & 5 & 5 \\ 0 & 0 & -8 & 1 & -20 \end{pmatrix} \sim \begin{pmatrix} 1 & 0 & 2 & -1 & 3 \\ 0 & 1 & 1 & 0 & 5 \\ 0 & 0 & -1 & 4 & 13 \\ 0 & 0 & 0 & -15 & -60 \\ 0 & 0 & 0 & -31 & -124 \end{pmatrix} \sim \begin{pmatrix} 1 & 0 & 2 & -1 & 3 \\ 0 & 1 & 1 & 0 & 5 \\ 0 & 0 & -1 & 4 & 13 \\ 0 & 0 & 0 & 1 & 4 \\ 0 & 0 & 0 & 0 & 0 \end{pmatrix}$

and so $r(A \mid H) = 4$. Because $r(A) = r(A \mid H) = 4$, there is a unique solution.

12. $A = \begin{pmatrix} 1 & -2 & 1 & -1 & 2 \\ 0 & 1 & 1 & 2 & -1 \\ 1 & -1 & 2 & 2 & 2 \end{pmatrix}$ and $(A \mid H) = \begin{pmatrix} 1 & -2 & 1 & -1 & 2 & -7 \\ 0 & 1 & 1 & 2 & -1 & 5 \\ 1 & -1 & 2 & 2 & 2 & -1 \end{pmatrix}$

$\begin{pmatrix} 1 & -2 & 1 & -1 & 2 \\ 0 & 1 & 1 & 2 & -1 \\ 1 & -1 & 2 & 2 & 2 \end{pmatrix} \sim \begin{pmatrix} 1 & -2 & 1 & -1 & 2 \\ 0 & 1 & 1 & 2 & -1 \\ 0 & 1 & 1 & 3 & 0 \end{pmatrix} \sim \begin{pmatrix} 1 & -2 & 1 & -1 & 2 \\ 0 & 1 & 1 & 2 & -1 \\ 0 & 0 & 0 & 1 & 1 \end{pmatrix}$

and so $r(A) = 3$.

$\begin{pmatrix} 1 & -2 & 1 & -1 & 2 & -7 \\ 0 & 1 & 1 & 2 & -1 & 5 \\ 1 & -1 & 2 & 2 & 2 & -1 \end{pmatrix} \sim \begin{pmatrix} 1 & -2 & 1 & -1 & 2 & -7 \\ 0 & 1 & 1 & 2 & -1 & -5 \\ 0 & 1 & 1 & 3 & 0 & 6 \end{pmatrix} \sim \begin{pmatrix} 1 & -2 & 1 & -1 & 2 & -7 \\ 0 & 1 & 1 & 2 & -1 & -5 \\ 0 & 0 & 0 & 1 & 1 & 11 \end{pmatrix}$

and so $r(A \mid H) = 3$.

Because $r(A) = r(A \mid H) = 3 < 5$, there is a two-parameter family of solutions.

9.5   Vector Spaces

1. All nine criteria on pages 436 and 437 are satisfied, particularly since the addition and scalar multiplication rules obey the laws of real numbers. In addition, the "zero vector" is $(0, 0, \dots, 0)$ and the "inverse vector" to $(u_1, u_2, \dots, u_n)$ is $(-u_1, -u_2, \dots, -u_n)$.

3. The general polynomial in this case is $a + bx + cx^2 + dx^3$. The first three criteria on page 436 are easily satisfied. The "zero vector" occurs when $a = b = c = d = 0$. The "inverse vector" is $-a - bx - cx^2 - dx^3$. The dimension of this vector space is 4.

6. Yes. Because $(1, 1, 0)$ and $(1, 0, 1)$ are linearly independent, they form a two-dimensional subspace of R^3.

9. To see if $(1, 0, 2)$ is in the set spanned by $(1, 1, 1)$, $(1, -1, -1)$, $(3, 1, 1)$, express it as $(1, 0, 2) = c_1(1, 1, 1) + c_2(1, -1, -1) + c_3(3, 1, 1)$, which gives $c_1 + c_2 + 3c_3 = 1$, $c_1 - c_2 + c_3 = 0$, $c_1 - c_2 + c_3 = 2$. In this case, the matrix of the coefficients of the c_j

is $A = \begin{pmatrix} 1 & 1 & 3 \\ 1 & -1 & 1 \\ 1 & -1 & 1 \end{pmatrix} \sim \begin{pmatrix} 1 & 1 & 3 \\ 1 & -1 & 1 \\ 0 & 0 & 0 \end{pmatrix}$ and so $r(A) = 2$. $A \mid H = \begin{pmatrix} 1 & 1 & 3 & 1 \\ 1 & -1 & 1 & 0 \\ 1 & -1 & 1 & 2 \end{pmatrix}$ and

so $r(A \mid H) = 3$. Because $r(\mathbf{A}) = 2 < r(\mathbf{A} \mid \mathbf{H}) = 3$, there is no solution, and the vector is not in the set spanned by the others.

12. Express **u** as $(1, 2, 3) = u_1(1, 1, 0) + u_2(1, 0, 1) + u_3(1, 1, 1)$, which gives

$$
\begin{aligned}
u_1 + u_2 + u_3 &= 1 \\
u_1 \quad\;\; + u_3 &= 2 \\
u_2 + u_3 &= 3
\end{aligned}
$$

The corresponding augmented matrix is

$$
\begin{pmatrix} 1 & 1 & 1 & 1 \\ 1 & 0 & 1 & 2 \\ 0 & 1 & 1 & 3 \end{pmatrix} \sim \begin{pmatrix} 1 & 1 & 1 & 1 \\ 0 & -1 & 0 & 1 \\ 0 & 1 & 1 & 3 \end{pmatrix} \sim \begin{pmatrix} 1 & 1 & 1 & 1 \\ 0 & -1 & 0 & 1 \\ 0 & 0 & 1 & 4 \end{pmatrix}
$$

from which we see that $u_3 = 4$, $u_2 = -1$, and $u_1 = -2$. Therefore,

$$
(1, 2, 3) = -2(1, 1, 0) - (1, 0, 1) + 4(1, 1, 1)
$$

15. $f_1(x) = x^2$ and $f_2(x) = x|x|$

$$
W(x) = \begin{vmatrix} x^2 & x|x| \\ 2x & |x| + x\dfrac{d|x|}{dx} \end{vmatrix} = \begin{cases} 0 & x > 0 \\ 0 & x = 0 \\ -2x^3 & x < 0 \end{cases}
$$

Therefore, $f_1(x)$ and $f_2(x)$ are linearly independent over $[-1, 1]$ because $W(x) \neq 0$ for $x < 0$. (See the figure on the left below.) For $(0, 1]$, the test is inconclusive, but $f_1(x)$ and $f_2(x)$ are identical for $x > 0$. (See the figure on the right below.)

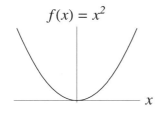

$f(x) = x^2$

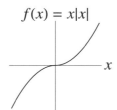

$f(x) = x|x|$

18. We first show that the two functions $f_1(x)$ and $f_2(x)$ are linearly independent over the interval $-\infty < x < \infty$. The two conditions

$$
\begin{aligned}
0c_1 - x^2 c_2 &= 0 & x < 0 \\
x^2 c_1 + 0 c_2 &= 0 & x \geq 0
\end{aligned}
$$

imply that $c_1 = c_2 = 0$, and so $f_1(x)$ and $f_2(x)$ are linearly indpependent over the interval $-\infty < x < \infty$. But $W = \begin{vmatrix} 0 & -x^2 \\ 0 & -2x \end{vmatrix}$ if $x < 0$ and $\begin{vmatrix} x^2 & 0 \\ 2x & 0 \end{vmatrix}$ if $x > 0$, and $W = 0$ for all x.

9.6   Inner Product Spaces

1. The three criteria on page 444 are satisfied by $\mathbf{u} \cdot \mathbf{v} = \|\mathbf{u}\| \|\mathbf{v}\| \cos \theta$:

1. $(\alpha \mathbf{u}_1 + \beta \mathbf{u}_2) \cdot \mathbf{v} = \alpha(\mathbf{u}_1 \cdot \mathbf{v}) + \beta(\mathbf{u}_2 \cdot \mathbf{v})$

2. $\mathbf{u} \cdot \mathbf{v} = \| \mathbf{u} \| \| \mathbf{v} \| \cos \theta$ is symmetric in $\mathbf{u}$ and $\mathbf{v}$.

3. $\mathbf{u} \cdot \mathbf{u} = \| \mathbf{u} \|^2 \geq 0$.

3. $\langle \mathbf{u}, \mathbf{v} \rangle = \int_0^1 (1+x)x \, dx = \dfrac{5}{6}, \quad \langle \mathbf{u}, \mathbf{u} \rangle = \int_0^1 x^2 \, dx = \dfrac{1}{3}, \quad \langle \mathbf{v}, \mathbf{v} \rangle = \int_0^1 (1+x)^2 \, dx = \dfrac{7}{3}$

$$\langle \mathbf{u}, \mathbf{v} \rangle = \frac{5}{6} < \langle \mathbf{u}, \mathbf{u} \rangle^{1/2} \langle \mathbf{v}, \mathbf{v} \rangle^{1/2} = \left(\frac{1}{3}\right)^{1/2} \left(\frac{7}{3}\right)^{1/2} = \frac{7^{1/2}}{3} = \frac{2 \cdot 7^{1/2}}{6} = \frac{5.29 \cdots}{6}$$

6. Use the formulas (n and m integers) $\displaystyle\int_{-\pi}^{\pi} \sin nx \sin mx \, dx = \int_{-\pi}^{\pi} \cos nx \sin mx \, dx = 0$ if $m \neq 0$ and $\displaystyle\int_{-\pi}^{\pi} \sin nx \cos mx \, dx = 0$.

9. Using the definition of an inner product used in Example 1

$$\langle \mathbf{u}, \mathbf{u} \rangle = u_1^2 + u_2^2 + \cdots + u_n^2$$

and so

$$\| \mathbf{u} \| = (u_1^2 + u_2^2 + \cdots + u_n^2)^{1/2}$$

We'll now show that this definition of a norm satsifies Equations 8 through 10.

1. $\| \mathbf{u} \| > 0$ unless $u_1 = u_2 + \cdots u_n = 0$.

2. $\| c\mathbf{u} \| = [(cu_1)^2 + (cu_2)^2 + \cdots + (cu_n)^2]^{1/2} = |c|(u_1^2 + u_2^2 + \cdots + u_n^2)^{1/2}$ upon taking the positive square root.

3. To show that Equation 10 is satisfied, start with

$$\begin{aligned}\| \mathbf{u} + \mathbf{v} \|^2 &= (u_1 + v_1)^2 + (u_2 + v_2)^2 + \cdots + (u_n + v_n)^2 \\ &= \| \mathbf{u} \|^2 + \| \mathbf{v} \|^2 + 2\mathbf{u} \cdot \mathbf{v}\end{aligned}$$

and

$$(\| \mathbf{u} \| + \| \mathbf{v} \|)^2 = \| \mathbf{u} \|^2 + \| \mathbf{v} \|^2 + 2\| \mathbf{u} \| \| \mathbf{v} \|$$

But the Schwartz inequality applied to this inner product says that $\mathbf{u} \cdot \mathbf{v} \leq \| \mathbf{u} \| \cdot \| \mathbf{u} \|$, so $\| \mathbf{u} + \mathbf{v} \| \leq \| \mathbf{u} \| + \| \mathbf{v} \|$.

12. (a) $\| (1, -2, 3) \| = |1| + |-2| + |3| = 6$ and (b) $\| (2, 4, -1) \| = |2| + |4| + |-1| = 7$

15. Start with $\mathbf{u}_3 = \mathbf{v}_3 + b_1 \mathbf{u}_1 + b_2 \mathbf{u}_2$ and require that

$$\langle \mathbf{u}_1 \cdot \mathbf{u}_3 \rangle = \langle \mathbf{u}_1 \cdot \mathbf{v}_3 \rangle + b_1 \langle \mathbf{u}_1 \cdot \mathbf{u}_1 \rangle + b_2 \langle \mathbf{u}_1 \cdot \mathbf{u}_2 \rangle = \langle \mathbf{u}_1 \cdot \mathbf{v}_3 \rangle + b_1 \langle \mathbf{u}_1 \cdot \mathbf{u}_1 \rangle = 0$$

and

$$\langle \mathbf{u}_2 \cdot \mathbf{u}_3 \rangle = \langle \mathbf{u}_2 \cdot \mathbf{v}_3 \rangle + b_2 \langle \mathbf{u}_2 \cdot \mathbf{u}_2 \rangle = 0$$

So

$$\mathbf{u}_3 = \mathbf{v}_3 + b_1 \mathbf{u}_1 + b_2 \mathbf{u}_2 = \mathbf{v}_3 - \frac{\langle \mathbf{u}_1 \cdot \mathbf{v}_3 \rangle}{\langle \mathbf{u}_1 \cdot \mathbf{u}_1 \rangle} \mathbf{u}_1 - \frac{\langle \mathbf{u}_2 \cdot \mathbf{v}_3 \rangle}{\langle \mathbf{u}_2 \cdot \mathbf{u}_2 \rangle} \mathbf{u}_2$$

9.7 Complex Inner Product Spaces

1. Let's write $(i, -1, -i) = c_1(1, i, -1) + c_2(1 + i, 0, 1 - i)$. This equation is equivalent to the three equations

$$
\begin{aligned}
c_1 + c_2 + c_2 i &= i \\
c_1 i &= -1 \\
-c_1 + c_2 - c_2 i &= -i
\end{aligned}
$$

The solution to these equations is $c_1 = i$ and $c_2 = 0$, and so we see that $(i, -1, -i) = i(1, i, -1)$.

3. Start with $\begin{pmatrix} i & 0 & 0 \\ i & i & 0 \\ i & i & i \end{pmatrix}$ and express it in echelon form

$$
\begin{pmatrix} i & 0 & 0 \\ i & i & 0 \\ i & i & i \end{pmatrix} \sim \begin{pmatrix} i & 0 & 0 \\ 0 & i & 0 \\ 0 & i & i \end{pmatrix} \sim \begin{pmatrix} i & 0 & 0 \\ 0 & i & 0 \\ 0 & 0 & i \end{pmatrix}
$$

There is no row of zeros, and so we see that the three vectors are linearly independent.

6. We need to determine if $c_1 \mathbf{I} + c_2 \boldsymbol{\sigma}_x + c_3 \boldsymbol{\sigma}_y + c_4 \boldsymbol{\sigma}_z = \mathbf{0}$ only if $c_1 = c_2 = c_3 = c_4 = 0$. This matrix equation translates into the algebraic equation

$$
\begin{aligned}
c_1 + c_4 &= 0 \\
c_2 - ic_3 &= 0 \\
c_2 + ic_3 &= 0 \\
c_1 - c_4 &= 0
\end{aligned}
$$

The corresponding determinant is

$$
\begin{vmatrix} 1 & 0 & 0 & 1 \\ 0 & 1 & -i & 0 \\ 0 & 1 & i & 0 \\ 1 & 0 & 0 & -1 \end{vmatrix} = -4i \neq 0
$$

The fact that there is a nonzero determinant says that only the solution $c_1 = c_2 = c_3 = c_4 = 0$ exists, and so the matrices are linearly independent.

9. 1. $\langle \mathbf{u} \, | \, a\, \mathbf{v} + b\, \mathbf{w} \rangle = u_1^* a v_1 + u_1^* b w_1 + u_2^* a v_2 + u_2^* b w_2 + \cdots = a\langle \mathbf{u} \, | \, \mathbf{v} \rangle + b\langle \mathbf{u} \, | \, \mathbf{w} \rangle$

2. $\langle \mathbf{u} \, | \, \mathbf{v} \rangle = u_1^* v_1 + u_2^* v_2 + \cdots = (u_1 v_1^* + u_2 v_2^* + \cdots)^* = \langle \mathbf{v} \, | \, \mathbf{u} \rangle^*$

3. $\langle \mathbf{u} \, | \, \mathbf{u} \rangle = u_1^* u_1 + u_2^* u_2 + \cdots \geq 0$

12. We are given $\mathbf{v}_1 = (1, i)$ and $\mathbf{v}_2 = (1, 1)$. Using The Gram-Schmidt orthogonalization procedure, we write $\mathbf{u}_1 = (1, i)$ and $\mathbf{u}_2 = \mathbf{u}_1 + a\, \mathbf{v}_2$. Requiring that $\mathbf{u}_1$ and $\mathbf{u}_2$ be orthogonal gives

$$
\langle \mathbf{u}_1 \, | \, \mathbf{u}_2 \rangle = \langle \mathbf{u}_1 \, | \, \mathbf{u}_1 \rangle + a\langle \mathbf{u}_1 \, | \, \mathbf{v}_2 \rangle = 0 = 2 + a(1 - i)
$$

or $a = -\dfrac{2}{1 - i} = -\dfrac{2}{2}(1 + i) = -(1 + i)$. Therefore,

$$
\mathbf{u}_2 = \mathbf{u}_1 - (1 + i)\, \mathbf{u}_2 = (1, i) - (1 + i, 1 + i) = (-i, -1)
$$

and so $\mathbf{u}_1 = (1, i)$ and $\mathbf{u}_2 = (-i, -1)$ are orthogonal.

Matrices and Eigenvalue Problems

10.1 Orthogonal and Unitary Transformations

1. $S\,\mathbf{u} = \begin{pmatrix} -1 & 0 \\ 0 & 1 \end{pmatrix} \begin{pmatrix} u_x \\ u_y \end{pmatrix} = \begin{pmatrix} -u_x \\ u_y \end{pmatrix}$.

3. (a) $\begin{pmatrix} -1 & 0 \\ 0 & -1 \end{pmatrix} \begin{pmatrix} u_x \\ u_y \end{pmatrix} = \begin{pmatrix} -u_x \\ -u_y \end{pmatrix}$. Inversion through the origin.

(b) $\begin{pmatrix} 1 & 0 \\ 0 & -1 \end{pmatrix} \begin{pmatrix} u_x \\ u_y \end{pmatrix} = \begin{pmatrix} u_x \\ -u_y \end{pmatrix}$. Reflection through the x axis.

(c) $\dfrac{1}{\sqrt{2}} \begin{pmatrix} 1 & -1 \\ 1 & 1 \end{pmatrix} \begin{pmatrix} u_x \\ u_y \end{pmatrix} = \dfrac{1}{\sqrt{2}} \begin{pmatrix} u_x - u_y \\ u_x + u_y \end{pmatrix}$. According to Equation 1, this represents a rotation by $45°$ in a counterclockwise direction.

6. We'll determine how A acts upon each of the standard basis vectors. To do this let $a\,\mathbf{u}_1 + b\,\mathbf{u}_2 + c\,\mathbf{u}_3 = \mathbf{e}_1$, which in augmented matrix form is

$$\begin{pmatrix} a & 0 & c & 1 \\ 0 & b & c & 0 \\ a & b & 0 & 0 \end{pmatrix} \sim \begin{pmatrix} a & 0 & c & 1 \\ 0 & b & c & 0 \\ 0 & 0 & -2c & -1 \end{pmatrix}$$

so $c = 1/2$, $b = -1/2$, and $a = 1/2$, and so $\mathbf{e}_1 = \frac{1}{2}\,\mathbf{u}_1 - \frac{1}{2}\,\mathbf{u}_2 + \frac{1}{2}\,\mathbf{u}_3$ and

$$A\,\mathbf{e}_1 = \frac{1}{2}\,\mathbf{v}_1 - \frac{1}{2}\,\mathbf{v}_2 + \frac{1}{2}\,\mathbf{v}_3 = (1/2, 1/2, 0)$$

Similarly, we find that

$$A\,\mathbf{e}_2 = \frac{1}{2}\,\mathbf{v}_1 + \frac{1}{2}\,\mathbf{v}_2 + \frac{1}{2}\,\mathbf{v}_3 = (-1/2, -3/2, 1)$$

and

$$A\,\mathbf{e}_3 = \frac{1}{2}\,\mathbf{v}_1 + \frac{1}{2}\,\mathbf{v}_2 - \frac{1}{2}\,\mathbf{v}_3 = (3/2, 1/2, -1)$$

Therefore, the matrix is $\mathsf{A} = \dfrac{1}{2}\begin{pmatrix} 1 & -1 & 3 \\ 1 & -3 & 1 \\ 0 & 2 & -2 \end{pmatrix}$.

Note that $\mathbf{A}\,\mathbf{u}_1 = \mathbf{v}_1$, $\mathbf{A}\,\mathbf{u}_2 = \mathbf{v}_2$, and $\mathbf{A}\,\mathbf{u}_3 = \mathbf{v}_3$.

9. Use $\mathsf{A}\mathsf{A}^\mathsf{T} = \mathsf{I}$ to write $\displaystyle\sum_{j=1}^{n} a_{ij}a_{jk}^\mathsf{T} = \delta_{ik} = \sum_{j=1}^{n} a_{ij}a_{kj}$, which says that the rows act as orthogonal vectors.

12.

$$\mathsf{A}^\mathsf{T}\mathsf{A} = \frac{1}{9}\begin{pmatrix} 1 & 2 & -2 \\ 2 & 1 & 2 \\ 2 & -2 & -1 \end{pmatrix}\begin{pmatrix} 1 & 2 & 2 \\ 2 & 1 & -2 \\ -2 & 2 & -1 \end{pmatrix} = \frac{1}{9}\begin{pmatrix} 9 & 0 & 0 \\ 0 & 9 & 0 \\ 0 & 0 & 9 \end{pmatrix}$$

$$\mathsf{A}\mathsf{A}^\mathsf{T} = \frac{1}{9}\begin{pmatrix} 1 & 2 & 2 \\ 2 & 1 & -2 \\ -2 & 2 & -1 \end{pmatrix}\begin{pmatrix} 1 & 2 & -2 \\ 2 & 1 & 2 \\ 2 & -2 & -1 \end{pmatrix} = \frac{1}{9}\begin{pmatrix} 9 & 0 & 0 \\ 0 & 9 & 0 \\ 0 & 0 & 9 \end{pmatrix}$$

15.

$$\mathsf{A}^\dagger\mathsf{A} = \frac{1}{36}\begin{pmatrix} 2+4i & 4i \\ -4i & -2+4i \end{pmatrix}\begin{pmatrix} 2-4i & 4i \\ -4i & -2-4i \end{pmatrix} = \frac{1}{36}\begin{pmatrix} 36 & 0 \\ 0 & 36 \end{pmatrix}$$

$$\mathsf{A}\mathsf{A}^\dagger = \begin{pmatrix} 2-4i & 4i \\ -4i & -2-4i \end{pmatrix}\begin{pmatrix} 2+4i & 4i \\ -4i & -2+4i \end{pmatrix} = \frac{1}{36}\begin{pmatrix} 36 & 0 \\ 0 & 36 \end{pmatrix}$$

10.2   Eigenvalues and Eigenvectors

1. The characteristic equation of A is $(1-\lambda)(2-\lambda)-2 = \lambda^2 - 3\lambda = 0$, and so $\lambda = 0$ and 3. For $\lambda = 0$, we have

$$x + y = 0$$
$$2x + 2y = 0$$

which gives $y = -x$, and $\mathbf{x}_1 = (a, -a)^\mathsf{T}$, or $(1, -1)^\mathsf{T}$.

For $\lambda = 3$, we have

$$x + y = 3x$$
$$2x + 2y = 3y$$

which gives $y = 2x$ and $\mathbf{x}_2 = (b, 2b)^\mathsf{T}$, or $(1, 2)^\mathsf{T}$.

There are two linearly independent eigenvectors.

3. The characteristic equation of A is $(1-\lambda)(3-\lambda)+1 = \lambda^2 - 4\lambda + 4 = 0$, which gives $\lambda = 2$ and 2. The corresponding equations for the eigenvectors are

$$x - y = 2x$$
$$x + 3y = 2y$$

The solution is $y = -x$, which gives $\mathbf{x}_1 = (a, -a)^\mathsf{T}$, or $(1, -1)^\mathsf{T}$. This is the *only* eigenvector obtained.

6. The characteristic equation for a general 3×3 matrix is

$$\begin{vmatrix} a_{11} - \lambda & a_{12} & a_{13} \\ a_{21} & a_{22} - \lambda & a_{23} \\ a_{31} & a_{32} & a_{33} - \lambda \end{vmatrix} = (a_{11} - \lambda) \begin{vmatrix} a_{22} - \lambda & a_{23} \\ a_{32} & a_{33} - \lambda \end{vmatrix} - a_{12} \begin{vmatrix} a_{21} & a_{23} \\ a_{31} & a_{33} - \lambda \end{vmatrix}$$

$$+ a_{13} \begin{vmatrix} a_{21} & a_{22} - \lambda \\ a_{31} & a_{32} \end{vmatrix}$$

$$= (a_{11} - \lambda)\left[(a_{22} - \lambda)(a_{33} - \lambda) - a_{23}a_{32}\right] - a_{12}\left[a_{21}(a_{33} - \lambda)\right.$$
$$\left. - a_{31}a_{23}\right] + a_{13}\left[a_{21}a_{31} - a_{31}(a_{22} - \lambda)\right] = 0$$

The resulting expression is $(a_{11} - \lambda)(a_{22} - \lambda)(a_{33} - \lambda) +$ terms of order λ^1 or λ^0. Thus the coefficient of $\lambda^2 = a_{11} + a_{22} + a_{33}$. But the coefficient of λ^2 for $(\lambda_1 - \lambda)(\lambda_2 - \lambda)(\lambda_3 - \lambda)$ is $\lambda_1 + \lambda_2 + \lambda_3$, so we see that $\operatorname{Tr} \mathsf{A} = \lambda_1 + \lambda_2 + \lambda_3$.

9. Because $\det(\mathsf{A} - \lambda\mathsf{I}) = \det(\mathsf{A} - \lambda\mathsf{I})^\mathsf{T} = \det(\mathsf{A}^\mathsf{T} - \lambda\mathsf{I})$, the characteristic of A and A^T and hence the eigenvalues are the same.

12. Start with $\mathsf{A}\mathbf{x} = \lambda\mathbf{x}$ and multiply both sides from the left by $\mathbf{x}^\dagger$ to obtain $\mathbf{x}^\dagger\mathsf{A}\mathbf{x} = \lambda\mathbf{x}^\dagger\mathbf{x}$. Now take the complex conjugate of $\mathsf{A}\mathbf{x} = \lambda\mathbf{x}$ to obtain $\mathsf{A}^*\mathbf{x}^* = \lambda^*\mathbf{x}^*$. Take the transpose of both sides to get $\mathbf{x}^\dagger\mathsf{A}^\dagger = \lambda^*\mathbf{x}^\dagger$ and multiply both sides from the right by $\mathbf{x}$ to get $\mathbf{x}^\dagger\mathsf{A}^\dagger\mathbf{x} = \lambda^*\mathbf{x}^\dagger\mathbf{x}$. But $\mathsf{A}^\dagger = -\mathsf{A}$, and so we see that $\lambda^* = -\lambda$, or that λ is pure imaginary.

15. $f(\mathsf{A})\mathbf{x} = \displaystyle\sum_{n=0}^{\infty} a_n \mathsf{A}^n \mathbf{x} = \sum_{n=0}^{\infty} a_n \lambda^n \mathbf{x} = f(\lambda)\mathbf{x}$

10.3   Some Applied Eigenvalue Problems

1. For $\mathbf{x}(t) = c_1, \mathbf{u}_1 e^{\lambda_1 t} + c_2 \mathbf{u}_2 e^{\lambda_2 t} + \cdots$, $\dfrac{d\mathbf{x}}{dt} = c_1\lambda_1 \mathbf{u}_1 e^{\lambda_1 t} + c_2\lambda_2 \mathbf{u}_2 e^{\lambda_2 t} + \cdots$. But

$$\mathsf{A}\mathbf{x}(x) = c_1\mathsf{A}\,\mathbf{u}_1 e^{\lambda_1 t} + c_2\mathsf{A}\,\mathbf{u}_2 e^{\lambda_2 t} + \cdots = c_1\lambda_1 \mathbf{u}_1 e^{\lambda_1 t} + c_2\lambda_2 \mathbf{u}_2 e^{\lambda_2 t} + \cdots$$

3. Let $x_j(t) = u_j e^{\lambda t}$. Then

$$\lambda u_1 = u_1 + u_2$$
$$\lambda u_2 = 4u_1 + u_2$$

The eigenvalues are given by

$$\begin{vmatrix} 1 - \lambda & 1 \\ 4 & 1 - \lambda \end{vmatrix} = \lambda^2 - 2\lambda - 3 = 0$$

which gives $\lambda = 3$ and -1. The linear algebraic equations for the eigenvectors are given by

$$\lambda = -1 \qquad \begin{matrix} 2u_1 + u_2 = 0 \\ 4u_1 + 2u_2 = 0 \end{matrix} \quad \longrightarrow \quad \begin{pmatrix} -1 \\ 2 \end{pmatrix}$$

$$\lambda = 3 \qquad \begin{matrix} -2u_1 + u_2 = 0 \\ 4u_1 - 2u_2 = 0 \end{matrix} \quad \longrightarrow \quad \begin{pmatrix} 1 \\ 2 \end{pmatrix}$$

The general solution is

$$\mathbf{x}(t) = \begin{pmatrix} x_1(t) \\ x_2(t) \end{pmatrix} = c_1 \begin{pmatrix} -1 \\ 2 \end{pmatrix} e^{-\lambda t} + c_2 \begin{pmatrix} 1 \\ 2 \end{pmatrix} e^{3\lambda t}$$

6. The eigenvalues and corresponding eigenvectors of $A = \begin{pmatrix} 2 & -1 \\ -6 & 1 \end{pmatrix}$ are $\begin{pmatrix} 1 \\ 3 \end{pmatrix}$ for $\lambda = -1$ and $\begin{pmatrix} 1 \\ -2 \end{pmatrix}$ for $\lambda = 4$. Therefore, the general solution is

$$\begin{pmatrix} x_1(t) \\ x_2(t) \end{pmatrix} = c_1 \begin{pmatrix} 1 \\ 3 \end{pmatrix} e^{-t} + c_2 \begin{pmatrix} 1 \\ -2 \end{pmatrix} e^{4t}$$

The initial condition $\mathbf{x}(0) = (1,0)^\mathsf{T}$ gives $c_1 + c_2 = 1$ and $3c_1 - 2c_2 = 0$, or $c_1 = 2/5$ and $c_2 = 3/5$.

9. First we'll present the solution to Problem 8. For small values of θ_1, $V_1(\theta_1) = mgl(1-\cos\theta_1) = \frac{1}{2}mgl\theta_1^2$, Similarly $V_2(\theta_2) = \frac{1}{2}mgl\theta_2^2$, and Hooke's law gives $\frac{1}{2}kl^2(\theta_2-\theta_1)^2$. Applying this result to Problem 9, we have that

$$V(s_1, s_2) = \frac{mg}{2l}s_1^2 + \frac{mg}{2l}s_2^2 + \frac{1}{2}k(s_2 - s_1)^2$$

where we have used the realtion $s = l\theta$. The forces associated with $V(s_1, s_2)$ are

$$-\frac{\partial V}{\partial s_1} = -\frac{mgs_1}{l} + k(s_2 - s_1)$$
$$-\frac{\partial V}{\partial s_2} = -\frac{mgs_2}{l} - k(s_2 - s_1)$$

which are used in Example 4.

12. For the in-phase motion $x_1(t) = A\cos\omega t$ and $x_2(t) = A\cos\omega t$ and

$$V = \frac{k}{2}x_1^2 + \frac{k}{2}(x_2 - x_1)^2 + \frac{k}{2}x_2^2$$
$$= \frac{kA^2}{2}\cos^2\omega t + \frac{kA^2}{2}(\cos\omega t - \cos\omega t)^2 + \frac{kA^2}{2}\cos^2\omega t = kA^2\cos^2\omega t$$

The kinetic energy is

$$K = \frac{m}{2}\dot{x}_1^2 + \frac{m}{2}\dot{x}_2^2 = mA^2\omega^2\sin^2\omega t$$

and so the total energy is

$$E = mA^2\omega^2\sin^2\omega t + kA^2\cos^2\omega t = kA^2$$

For the out-of-phase motion, $x_1(t) = A\cos 3^{1/2}\omega t$ and $x_2(t) = -A\cos 3^{1/2}\omega t$ and

$$V = \frac{k}{2}x_1^2 + \frac{k}{2}(x_2 - x_1)^2 + \frac{k}{2}x_2^2$$
$$= \frac{kA^2}{2}\cos^2 3^{1/2}\omega t + 2kA^2\cos^2 3^{1/2}\omega t + \frac{kA^2}{2}\cos^2 3^{1/2}\omega t = 3kA^2\cos^2 3^{1/2}\omega t$$

The kinetic energy is

$$
\begin{aligned}
K &= \frac{m}{2}\dot{x}_1^2 + \frac{m}{2}\dot{x}_2^2 \\
&= \frac{3m}{2}A^2\omega^2\sin^2 3^{1/2}\omega t + \frac{3m}{2}A^2\omega^2\sin^2 3^{1/2}\omega t = 3mA^2\omega^2\sin^2 3^{1/2}\omega t
\end{aligned}
$$

and so the total energy is

$$
E = 3kA^2\cos^2 3^{1/2}\omega t + 3mA^2\omega^2\sin^2 3^{1/2}\omega t = 3kA^2
$$

15. The potential energy of the system is given by

$$
\begin{aligned}
V &= \frac{k_1}{2}x_1^2 + \frac{k_2}{2}(x_2 - x_1)^2 + \frac{k_3}{2}(x_3 - x_2)^2 + \frac{k_4}{2}x_3^2 \\
&= 3x_1^2 + (x_2 - x_1)^2 + \frac{1}{2}(x_3 - x_2)^2 + \frac{1}{2}x_3^2
\end{aligned}
$$

and so the equations of motion are

$$
\begin{aligned}
m_1\ddot{x}_1 &= -\frac{\partial V}{\partial x_1} = -6x_1 + 2(x_2 - x_1) = -8x_1 + 2x_2 \\
m_2\ddot{x}_2 &= -\frac{\partial V}{\partial x_2} = -2(x_2 - x_1) + (x_3 - x_2) = 2x_1 + x_3 - 3x_2 \\
m_3\ddot{x}_3 &= -\frac{\partial V}{\partial x_3} = -(x_3 - x_2) - x_3 = -2x_3 + x_2
\end{aligned}
$$

18. First we will calculate the three fundamental frequencies (Problem 16). Substitute $x_j(t) = c_j e^{i\omega t}$ into the equations of motion to get

$$
\begin{pmatrix} 8 & -2 & 0 \\ -2 & 3 & -1 \\ 0 & -1 & 2 \end{pmatrix} \begin{pmatrix} c_1 \\ c_2 \\ c_3 \end{pmatrix} = \omega^2 \begin{pmatrix} 8c_1 \\ 2c_2 \\ 2c_3 \end{pmatrix}
$$

Multiply the second and third rows by 4 to get

$$
\begin{pmatrix} 8 & -2 & 0 \\ -8 & 12 & -4 \\ 0 & -4 & 8 \end{pmatrix} \begin{pmatrix} c_1 \\ c_2 \\ c_3 \end{pmatrix} = 8\omega^2 \begin{pmatrix} c_1 \\ c_2 \\ c_3 \end{pmatrix}
$$

Let $\lambda = 8\omega^2$. Then the characteristic equation is

$$
\begin{vmatrix} 8 - \lambda & -2 & 0 \\ -8 & 12 - \lambda & -4 \\ 0 & -4 & 8 - \lambda \end{vmatrix} = 0
$$

or $(8 - \lambda)^2(12 - \lambda) - 32(8 - \lambda) = 0$. Thus $\lambda = 8$, 16, and 4, and so $\omega_1^2 = \frac{1}{2}$, $\omega_2^2 = 1$, and $\omega_3^2 = 2$. The eigenvector corresponding to $\omega^2 = 1/2$ is $(1, 2, 2)$, and so using $x_1 = \cos\omega t$, $x_2 = 2\cos\omega t$, and $x_3 = 2\cos\omega t$ in the potential energy, (see the solution to Problem 15)

$$
\begin{aligned}
V &= 3x_1^2 + (x_2 - x_1)^2 + \frac{1}{2}(x_3 - x_2)^2 + \frac{1}{2}x_3^2 \\
&= 3a^2\cos^2\omega t + a^2\cos^2\omega t + 2a^2\cos^2\omega t = 6a^2\cos^2\omega t
\end{aligned}
$$

The kinetic energy is given by

$$\begin{aligned}
K &= \frac{8}{2}\dot{x}_1^2 + \frac{2}{2}\dot{x}_2^2 + \frac{2}{2}\dot{x}_3^2 \\
&= 4a^2\omega^2\sin^2\omega t + 4a^2\omega^2\sin^2\omega t + 4a^2\omega^2\sin^2\omega t = 12a^2\omega^2\sin^2\omega t
\end{aligned}$$

But $\omega^2 = 1/2$, so the total energy $E = K + V = 6a^2$.

10.4   Change of Basis

1. First write $f(x,y)$ as $f(x,y) = (x\ \ y)\begin{pmatrix} 2 & 1 \\ 1 & 2 \end{pmatrix}\begin{pmatrix} x \\ y \end{pmatrix} = \mathbf{x}^\mathsf{T}\mathbf{A}\mathbf{x}$. Now use Equation 6, where

$$\begin{aligned}
\mathsf{A}' = \mathsf{R}^\mathsf{T}\mathsf{A}\mathsf{R} &= \begin{pmatrix} \cos\theta & \sin\theta \\ -\sin\theta & \cos\theta \end{pmatrix}\begin{pmatrix} 2 & 1 \\ 1 & 2 \end{pmatrix}\begin{pmatrix} \cos\theta & -\sin\theta \\ \sin\theta & \cos\theta \end{pmatrix} \\
&= \begin{pmatrix} 2\cos^2\theta + \cos\theta\sin\theta + 2\sin^2\theta + \cos\theta\sin\theta & \cos^2\theta - \sin^2\theta \\ \cos^2\theta - \sin^2\theta & 2\sin^2\theta - 2\sin\theta\cos\theta + 2\cos^2\theta \end{pmatrix} \\
&= \begin{pmatrix} 2 + \sin 2\theta & \cos 2\theta \\ \cos 2\theta & 2 - \sin 2\theta \end{pmatrix}
\end{aligned}$$

to get

$$\begin{aligned}
f(x',y') &= (x'\ \ y')\begin{pmatrix} 2 + \sin 2\theta & \cos 2\theta \\ \cos 2\theta & 2 - \sin 2\theta \end{pmatrix}\begin{pmatrix} x' \\ y' \end{pmatrix} \\
&= (2 + \sin 2\theta)x'^2 + (2\cos 2\theta)x'y' + (2 - \sin 2\theta)y'^2 = 1
\end{aligned}$$

Thus if $\cos 2\theta = 0$ or $\theta = \pi/4$, then the cross term is eliminated.

3. We have $f(x,y) = (x\ \ y)\begin{pmatrix} a & b \\ b & c \end{pmatrix}\begin{pmatrix} x \\ y \end{pmatrix} = \mathbf{x}^\mathsf{T}\mathbf{A}\mathbf{x}$. Equation 6 gives

$$\begin{aligned}
\mathsf{A}' = \mathsf{R}^\mathsf{T}\mathsf{A}\mathsf{R} &= \begin{pmatrix} a\cos^2\theta + c\sin^2\theta + 2b\sin\theta\cos\theta & b(\cos^2\theta - \sin^2\theta) + (c-a)\cos\theta\sin\theta \\ b(\cos^2\theta - \sin^2\theta) + (c-a)\cos\theta\sin\theta & a\sin^2\theta + c\cos^2\theta - 2b\sin\theta\cos\theta \end{pmatrix} \\
&= \begin{pmatrix} a\cos^2\theta + c\sin^2\theta + b\sin 2\theta & b\cos 2\theta + \dfrac{c-a}{2}\sin 2\theta \\ b\cos 2\theta + \dfrac{c-a}{2}\sin 2\theta & a\sin^2\theta + c\cos^2\theta - b\sin 2\theta \end{pmatrix}
\end{aligned}$$

Set $b\cos 2\theta + \dfrac{c-a}{2}\sin 2\theta = 0$ or $\tan 2\theta = \dfrac{2b}{a-c}$.

6. From Figure 10.18,

$$\begin{aligned}
\mathbf{u}_E &= u\cos\alpha\,\mathbf{e}_1 + u\sin\alpha\,\mathbf{e}_2 = u_{E_1}\,\mathbf{e}_1 + u_{E_2}\,\mathbf{e}_2 \\
\mathbf{u}_z &= u\cos(60° - \alpha)\,\mathbf{z}_1 - u\sin(60° - \alpha)\,\mathbf{z}_2 = u_{z_1}\,\mathbf{z}_1 + u_{z_2}\,\mathbf{z}_2
\end{aligned}$$

where

$$\begin{aligned}
u_{z_1} &= u\cos(60° - \alpha) = \frac{1}{2}u\cos\alpha + \frac{\sqrt{3}}{2}u\sin\alpha = \frac{1}{2}u_{E_1} + \frac{\sqrt{3}}{2}u_{E_2} \\
u_{z_2} &= -u\sin(60° - \alpha) = -\frac{\sqrt{3}}{2}u\cos\alpha + \frac{1}{2}u\sin\alpha = -\frac{\sqrt{3}}{2}u_{E_1} + \frac{1}{2}u_{E_2}
\end{aligned}$$

Therefore, the components of $\mathbf{u}_z$ are related to those of $\mathbf{u}_E$ by

$$
\begin{pmatrix} u_{z_1} \\ u_{z_2} \end{pmatrix} = \begin{pmatrix} \dfrac{1}{2} & \dfrac{\sqrt{3}}{2} \\ -\dfrac{\sqrt{3}}{2} & \dfrac{1}{2} \end{pmatrix} \begin{pmatrix} u_{E_1} \\ u_{E_2} \end{pmatrix}
$$

9. We must first do Problem 7. We construct $Z = \begin{pmatrix} 1 & -1 & 0 \\ 0 & 1 & 1 \\ 2 & 2 & 0 \end{pmatrix}$ from $\mathbf{z}_1$, $\mathbf{z}_2$, and $\mathbf{z}_3$ and

then use the relation $\mathbf{u}_E = Z\,\mathbf{u}_Z$ to write

$$
\mathbf{u}_E = \begin{pmatrix} 1 & -1 & 0 \\ 0 & 1 & 1 \\ 2 & 2 & 0 \end{pmatrix} \begin{pmatrix} -1 \\ 2 \\ 1 \end{pmatrix} = \begin{pmatrix} -3 \\ 3 \\ 2 \end{pmatrix}
$$

Now, to calculate $\mathbf{u}_Z$ from $\mathbf{u}_E = (-3, 3, 2)^{\mathsf{T}}$, we need Z^{-1}, which is $Z^{-1} = \dfrac{1}{4}\begin{pmatrix} 2 & 0 & 1 \\ -2 & 0 & 1 \\ 2 & 4 & -1 \end{pmatrix}$.

Using $\mathbf{u}_Z = Z^{-1}\mathbf{u}_E$ gives

$$
\mathbf{u}_z = \frac{1}{4}\begin{pmatrix} 2 & 0 & 1 \\ -2 & 0 & 1 \\ 2 & 4 & -1 \end{pmatrix}\begin{pmatrix} -3 \\ 3 \\ 2 \end{pmatrix} = \begin{pmatrix} -1 \\ 2 \\ 1 \end{pmatrix}
$$

12. First we must do Problem 11, which is very similar to Problems 8 and 9. We first construct $Z = \begin{pmatrix} 2 & 3 \\ -4 & 8 \end{pmatrix}$ and use $\mathbf{u}_E = Z\,\mathbf{u}_Z$ or

$$
\mathbf{u}_Z = Z^{-1}\mathbf{u}_E = \frac{1}{28}\begin{pmatrix} 8 & -3 \\ 4 & 2 \end{pmatrix}\begin{pmatrix} 1 \\ 1 \end{pmatrix} = \frac{1}{28}\begin{pmatrix} 5 \\ 6 \end{pmatrix}
$$

Now $\| \mathbf{u}_E \| = (1^2 + 1^2)^{1/2} = \sqrt{2}$ and

$$
\begin{aligned}
\| \mathbf{u}_z \| &= \left[\left(\frac{5}{28}\mathbf{z}_1 + \frac{6}{28}\mathbf{z}_2 \right) \cdot \left(\frac{5}{28}\mathbf{z}_1 + \frac{6}{28}\mathbf{z}_2 \right) \right]^{1/2} \\
&= \frac{1}{28}\left[25z_1^2 + 2(30)\mathbf{z}_1 \cdot \mathbf{z}_2 + 36z_2^2 \right] \\
&= \frac{1}{28}\left[(25)(20) + 2(30)(-26) + (36)(73) \right]^{1/2} = \sqrt{2}
\end{aligned}
$$

15. We use the relation $W\,\mathbf{u}_w = Z u_z$, or $\mathbf{u}_w = W^{-1}Z u_z$. Now $W = \begin{pmatrix} 1 & 1 & 1 \\ 0 & 1 & 1 \\ 0 & 0 & 1 \end{pmatrix}$ and

$Z = \begin{pmatrix} 1 & 1 & 1 \\ 1 & 0 & 1 \\ 0 & 1 & 1 \end{pmatrix}$ and $W^{-1} = \begin{pmatrix} 1 & -1 & 0 \\ 0 & 1 & -1 \\ 0 & 0 & 1 \end{pmatrix}$. Therefore,

$$
W^{-1}Z = \begin{pmatrix} 1 & -1 & 0 \\ 0 & 1 & -1 \\ 0 & 0 & 1 \end{pmatrix}\begin{pmatrix} 1 & 1 & 1 \\ 1 & 0 & 1 \\ 0 & 1 & 1 \end{pmatrix} = \begin{pmatrix} 0 & 1 & 0 \\ 1 & -1 & 0 \\ 0 & 1 & 1 \end{pmatrix}
$$

transforms $\mathbf{u}_z$ to $\mathbf{u}_w$.

18. The vector $\mathbf{v}_E = \mathsf{A}\mathbf{u}_E$ in the standard basis is

$$\mathbf{v}_E = \mathsf{A}\mathbf{u}_E = \begin{pmatrix} 1 & 0 & 3 \\ 2 & 1 & -2 \\ 1 & 2 & 1 \end{pmatrix} \begin{pmatrix} 3 \\ 0 \\ 2 \end{pmatrix} = \begin{pmatrix} 9 \\ 2 \\ 5 \end{pmatrix}$$

Now use $\mathbf{v}_E = \mathsf{W}\mathbf{v}_w$, or $\mathbf{v}_w = \mathsf{W}^{-1}\mathbf{v}_E$. Now W is $\mathsf{W} = \begin{pmatrix} 1 & 0 & 0 \\ 0 & 1 & 0 \\ 0 & 2 & 1 \end{pmatrix}$ and so $\mathsf{W}^{-1} = \begin{pmatrix} 1 & 0 & 0 \\ 0 & 1 & 0 \\ 0 & -2 & 1 \end{pmatrix}$

Therefore, $\mathbf{v}_w$ is given by

$$\mathbf{v}_w = \mathsf{W}^{-1}\mathbf{v}_E = \begin{pmatrix} 1 & 0 & 0 \\ 0 & 1 & 0 \\ 0 & -2 & 1 \end{pmatrix} \begin{pmatrix} 9 \\ 2 \\ 5 \end{pmatrix} = \begin{pmatrix} 9 \\ 2 \\ 1 \end{pmatrix}$$

21. From the statement of Problem 20, we have the relation $\mathsf{A}_W = \mathsf{W}^{-1}\mathsf{Z}\mathsf{A}_Z\mathsf{Z}^{-1}\mathsf{W} = \mathsf{W}^{-1}\mathsf{Z}\mathsf{A}_Z(\mathsf{W}^{-1}\mathsf{Z})^{-1}$. From Problem 15, we have $\mathsf{W}^{-1}\mathsf{Z} = \begin{pmatrix} 0 & 1 & 0 \\ 1 & -1 & 0 \\ 0 & 1 & 1 \end{pmatrix}$ and so

$$\mathsf{A}_W = \begin{pmatrix} 0 & 1 & 0 \\ 1 & -1 & 0 \\ 0 & 1 & 1 \end{pmatrix} \begin{pmatrix} 1 & 1 & 2 \\ 2 & 2 & 1 \\ 3 & 1 & 2 \end{pmatrix} \begin{pmatrix} 1 & 1 & 0 \\ 1 & 0 & 0 \\ -1 & 0 & 1 \end{pmatrix} = \begin{pmatrix} 3 & 2 & 1 \\ -3 & -1 & 1 \\ 5 & 5 & 3 \end{pmatrix}$$

10.5   Diagonalization of Matrices

1.

$$\begin{pmatrix} a_{11} & a_{12} & a_{13} & \cdots \\ a_{21} & a_{22} & a_{23} & \cdots \\ a_{31} & a_{32} & a_{33} & \cdots \\ \vdots & \vdots & \vdots & \ddots \end{pmatrix} \begin{pmatrix} u_{11} & u_{21} & u_{31} & \cdots \\ u_{12} & u_{22} & u_{32} & \cdots \\ u_{13} & u_{23} & u_{33} & \cdots \\ \vdots & \vdots & \vdots & \ddots \end{pmatrix}$$
$$= \begin{pmatrix} a_{11}u_{11} + a_{12}u_{12} + \cdots & a_{11}u_{21} + \cdots & \cdots \\ a_{21}u_{11} + a_{22}u_{12} + \cdots & a_{21}u_{21} + \cdots & \cdots \\ a_{31}u_{11} + a_{32}u_{12} + \cdots & & \end{pmatrix}$$
$$= \left(\mathsf{A}\mathbf{u}_1, \ \mathsf{A}\mathbf{u}_2, \ \cdots \right) = \left(\lambda_1\mathbf{u}_1, \ \lambda_2\mathbf{u}_2, \ \cdots \right)$$

3. $\mathsf{A} = \begin{pmatrix} 1 & 0 \\ 6 & -1 \end{pmatrix}$. The eigenvalues of A are $\lambda = \pm 1$, and the eigenvectors are

$$\mathbf{u}_1 = \frac{1}{\sqrt{10}} \begin{pmatrix} 1 \\ 3 \end{pmatrix} \qquad \text{and} \qquad \mathbf{u}_2 = \begin{pmatrix} 0 \\ 1 \end{pmatrix}$$

The (normalized) modal matrix is $S = \begin{pmatrix} \dfrac{1}{\sqrt{10}} & 0 \\[2mm] \dfrac{3}{\sqrt{10}} & 1 \end{pmatrix}$, and $S^{-1} = \sqrt{10}\begin{pmatrix} 1 & 0 \\[2mm] -\dfrac{3}{\sqrt{10}} & \dfrac{1}{\sqrt{10}} \end{pmatrix}$.

Then

$$D = S^{-1}AS = \sqrt{10}\begin{pmatrix} 1 & 0 \\[2mm] -\dfrac{3}{\sqrt{10}} & \dfrac{1}{\sqrt{10}} \end{pmatrix}\begin{pmatrix} 1 & 0 \\ 6 & -1 \end{pmatrix}\begin{pmatrix} \dfrac{1}{\sqrt{10}} & 0 \\[2mm] \dfrac{3}{\sqrt{10}} & 1 \end{pmatrix} = \begin{pmatrix} 1 & 0 \\ 0 & -1 \end{pmatrix}$$

6. (a) No. The eigenvalues are repeated $(2, 2)$ and there is only one linearly independent eigenvector $(0, 1)$.

(b) No. The eigenvalues are repeated $(2, 2, 3)$ and there are only two linearly independent eigenvectors.

(c) Yes. Symmetric matrix

9. $A = \begin{pmatrix} -2 & 3+3i \\ 3-3i & 1 \end{pmatrix}$. The eigenvalues of A are $\lambda = -5$ and 4, and the eigenvectors are $\mathbf{u}_1 = (-1-i, 1)^\mathsf{T}/\sqrt{3}$ and $\mathbf{u}_2 = (\frac{1}{2}+\frac{i}{2}, 1)^\mathsf{T}/\sqrt{3/2}$. The (normalized) modal matrix is

$$S = \begin{pmatrix} \dfrac{-1-i}{\sqrt{3}} & \dfrac{1+i}{\sqrt{6}} \\[3mm] \dfrac{1}{\sqrt{3}} & \sqrt{\dfrac{2}{3}} \end{pmatrix} \qquad \text{and} \qquad S^\dagger = \begin{pmatrix} \dfrac{-1+i}{\sqrt{3}} & \dfrac{1}{\sqrt{3}} \\[3mm] \dfrac{1-i}{\sqrt{6}} & \sqrt{\dfrac{2}{3}} \end{pmatrix}$$

Thus $D = S^\dagger AS = \begin{pmatrix} -5 & 0 \\ 0 & 4 \end{pmatrix}$

12. The eigenvalues and corresponding eigenvectors of $A = \begin{pmatrix} -3 & 2 \\ -1 & 0 \end{pmatrix}$ are $\lambda_1 = -2$ and $\lambda_2 = -1$ with $\mathbf{u}_1 = (2, 1)^\mathsf{T}$ and $\mathbf{u}_2 = (1, 1)^\mathsf{T}$. The normalized modal matrix S is $S = \begin{pmatrix} 2/\sqrt{5} & 1/\sqrt{2} \\ 1/\sqrt{5} & 1/\sqrt{2} \end{pmatrix}$ with $S^{-1} = \begin{pmatrix} \sqrt{5} & -\sqrt{5} \\ -\sqrt{2} & 2\sqrt{2} \end{pmatrix}$. The diagonalized form of A is $D = S^{-1}AS = \begin{pmatrix} -2 & 0 \\ 0 & -1 \end{pmatrix}$. Therefore, $e^D = \begin{pmatrix} e^{-2} & 0 \\ 0 & e^{-1} \end{pmatrix}$ and

$$e^A = Se^DS^{-1} = \frac{1}{e^2}\begin{pmatrix} 2-e & 2e-2 \\ 1-e & 2e-1 \end{pmatrix}$$

15. The eigenvalues and corresponding normalized eigenvectors of $A = \begin{pmatrix} 0 & 1 \\ -1 & 0 \end{pmatrix}$ are $\lambda_1 = i$ and $\lambda_2 = -i$ with $\mathbf{u}_1 = (1, i)^\mathsf{T}/\sqrt{2}$ and $\mathbf{u}_2 = (1, -i)^\mathsf{T}/\sqrt{2}$. The normalized modal matrix S is $S = \dfrac{1}{\sqrt{2}}\begin{pmatrix} 1 & 1 \\ i & -i \end{pmatrix}$ with $S^{-1} = \dfrac{1}{\sqrt{2}}\begin{pmatrix} 1 & -i \\ 1 & i \end{pmatrix}$. The diagonalized form of A is

$D = S^{-1}AS = \begin{pmatrix} i & 0 \\ 0 & -i \end{pmatrix}$, Therefore, the diagonal form of e^{Ax} is $e^{Ax} = \begin{pmatrix} e^{ix} & 0 \\ 0 & e^{-ix} \end{pmatrix}$ and

$$e^{Ax} = Se^{Dx}S^{-1} = \begin{pmatrix} \dfrac{e^{ix}+e^{-ix}}{2} & -i\dfrac{e^{ix}-e^{-ix}}{2} \\ i\dfrac{e^{ix}+e^{-ix}}{2} & \dfrac{e^{ix}+e^{-ix}}{2} \end{pmatrix} = \begin{pmatrix} \cos x & \sin x \\ -\sin x & \cos x \end{pmatrix}$$

18. $SS^T = \begin{pmatrix} 5 & 0 \\ 0 & 5 \end{pmatrix} = S^TS$

10.6   Quadratic Forms

1. (a) $\begin{pmatrix} 3 & -\frac{1}{2} & 1 \\ -\frac{1}{2} & 0 & 0 \\ 1 & 0 & 1 \end{pmatrix}$

(b) $\begin{pmatrix} 1 & -\frac{3}{2} \\ -\frac{3}{2} & 6 \end{pmatrix}$

(c) $\begin{pmatrix} 0 & 0 & 3 \\ 0 & 1 & 0 \\ 3 & 0 & 0 \end{pmatrix}$

3. The eigenvalues and corresponding normalized eigenvectors of $A = \begin{pmatrix} 5 & 2 \\ 2 & 2 \end{pmatrix}$ are $\lambda_1 = 1$ and $\lambda_2 = 6$ with $\mathbf{u}_1 = \dfrac{1}{\sqrt{5}}\begin{pmatrix} 1 \\ -2 \end{pmatrix}$ and $\mathbf{u}_2 = \dfrac{1}{\sqrt{5}}\begin{pmatrix} 2 \\ 1 \end{pmatrix}$. The normalized modal matrix is $S = \dfrac{1}{\sqrt{5}}\begin{pmatrix} 1 & 2 \\ -2 & 1 \end{pmatrix}$ and its inverse is $S^{-1} = \dfrac{1}{\sqrt{5}}\begin{pmatrix} 1 & -2 \\ 2 & 1 \end{pmatrix}$. Consequently,

$$\begin{pmatrix} x \\ y \end{pmatrix} = \dfrac{1}{\sqrt{5}}\begin{pmatrix} 1 & 2 \\ -2 & 1 \end{pmatrix}\begin{pmatrix} x' \\ y' \end{pmatrix} = \dfrac{1}{\sqrt{5}}\begin{pmatrix} x'+2y' \\ -2x'+y' \end{pmatrix}$$

Now the standard form of $5x^2 + 4xy + 2y^2 = 1$ is given by

$$5(x'+2y')^2 + \frac{4}{5}(x'+2y')(-2x'+y') + \frac{2}{5}(-2x'+y')^2$$
$$= (x'^2 + 4x'y' + 4y'^2) + \frac{4}{5}(-2x'^2 - 3x'y' + 2y'^2) + \frac{2}{5}(4x'^2 - 4x'y' + y'^2)$$
$$= x'^2 + 6y'^2$$

6. The associated matrix for the quadratic form is $A = \begin{pmatrix} 2 & 2 & -2 \\ 2 & 5 & -4 \\ -2 & -4 & 5 \end{pmatrix}$. The eigenvalues are $\lambda = 1, 1,$ and 10, and so the standard form $x'^2 + y'^2 + 10z'^2 = 10$ shows that the surface is of an ellipsoid.

This is a good problem to carry through to the end because there are repeated eigenvalues. The three eigenvectors come out to be

$$\mathbf{v}_1 = \begin{pmatrix} 2 \\ 0 \\ 1 \end{pmatrix} \qquad \mathbf{v}_2 = \begin{pmatrix} -2 \\ 1 \\ 0 \end{pmatrix} \qquad \mathbf{v}_3 = \begin{pmatrix} -1 \\ -2 \\ 2 \end{pmatrix}$$

Because two of the eigenvalues ($\lambda = 1$) are repeated, the eigenvectors are not necessarily orthogonal. We shall use the Gram-Schmidt procedure to form an orthogonal set from the above eigenvectors. Choosing $\mathbf{u}_1 = \mathbf{v}_1$, we have (see Equations 9.6.16 and 17)

$$\mathbf{u}_2 = \mathbf{v}_2 + a\,\mathbf{u}_1 = \begin{pmatrix} -2 \\ 1 \\ 0 \end{pmatrix} + a_1 \begin{pmatrix} 2 \\ 0 \\ 1 \end{pmatrix} = \begin{pmatrix} -2 \\ 1 \\ 0 \end{pmatrix} + \frac{4}{5} \begin{pmatrix} 2 \\ 0 \\ 1 \end{pmatrix} = \begin{pmatrix} -2/5 \\ 1 \\ 4/5 \end{pmatrix}$$

and $\mathbf{u}_3$ turns out to be $\mathbf{u}_3 = (-1 \ -2 \ 2)^\mathsf{T}$. Note that

$$\mathbf{u}_1 = \begin{pmatrix} 2 \\ 0 \\ 1 \end{pmatrix} \qquad \mathbf{u}_2 = \begin{pmatrix} -2/5 \\ 1 \\ 4/5 \end{pmatrix} \qquad \mathbf{u}_3 = \begin{pmatrix} -1 \\ -2 \\ 2 \end{pmatrix}$$

are orthogonal. The normalized modal matrix is $\mathsf{S} = \begin{pmatrix} 2/\sqrt{5} & -2\sqrt{5}/15 & -1/3 \\ 0 & 5\sqrt{5}/15 & -2/3 \\ 1/\sqrt{5} & 4\sqrt{5}/15 & 2/3 \end{pmatrix}$. You can readily show that $\mathsf{D} = \mathsf{S}^{-1}\mathsf{A}\,\mathsf{S} = \mathsf{S}^\mathsf{T}\mathsf{A}\,\mathsf{S}$, where the diagonal elements of D are 1, 1, and 10.

Now, Equation 4, $\mathbf{x} = \mathsf{S}\,\mathbf{x}'$, gives

$$x = \frac{2x'}{\sqrt{5}} - \frac{2\,y'}{3\sqrt{5}} - \frac{z'}{3}$$

$$y = \frac{\sqrt{5}\,y'}{3} - \frac{2z'}{3}$$

$$z = \frac{x'}{\sqrt{5}} + \frac{4y'}{3\sqrt{5}} + \frac{2z'}{3}$$

Substituting these into $2x^2 + 5y^2 + 5z^2 + 4xy - 4xz - 8yz = 10$ yields $x'^2 + y'^2 + 10z'^2 = 10$.

9. We use the vector relation $\mathbf{v}_1 \times (\mathbf{v}_2 \times \mathbf{v}_3) = (\mathbf{v}_1 \cdot \mathbf{v}_3)\,\mathbf{v}_2 - (\mathbf{v}_1 \cdot \mathbf{v}_2)\,\mathbf{v}_3$ to write $\mathbf{r} \times (\mathbf{w} \times \mathbf{r})$ as

$$
\begin{aligned}
r^2\,\mathbf{w} - (\mathbf{w} \cdot \mathbf{r})\,\mathbf{r} &= (x^2 + y^2 + z^2)(w_x\,\mathbf{i} + w_y\,\mathbf{j} + w_z\,\mathbf{k}) - (xw_x + yw_y + zw_z)(x\,\mathbf{i} + y\,\mathbf{j} + z\,\mathbf{k}) \\
&= [\,(x^2 + y^2 + z^2)w_x - x(xw_x + yw_y + zw_z)\,]\,\mathbf{i} + \text{like terms in } \mathbf{j} \text{ and } \mathbf{k} \\
&= [\,(y^2 + z^2)w_x - xyw_y - xzw_z\,]\,\mathbf{i} + \text{like terms in } \mathbf{j} \text{ and } \mathbf{k} \\
&\Rightarrow [\,I_{xx}w_x + I_{xy}w_y + I_{xz}w_z\,]\,\mathbf{i} + \text{like terms in } \mathbf{j} \text{ and } \mathbf{k}
\end{aligned}
$$

12. Start with det $(\mathsf{AB}) = $ det A det B and let $\mathsf{B} = \mathsf{A}^{-1}$.

15.

$$I = \int_{-\infty}^{\infty} e^{itx - \frac{1}{2}hx^2}\, dx = \sum_{n=0}^{\infty} \frac{(it)^n}{n!} \int_{-\infty}^{\infty} x^n e^{-hx^2/2}\, dx$$

$$= \sum_{n=0}^{\infty} \frac{(it)^n}{(2n)!} \frac{(2n)!}{2^n n!} \frac{1}{h^n} \left(\frac{2\pi}{h}\right)^{1/2} = \left(\frac{2\pi}{h}\right)^{1/2} \sum_{n=0}^{\infty} \frac{1}{n!} \left(-\frac{t^2}{2h}\right)^n$$

$$= \left(\frac{2\pi}{h}\right)^{1/2} e^{-t^2/2h}$$

18. (a) The eigenvalues of $S = \begin{pmatrix} 1 & -2 \\ -2 & 5 \end{pmatrix}$ are $\lambda_1 = 3 - 2\sqrt{2}$ and $\lambda_2 = 3 + 2\sqrt{2}$. Both eigenvalues are positive, so the quadratic form is positive definite.

(b) The eigenvalues of $S = \begin{pmatrix} 1 & 3 \\ 3 & 3 \end{pmatrix}$ are $\lambda_1 = 2 - \sqrt{10}$ and $\lambda_2 = 2 + \sqrt{10}$, and so the quadratic form is not positive definite.

(c) The eigenvalues of $S = \begin{pmatrix} 2 & -6 \\ -6 & 5 \end{pmatrix}$ are $\lambda_1 = \dfrac{7 - 3\sqrt{17}}{3}$ and $\lambda_2 = \dfrac{7 + 3\sqrt{17}}{3}$, and so the quadratic form is not positive definite.

CHAPTER 11

Ordinary Differential Equations

11.1   Differential Equations of First Order and First Degree

1. (a) Substituting $y = x^2 + cx$ into $xy' = x^2 + c$ gives $xy' = 2x^2 + cx = x^2 + y$.

(b) Substituting $y = c_1 \cos x + c_2 \sin x$ into $y'' + y = 0$ gives $-c_1 \cos x - c_2 \sin x + c_1 \cos x + c_2 \sin x = 0$.

(c) Substituting $y = c_1 e^{2x} + c_2 e^{-3x}$ into $y'' + y' - 6y = 0$ gives

$$(4c_1 e^{2x} + 9c_2 e^{-3x}) + (2c_1 e^{2x} - 3c_2 e^{-3x}) - 6(c_1 e^{2x} + c_2 e^{-3x}) = 0$$

(d) Substituting $y = cx + c^2$ into $y = xy' + y'^2$ gives $xc + c^2 = xc + c^2$.

3. Solve the equation for dy/dx to get $\left(\dfrac{dy}{dx}\right)^2 = \dfrac{y^2}{x^2}$, or $\dfrac{dy}{dx} = \pm\dfrac{y}{x}$. Then integrate to get $\ln y = \pm \ln x + c$, or $y = c_1 x$, or $y = c_2/x$.

6. (a) The equation is exact because $\partial(2xy)/\partial y = \partial(x^2 + y^2)/\partial x$. If we let the solution be $f(x,y)$, then $\dfrac{\partial f}{\partial x} = 2xy$ gives $f = x^2 y + g(y)$. Now $\dfrac{\partial f}{\partial y} = x^2 + \dfrac{dg}{dy} = x^2 + y^2$, and so $g(y) = \dfrac{y^3}{3} + c$. Therefore, $f(x,y) = x^2 y + \dfrac{y^3}{3} + c$.

(b) The equation is not exact because $\partial(2x + y)/\partial y \neq \partial(y - x)/\partial x$.

(c) The equation is exact because $\partial(4 + xy^2)/\partial y = \partial(yx^2)/\partial x$. If we let $f(x,y)$ be the solution, then $\dfrac{\partial f}{\partial y} = yx^2$ gives $f = \dfrac{x^2 y^2}{2} + h(x)$. Now $\dfrac{\partial f}{\partial x} = xy^2 + \dfrac{dh}{dx} = 4 + xy^2$ and so $h(x) = 4x + c$. Therefore, $f(x,y) = \dfrac{x^2 y^2}{2} + 4x + c$.

(d) The equation is exact because $\partial(\sin x + y)/\partial y = \partial(x - \cos y)/\partial x$. If we let $f(x,y)$ be the solution, then $\dfrac{\partial f}{\partial x} = \sin x + y$ gives $f = -\cos x + xy + g(y)$. Now $\dfrac{\partial f}{\partial y} = x + \dfrac{dg}{dy} = x - 2\cos y$ and so $g = -2\sin y + c$. Therefore, $f(x,y) = xy - \cos x - 2\sin y + c$.

9. The equation of the tangent line is given by $y = x\dfrac{dy}{dx} + 2xy^2$. If we write this equation as $\dfrac{y\,dx - x\,dy}{y^2} = 2x\,dx$, then we can see that it takes the form $d(x/y) = 2x\,dx$, whose solution is $\dfrac{x}{y} = x^2 + c$ or $x - x^2y = cy$.

12. Equation 8.1.3 gives us the slope of a cardioid

$$\frac{dy}{dx} = \frac{r'(\theta)\sin\theta + r(\theta)\cos\theta}{r'(\theta)\cos\theta - r(\theta)\sin\theta} = \frac{c\sin\theta\cos\theta + c(1+\sin\theta)\cos\theta}{c\cos^2\theta - c(1+\sin\theta)\sin\theta}$$

$$= \frac{\cos\theta + 2\cos\theta\sin\theta}{\cos^2\theta - \sin^2\theta - \sin\theta}$$

The slope of a curve orthogonal to this is -1 times this result, or

$$\frac{\dfrac{r'(\theta)}{r(\theta)}\sin\theta + \cos\theta}{\dfrac{r'(\theta)}{r(\theta)}\cos\theta - \sin\theta} = \frac{(1+\sin\theta)\sin\theta - \cos^2\theta}{\cos\theta + 2\cos\theta\sin\theta}$$

$$= \text{ slope of curves orthogonal to } r(\theta) = c(1 + \sin\theta)$$

After some amount of straightforward algebra, this equation becomes
$\dfrac{r'(\theta)}{r(\theta)} = -\dfrac{1+\sin\theta}{\cos\theta} = -\sec\theta - \tan\theta$. Integration gives $\ln r + \ln(\sec\theta + \tan\theta) - \ln\cos\theta = c$
or $r\left(\dfrac{\sec\theta + \tan\theta}{\cos\theta}\right) = c_1$, or $r = c_1\left(\dfrac{\cos^2\theta}{1+\sin\theta}\right) = c_1(1 - \sin\theta)$. The accompanying figure shows this family of curves.

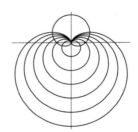

15. We first must derive an expression for $A(h)$, the area of the surface of the water, as a function of its height. The accompanying figure shows that $(1-h)^2 + r^2 = 1$, or that

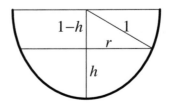

$$A(h) = \pi r^2 = \pi\left[1 - (1-h)^2\right] = \pi(2h - h^2) = \pi h(2-h)$$

Substituting this into the result of Problem 14 gives $\dfrac{dh}{dt} = -\dfrac{ac(2g)^{1/2}h^{1/2}}{\pi h(2-h)}$ or

$$h^{1/2}(2-h)\,dh = -\frac{ac}{\pi}(2g)^{1/2}dt$$

Solving this equation for h gives

$$\frac{4}{3}(1.00\text{ m})h^{3/2} - \frac{2}{5}h^{5/2} = -\frac{ac}{\pi}(2g)^{1/2}\,t + \text{constant}$$

Using $c = 0.6$, $a = 1.00$ cm^2, and $g = 9.81$ m^2·s^{-1} gives

$$\frac{4}{3}(1.00\text{ m})h^{3/2} - \frac{2}{5}h^{5/2} = -\frac{(1.00\text{ cm}^2)(0.6)(19.62\text{ m·s}^{-1})^{1/2}t}{\pi} + \text{constant}$$

$$= -(8.46 \times 10^{-5}\text{ m}^{5/2}\text{·s}^{-1})\,t + \text{constant}$$

Using the initial conditions $t = 0$ and $h = 1.00$ m, we find that constant $= 0.933$ m$^{5/2}$. When $h = 0$, $t = 0.933\,\text{m}^{5/2}/(8.46 \times 10^{-5}\,\text{m}^{5/2}\text{·s}^{-1}) = 1.10 \times 10^4\text{ s} = 3.06$ h. The accompanying figure shows t plotted against h.

18. Start with

$$M(x,y)\,dx + N(x,y)\,dy = x^n\phi_1\left(\frac{y}{x}\right)dx + x^n\phi_2\left(\frac{y}{x}\right)dy$$

Now let $u = y/x$, or $y = ux$ to get

$$M(x,y)\,dx + N(x,y)\,dy = x^n\phi_1(u)\,dx + x^n\phi_2(u)\,[u\,dx + x\,du] = 0$$

or $[x^n\phi_1(u) + x^n u\phi_2(u)]\,dx + x^n\phi_2(u)x\,du = 0$ or

$$\frac{dx}{x} + \frac{\phi_2(u)\,du}{\phi_1(u) + u\phi_2(u)} = 0$$

21. Use $a = 2$, $b = 1$, $c = -4$, $\alpha = 1$, $\beta = -1$, and $\gamma = 1$ in the formula of the previous problem to write $2x_0 + y_0 - 4 = 0$ and $x_0 - y_0 + 1 = 0$, which gives $x_0 = 1$ and $y_0 = 2$. Substituting $x = u+1$ and $y = v+2$ into the differential equation gives $\dfrac{dv}{du} = \dfrac{2u+v}{u-v}$. Now let $v = wu$ to obtain $\dfrac{1-w}{2+w^2}\,dw = \dfrac{du}{u}$, which integrates to give

$$\frac{1}{\sqrt{2}}\tan^{-1}\left(\frac{w}{\sqrt{2}}\right) - \frac{1}{2}\ln(2+w^2) = \ln u + c$$

or

$$\frac{1}{\sqrt{2}}\tan^{-1}\left(\frac{v}{\sqrt{2}\,u}\right) = \ln(2u^2 + v^2)^{1/2} + c$$

11.2   Linear First-Order Differential Equations

1. (a) The equation is a first-order linear differential equation and so we use Equation 4 with $\int p\,dx = \int 3x^2\,dx = x^3$. Therefore,

$$y(x)e^{x^3} = \int x^2 e^{x^3}\,dx + c = \frac{1}{3}e^{x^3} + c$$

so that $y(x) = \frac{1}{3} + ce^{-x^3}$.

(b) The equation is a first-order linear differential equation and so we use Equation 4 with $\int p\,dx = 2\ln x = \ln x^2$. Therefore,

$$yx^2 = \int (x^4 + 2x^2)\,dx + c = \frac{x^5}{5} + \frac{2}{3}x^3 + c$$

so that $y(x) = \frac{x^3}{5} + \frac{2x}{3} + \frac{c}{x^2}$.

3. (a) The equation is a first-order linear differential equation and so we use Equation 4 with $\int p\,dx = \ln x$ and so $e^{\int p\,dx} = x$. Therefore,

$$xy = \int 2x\,dx + c = x^2 + c$$

or $y(x) = x + \frac{c}{x}$. Using the condition $y(2) = 2$ gives $2 = 2 + \frac{c}{x}$ which gives $c = 0$, so that $y(x) = x$.

(b) The equation is a first-order linear differential equation and so we use Equation 4 with $\int p\,dx = \int \tan x\,dx = -\ln \cos x$, or $e^{\int p\,dx} = 1/\cos x$. Therefore,

$$\frac{y(x)}{\cos x} = \int \cos x\,dx + c = \sin x + c$$

or $y(x) = \sin x \cos x + c \cos x$. Using the condition $y(0) = -1$ gives $c = -1$, and so we have $y(x) = \cos x(\sin x - 1)$.

6. For $0 \le t < 1$, the solution is the same as that in Example 1 (with $i_0 = 0$) $i(t) = \frac{E_0}{R}(1 - e^{-Rt/L}$ for $0 \le t < 1$.

For $t > 1$, the differential equation is $L\frac{di}{dt} + Ri = 0$, whose solution is $i(t) = ce^{-Rt/L}$ for $t > 1$. Assuming that the current is a continuous function of time,

$$i(t = 1) = ce^{-R/L} = \frac{E_0}{R}(1 - e^{-R/L})$$

or $c = \frac{E_0}{R}(e^{R/L} - 1)$. So

$$i(t) = \frac{E_0}{R}(e^{R/L} - 1)e^{-Rt/L} \qquad t > 1$$

9. First we write $y(x) = u(x)e^{-\int p\,dx}$ from which we obtain

$$\frac{dy}{dx} = \frac{du}{dx}e^{-\int p\,dx} - p(x)u(x)e^{-\int p\,dx}$$

or

$$\frac{dy}{dx} + py = \frac{du}{dx}e^{-\int p\,dx} = q(x)$$

Solving this for $u(x)$ gives $u(x) = \int q(x)e^{\int p\,dx}dx + c$, which agrees with Equation 3.

12. (a) The equation is Bernoulli's equation with $n = 2$, so we let $u = 1/y$ or $\frac{dy}{dx} = -\frac{1}{u^2}\frac{du}{dx}$. The differential equation to be solved becomes

$$-\frac{x}{u^2}\frac{du}{dx} + \frac{1}{u} = \frac{3x^3}{u^2}$$

or $\frac{du}{dx} - \frac{u}{x} = -3x^2$. This equation is a linear first-order differential equation with $e^{\int p\,dx} = e^{-\ln x} = 1/x$, and so its solution is given by

$$\frac{u}{x} = -\int 3x\,dx + c = -\frac{3x^2}{2} + c$$

or $u(x) = cx - \frac{3x^3}{2}$. The solution in terms of $y(x) = 1/u(x)$ is $y(x) = 1/(cx - \frac{3}{2}x^3)$.

(b) The equation is Bernoulli's equation with $n = 3$. Therefore, we write $u(y) = 1/x^2(y)$. Differentiating $x(y) = u(y)^{1/2}$ gives $\frac{dx}{dy} = \frac{-1}{2u^{3/2}}\frac{du}{dy}$. The differential equation to be solved becomes

$$-\frac{1}{2u^{3/2}}\frac{du}{dy} + \frac{2}{yu^{1/2}} = \frac{2y}{u^{3/2}}$$

or $\frac{du}{dy} - \frac{4u}{y} = -4y$. This equation is a first-order linear differential equation whose solution is

$$\frac{u(y)}{y^4} = \int -\frac{4}{y^3}dy + c = \frac{2}{y^2} + c$$

or $u(y) = cy^4 + 2y^2$. Expressing the solution in terms of $x(y)$ gives $x(y) = (cy^4 + 2y^2)^{1/2}$.

15. (a) The equation is first-order linear with $\int p(t)\,dt = \int 2t\,dt = t^2$ and $e^{\int p\,dt} = e^{t^2}$. Therefore,

$$u(t)e^{t^2} = \int 4te^{t^2}\,dt = 2e^{t^2} + c$$

or $u(t) = 2 + ce^{-t^2}$.

(b) This is Bernoulli's equation with $n = 2$, so we let $u = y^{-3}$, which reduces the equation to $\dfrac{du}{dx} - u = 2x - 1$. Now, $\displaystyle\int p\,dx = -x$, and so

$$u(x)e^{-x} = \int (2x - 1)e^{-x}\,dx = -2xe^{-x} - e^{-x} + c$$

In terms of $y(x)$, we have $\dfrac{1}{y^3} = -1 - 2x + ce^x$

18. The integrating factor is $e^{\int p\,dx} = e^{-\int 2x\,dx} = e^{-x^2}$, so $\dfrac{d}{dx}(ye^{-x^2}) = e^{-x^2}$, Thus

$$y(x) = e^x \int_0^x e^{-u^2}\,du + ce^{x^2} = e^{x^2}\left[c + \frac{\pi^{1/2}}{2}\operatorname{erf}(x)\right]$$

21. Write the equation to be solved as $\dfrac{dy}{dx} + \dfrac{y}{x} = \dfrac{\sin x}{x^2}$. Now, $\displaystyle\int p\,dx = \ln x$ and so $e^{\int p\,dx} = x$. The solution is $xy(x) = \displaystyle\int \frac{\sin x}{x}\,dx + c$. Using the fact that $y(1) = 2$ gives $c = 2 - \displaystyle\int_0^1 \frac{\sin x}{x}\,dx$ and so the solution is

$$xy(x) = 2 + \int_0^1 \frac{\sin u}{u}\,du \qquad x \geq 1$$

11.3   Homogeneous Linear Differential Equations with Constant Coefficients

1. The Wronskian determinant is

$$\begin{vmatrix} 1 & x & x^2 & x^3 \\ 0 & 1 & 2x & 3x^2 \\ 0 & 0 & 2 & 6x \\ 0 & 0 & 0 & 6 \end{vmatrix} = 6\begin{vmatrix} 1 & x & x^2 \\ 0 & 1 & 2x \\ 0 & 0 & 2 \end{vmatrix} = 6 \times 2 \times \begin{vmatrix} 1 & x \\ 0 & 1 \end{vmatrix} = 12 \neq 0$$

Therefore, 1, x, x^2, and x^3 are linearly indpenedent.

The Wronskian determinant for 1, $1 + x$, $1 + x^2$, and $1 + x^3$ is

$$\begin{vmatrix} 1 & 1+x & 1+x^2 & 1+x^3 \\ 0 & 1 & 2x & 3x^2 \\ 0 & 0 & 2 & 6x \\ 0 & 0 & 0 & 6 \end{vmatrix} = 6\begin{vmatrix} 1 & 1+x & 1+x^2 \\ 0 & 1 & 2x \\ 0 & 0 & 2 \end{vmatrix} = 6 \times 2 \times \begin{vmatrix} 1 & 1+x \\ 0 & 1 \end{vmatrix} = 12 \neq 0$$

Therefore, 1, $1 + x$, $1 + x^2$, and $1 + x^3$ are linearly independent.

3. The Wronskian determinant in this case is

$$\begin{vmatrix} 1+x & 1-x & x^2 \\ 1 & -1 & 2x \\ 0 & 0 & 2 \end{vmatrix} = 2\begin{vmatrix} 1+x & 1-x \\ 1 & -1 \end{vmatrix} = 2(-1 - x - 1 + x) = -4 \neq 0$$

Therefore, $1 + x$, $1 - x$, and x^2 are linearly independent.

6. (a) The auxiliary equation is $\alpha^2 + 6\alpha = 0$ which gives $\alpha = 0$, -6. Therefore, the general solution is $y(x) = c_1 + c_2 e^{-6x}$.

(b) The auxiliary equation is $\alpha^2 - 4\alpha + 3 = 0$ or $(\alpha - 3)(\alpha - 1) = 0$, which gives $\alpha = 1$, 3. Therefore, the general solution is $y(x) = c_1 e^x + c_2 e^{3x}$.

(c) The auxiliary equation is $\alpha^2 + 3 = 0$, which gives $\alpha = \pm \sqrt{3}\, i$, Therefore, the general solution is $y(x) = c_1 e^{\sqrt{3}\, ix} + c_2 e^{-\sqrt{3}\, ix} = c_3 \cos \sqrt{3}\, x + c_4 \sin \sqrt{3}\, x$.

9. (a) The auxiliary equation is $\alpha^2 - 4 = 0$ and so the general solution is $y(x) = c_1 e^{2x} + c_2 e^{-2x}$. Using $y(0) = 2$ and $y'(0) = 4$ gives $c_1 + c_2 = 2$ and $2c_1 - 2c_2 = 4$ or $c_1 = 2$ and $c_2 = 0$. Therefore, the particular solution is $y(x) = 2e^{2x}$.

(b) The auxiliary equation is $\alpha^2 - 5\alpha + 6 = 0$ and so the general solution is $y(x) = c_1 e^{2x} + c_2 e^{3x}$. Using $y(0) = -1$ and $y'(0) = 0$ gives $c_1 + c_2 = -1$ and $2c_2 + 3c_2 = 0$ or $c_1 = -3$ and $c_2 = 2$. Therefore, the particular solution is $y(x) = 2e^{3x} - 3e^{2x} = e^{2x}(2e^x - 3)$.

(c) The auxiliary equation is $\alpha - 2 = 0$ and so the general solution is $y(x) = ce^{2x}$. Using $y(0) = 2$ gives $y(x) = 2e^{2x}$.

12. Let $y(x) = x^2 u(x)$. Then $y'(x) = 2xu(x) + x^2 u'(x)$ and $y''(x) = 2u(x) + 4xu'(x) + x^2 u''(x)$. The differential equation to be solved becomes

$$2x^2 u(x) + 4x^3 u'(x) + x^4 u''(x) + 2x^2 u(x) + x^3 u'(x) - 4x^2 u(x) = 0$$

or $xu''(x) + 5u'(x) = 0$. Let $u'(x) = z(x)$ to get $z'(x) + \dfrac{5z(x)}{x} = 0$ or $z(x) = cx^{-5}$ and $u(x) = -x^{-4}/4$. Therefore, $y(x) = c/x^2$ is a second solution to the differential equation.

15. (a) Use the formula in Problem 14 with $a_2(x) = x^2$, $a_1(x) = x$, and $y_1(x) = x^2$ to write

$$y_2(x) = x^2 \int \frac{e^{-\ln x}}{x^4}\, dx = x^2 \int \frac{dx}{x^5} = x^2 \frac{x^{-4}}{(-4)} = -\frac{1}{4x^2}$$

or generally $y_2(x) = cx^{-2}$.

(b) In this case, use $a_2(x) = x^2$, $a_1(x) = -x$, and $y_1(x) = x$ to write

$$y_2(x) = x \int \frac{e^{\ln x}}{x^2}\, dx = x \ln x$$

18. $C \cos(t + \psi) = C \cos t \cos \psi - C \sin t \sin \psi$. Now let $A = C \cos \psi$ and $B = -C \sin \psi$, so that $\tan \psi = -\dfrac{B}{A}$ and $A^2 + B^2 = C^2$.

21. First note that $\theta(0) = \theta_0 = c_3$ in Equation 25. Now

$$\begin{aligned} \dot{\theta}(t) &= -\delta \omega_0 e^{-\delta \omega_0 t} \left[c_3 \cosh(\delta^2 - 1)^{1/2} \omega_0 t + c_4 \sin(\delta^2 - 1)^{1/2} \omega_0 t \right] \\ &\quad + e^{-\delta \omega_0 t} \left[c_3 (\delta^2 - 1)^{1/2} \omega_0 \sinh(\delta^2 - 1)^{1/2} \omega_0 t + c_4 (\delta^2 - 1)^{1/2} \omega_0 \cosh(\delta^2 - 1)^{1/2} \omega_0 t \right] \end{aligned}$$

and so $\dot{\theta}(0) = 0 = -\delta\omega_0 c_3 + c_4(\delta^2 - 1)^{1/2}\omega_0 = 0$ or $c_4 = c_3\dfrac{\delta}{(\delta^2 - 1)^{1/2}} = \dfrac{\theta_0\delta}{(\delta^2 - 1)^{1/2}}$. Thus,

$$\theta(t) = \frac{\theta_0 e^{-\delta\omega_0 t}}{(\delta^2 - 1)^{1/2}}[\delta\sinh(\delta^2 - 1)^{1/2}\omega_0 t + (\delta^2 - 1)^{1/2}\cosh(\delta^2 - 1)^{1/2}\omega_0 t]$$

24. Because $W = ce^{-\int p\,dx}$ and the exponential function never equals zero, W will equal zero only if $c = 0$, in which case W is identically zero over the entire interval. Thus, if $W \neq 0$ at any point in the interval, then $y_1(x)$ and $y_2(x)$ are linearly independent.

11.4   Nonhomogeneous Linear Differential Equations with Constant Coefficients

1. Because e^x is part of the complementary solution.

3. All three solutions contain x, which, therefore, must be the particular solution. Thus, the general solution is $y(x) = c_1 + c_2\sin x + x$.

6. (a) The complementary solution is $y_c(x) = c_1\cos x + c_2\sin x$. The particular solution is of the form $y_p(x) = \alpha e^x$. Substituting $y_p(x) = \alpha e^x$ into the differential equation gives $(\alpha + \alpha)e^x = e^x$ or $\alpha = 1/2$. Thus, $y_p(x) = \dfrac{1}{2}e^x$.

(b) The complementary solution is $y_c(x) = c_1 e^x + c_2 e^{2x}$. The particular solution is of the form $y_p(x) = \alpha + \beta x + \gamma x^2 + \delta x^3$. Substituting $y_p(x) = \alpha + \beta x + \gamma x^2 + \delta x^3$ into the differential equation gives

$$2\gamma + 6\delta x - 3\beta - 6\gamma x - 9\delta x^2 + 2\alpha + 2\beta x + 2\gamma x^2 + 2\delta x^3 = x^3$$

from which we find that $2\delta = 1$, $2\gamma - 9\delta = 0$ $(\gamma = 9/4)$, $6\delta - 6\gamma + 2\beta = 0$ $(\beta = 21/4)$, and $2\gamma - 3\beta + 2\alpha = 0$ $(\alpha = 45/8)$. Thus, $y_p(x) = \dfrac{45}{8} + \dfrac{21}{4}x + \dfrac{9}{4}x^2 + \dfrac{1}{2}x^3$.

9. The complementary solution is $y_c(x) = c_1\cos 2x + c_2\sin 2x$. The particular solution is of the form $y_p(x) = \alpha\cos 3x + \beta\sin 3x$. Substituting this into the differential equation gives

$$-9\alpha\cos 3x - 9\beta\sin 3x + 4\alpha\cos 3x + 4\beta\sin 3x = 6\cos 3x + 20\sin 3x$$

from which we find $\alpha = -6/5$ and $\beta = -4$. Thus, the general solution is

$$y(x) = c_1\cos 2x + c_2\sin 2x - \frac{6}{5}\cos 3x - 4\sin 3x$$

12.

$$a = \frac{E_0}{Z} = \frac{E_0}{\left[R^2 + \left(\omega L - \dfrac{1}{\omega c}\right)^2\right]^{1/2}} = \frac{E_0}{\left[R^2 + L^2\left(\omega - \dfrac{1}{\omega c L}\right)^2\right]^{1/2}}$$

$$= \frac{E_0}{\left[R^2 + L^2\left(\omega - \dfrac{\omega_0^2}{\omega}\right)^2\right]^{1/2}} = \frac{E_0}{R\left[1 + \dfrac{L^2\omega_0^2}{R^2}\left(\dfrac{\omega}{\omega_0} - \dfrac{\omega_0}{\omega}\right)^2\right]^{1/2}}$$

15. Start with the general solution to the differential equation in Problem 13

$$x(t) = c_1 \cos \omega_0 t + c_2 \sin \omega_0 t + \frac{A}{\omega_0^2 - \omega^2} \sin \omega t$$

Suppose initially the system is in its equilibrium state $[x(0) = 0, \dot{x}(0) = 0]$ and then the driving force is imposed. In this case $c_1 = 0$ and $c_2 = -\dfrac{A\omega/\omega_0}{\omega_0^2 - \omega}$ and

$$x(t) = \frac{A}{\omega_0(\omega_0^2 - \omega^2)} (\omega_0 \sin \omega t - \omega \sin \omega_0 t)$$

Now, if $\omega \approx \omega_0$, then we let $\omega = \omega_0 + \epsilon$, and write this equation as

$$x(t) \approx \frac{A}{2\omega_0\epsilon} [\sin(\omega_0 + \epsilon)t - \sin \omega_0 t]$$

We can use the trigonometric formula $\sin \alpha - \sin \beta = 2 \cos \dfrac{\alpha + \beta}{2} \sin \dfrac{\alpha - \beta}{2}$ to write $x(t) \approx \dfrac{A}{\omega_0\epsilon} \sin \dfrac{\epsilon t}{2} \cos \omega_0 t$, which is analogous to Equation 22.

18. As $t \to \infty$, the first term in $x(t)$ in Problem 17 becomes negligible compared to the second term, which we write as

$$x_{ss}(t) = -\frac{\gamma\Omega A \cos \Omega t + (\Omega^2 - \omega_0^2) A \sin \Omega t}{\gamma^2\Omega^2 + (\Omega^2 - \omega_0^2)^2}$$

Now use the relation $A \cos z + B \sin z = C \sin(t + \phi)$ where $C = (A^2 + B^2)^{1/2}$ and $\phi = \tan^{-1}(A/B)$ to write

$$x_{ss}(t) = \frac{A \sin(\Omega t + \phi)}{[\gamma^2\Omega^2 + (\Omega^2 - \omega_0^2)^2]^{1/2}}$$

where $\phi = \tan^{-1}[\gamma\Omega/(\Omega^2 - \omega_0^2)]^{1/2}$.

21. Start with the $x_{ss}(t)$ given in Problem 17 to write

$$\frac{dx_{ss}}{dt} = \dot{x}_{ss}(t) = \frac{\gamma\Omega^2 A \sin \Omega t}{\gamma^2\Omega^2 + (\Omega^2 - \omega_0^2)^2} + a \ \cos \Omega t \ \text{term}$$

Now

$$[\dot{x}_{ss}(t) A \sin \Omega t]_{ave} = \frac{\gamma \Omega^2 A^2}{\gamma^2\Omega^2 + (\Omega^2 - \omega_0^2)^2} \frac{1}{(2\pi/\Omega)} \int_0^{2\pi/\Omega} \sin^2 \Omega t \, dt$$

$$+ \text{a term containing} \int_0^{2\pi/\Omega} \sin \Omega t \cos \Omega t \, dt$$

The last integral equals zero and so we have

$$[\dot{x}_{ss}(t) A \sin \Omega t]_{ave} = \frac{\gamma \Omega^2 A^2}{2[\gamma^2\Omega^2 + (\Omega^2 - \omega_0^2)^2]}$$

24. We'll use the method of variation of parameters. The complementary solution is $y_c(x) = c_1 \cos x + c_2 \sin x = c_1 y_1(x) + c_2 y_2(x)$, from which we find that $W(y_1, y_2) = 1$. Equations 41 and 42 give $u_1'(x) = -\dfrac{\sin x \cdot f(x)}{1} = -f(x)\sin x$ and $u_2'(x) = f(x)\cos x$. Therefore, according to Equation 36,

$$
\begin{aligned}
y_p(x) &= \cos x \int_0^x (-f(u)\sin u)\, du + \sin x \int_0^x f(u)\cos u\, du \\
&= \int_0^x f(u)(\sin x \cos u - \sin u \cos x)\, du \\
&= \int_0^x f(u) \sin(x - u)\, du
\end{aligned}
$$

27. The complementary solution is $y_c = c_1 \cos x + c_2 \sin x$, from which we find that $W(x, y) = 1$. Thus, we have $u_1' = -\dfrac{\csc x \cdot \sin x}{1} = -1$ and $u_2' = \dfrac{\csc x \cdot \cos x}{1} = \cot x$, which integrates to give $u_1 = -x$ and $u_2 = \ln \sin x$. Thus, the general solution is

$$
y(x) = c_1 \cos x + c_2 \sin x - x \cos x + \sin x \ln \sin x
$$

11.5   Some Other Types of Higher-Order Differential Equations

1. The differential equation is of the Euler-Cauchy type. Substituting $y(x) = x^m$ gives $m^2 - 3m + 2 = 0$, which gives $m_1 = 1$ and $m_2 = 2$. Therefore, the solution is $y(x) = c_1 x + c_2 x^2$.

3. Substituting $y(x) = x^m$ into the differential equation gives $m(m - 1) - 3m + 4 = 0 = m^2 - 4m + 4 = 0$, from which we find the repeated root $m = 2$. We now use the technique of reduction of order and write $y(x) = u(x)x^2$, which gives $y'(x) = 2xu + u'x^2$ and $y''(x) = 2u + 4xu' + u''x^2$. Substituting into the differential equation gives $xu'' + u' = 0$. Let $z(x) = u'(x)$ to write $x\dfrac{dz}{dx} + z = 0$, whose solution is $xz = xu'(x) = $ constant. Integrating once more gives $u = \ln x$, and so the general solution is $y(x) = c_1 x^2 + c_2 x^2 \ln x$.

6. Substituting $y(x) = x^m$ into the differential equation gives $m(m - 1) + 3m + 5 = m^2 + 2m + 5 = 0$, which gives $m = -1 \pm 2i$. Thus the general solution is

$$
y(x) = \frac{c_1 \cos(2\ln x) + c_2 \sin(2\ln x)}{x}
$$

9. Substituting $y(x) = x^m$ into the differential equation gives

$$
m(m-1)(m-2)(m-2) - 3m(m-1) + 6m - 6 = m^3 - 6m^2 + 11m - 6 = 0
$$

from which we find that $m = 1$, 2, and 3. Therefore, the general solution is

$$
y(x) = c_1 x + c_2 x^2 + c_3 x^3
$$

12. Substituting $y_c(x) = x^m$ into the homogeneous equation gives $m(m-1) + 2m - 2 = m^2 + m - 2 = 0$, $m = 1$ and -2. Therefore, $y_c(x) = c_1 x + \dfrac{c_2}{x^2}$. We shall now use the method of variation of parameters. First, we write $W(y_1, y_2) = \begin{vmatrix} x & x^{-2} \\ 1 & -2x^{-3} \end{vmatrix} = -\dfrac{3}{x^2}$ and

so $u_1' = -\dfrac{6x \cdot x^{-2}}{-3x^{-2} \cdot x^2} = \dfrac{2}{x}$ and $u_2' = \dfrac{6x \cdot x^{-2}}{-3x^{-2} \cdot x^2} = -2x^2$. Integration of these two equations gives $u_1 = 2 \ln x$ and $u_2 = -\dfrac{2}{3} x^3$. The general solution is

$$y(x) = c_1 x + \frac{c_2}{x^2} - \frac{2}{3} x + 2x \ln x$$

15. The independent variable is missing in $yy'' = (y')^2$ and so we use the substitution $y' = p$ and write $y'' = p\dfrac{dp}{dy}$. Substituting these results into the differential equation gives $py\dfrac{dp}{dy} = p^2$, from which we find that $p = 0$ ($y = $ constant), or $\ln p = \ln y + c_1$ or $p = \alpha y$, or $\dfrac{dy}{dx} = \alpha y$. The solution to this equation is $y = Ae^{\alpha x}$, so we see that $y = Ae^{\alpha x}$ or $y = B$.

18. Multiply Newton's equation by dx/dt and use the identity $\dfrac{d}{dx}\left(\dfrac{dx}{dt}\right)^2 = 2\dfrac{dx}{dt}\dfrac{d^2x}{dt^2}$ to obtain $\dfrac{m}{2}\dfrac{d}{dt}\left(\dfrac{dx}{dt}\right)^2 = -f(x)\dfrac{dx}{dt}$. Now integrate with respect to t from A (where $\dot{x}(t) = 0$) to x to get $\dfrac{m}{2}\left(\dfrac{dx}{dt}\right)^2 = -\displaystyle\int_A^x f(u)\, du$. Solving for dx/dt gives

$$\frac{dx}{dt} = \left[\frac{2}{m}\int_x^A f(x)\, dx\right]^{1/2} = \left(\frac{2}{m}\right)^{1/2}[F(A) - F(x)]^{1/2}$$

where $F(x)$ is the antiderivative of $f(x)$. The displacement x goes from 0 to A in a time $\tau_0/4$, and so

$$\int_0^A \frac{dx}{[F(A) - F(x)]^{1/2}} = \left(\frac{2}{m}\right)^{1/2}\int_0^{\tau_0/4} dt = \left(\frac{2}{m}\right)^{1/2}\frac{\tau_0}{4} = \left(\frac{1}{8m}\right)^{1/2}\tau_0$$

or $\tau_0 = (8m)^{1/2}\displaystyle\int_0^A \frac{dx}{[F(A) - F(x)]^{1/2}}$. For a harmonic oscillator, $F(A) = kA^2/2$ and $F(x) = kx^2/2$, so

$$\tau_0 = 4\left(\frac{m}{k}\right)^{1/2}\int_0^A \frac{dx}{(A^2 - x^2)^{1/2}} = 4\left(\frac{m}{k}\right)^{1/2}\left[\sin^{-1}\frac{x}{A}\right]_0^A = 2\pi\left(\frac{m}{k}\right)^{1/2}$$

11.6 Systems of Linear Differential Equations

1. We are given that $\mathbf{y}_1' = A(x)\mathbf{y}_1$ and $\mathbf{y}_2' = A(x)\mathbf{y}_2(x)$. We see that $c_1\mathbf{y}_1 + c_2\mathbf{y}_2$ is also a solution because the equations are linear in $\mathbf{y}_1$ and $\mathbf{y}_2$.

3. The matrix A that describes the two differential equation is $\mathsf{A} = \begin{pmatrix} 5 & 4 \\ -1 & 0 \end{pmatrix}$. The eigenvalues and corresponding eigenvectors are $\lambda_1 = 1$ and $\lambda_2 = 4$ with $\mathbf{u}_1 = (1, -1)^\mathsf{T}$ and $\mathbf{u}_2 = (4, -1)^\mathsf{T}$. Thus the general solution is

$$\mathbf{y}(x) = \begin{pmatrix} y_1(x) \\ y_2(x) \end{pmatrix} = c_1 \begin{pmatrix} 1 \\ -1 \end{pmatrix} e^x + c_2 \begin{pmatrix} 4 \\ -1 \end{pmatrix} e^{4x}$$

6. The matrix A that describes the two differential equations is $\mathsf{A} = \begin{pmatrix} 3 & 2 \\ -1 & 1 \end{pmatrix}$. The eigenvalues and corresponding eigenvectors are $\lambda_1 = 2+i$ and $\lambda_2 = 2-i$ with $\mathbf{u}_1 = (-1-i, 1)$ and $\mathbf{u}_2 = (-1+i, 1)$. Thus the general solution is

$$\mathbf{y}(x) = c_1 \begin{pmatrix} -1+i \\ 1 \end{pmatrix} e^{(2-i)x} + c_2 \begin{pmatrix} -1-i \\ 1 \end{pmatrix} e^{(2+i)x}$$

Thus,

$$\begin{aligned}
y_1(x) &= e^{2x}\left[c_1(-1+i)e^{-ix} - c_2(1+i)e^{ix}\right] \\
&= e^{2x}\left[-c_1(\cos x - i\sin x) + ic_1(\cos x - i\sin x) - c_2(\cos x + i\sin x) - ic_2(\cos x + i\sin x)\right] \\
&= e^{2x}\left[(ic_1 - c_1 - c_2 - ic_2)\cos x + (ic_1 + c_1 - ic_2 + c_2)\sin x\right] \\
&= e^{2x}\{[i(c_1 - c_2) - (c_1 + c_2)]\cos x + [i(c_1 - c_2) + (c_1 + c_2)]\sin x\} \\
&= e^{2x}\left[(\beta - \alpha)\cos x + (\beta + \alpha)\sin x\right]
\end{aligned}$$

and

$$\begin{aligned}
y_2(x) &= e^{2x}\left[c_1(\cos x - i\sin x) + c_2(\cos x + i\sin x)\right] \\
&= e^{2x}\left[(c_1 + c_2)\cos x + i(c_2 - c_1)\sin x\right] \\
&= e^{2x}(\alpha \cos x - \beta \sin x)
\end{aligned}$$

where we have let $\alpha = c_1 + c_2$ and $\beta = i(c_1 - c_2)$.

A better way is to start with one of the eigenvalues (we choose $\lambda = 2-i$) and write

$$\begin{aligned}
y_+(x) &= \begin{pmatrix} -1+i \\ 1 \end{pmatrix} e^{(2-i)x} = e^{2x}\begin{pmatrix} -1+i \\ 1 \end{pmatrix}(\cos x - i\sin x) \\
&= e^{2x}\begin{pmatrix} -\cos x + \sin x \\ \cos x \end{pmatrix} + ie^{2x}\begin{pmatrix} \cos x + \sin x \\ -\sin x \end{pmatrix}
\end{aligned}$$

The real and imaginary parts are independent, so we can write

$$\mathbf{y}(x) = \begin{pmatrix} y_1(x) \\ y_2(x) \end{pmatrix} = c_1 e^{2x}\begin{pmatrix} \sin x - \cos x \\ \cos x \end{pmatrix} + c_2 e^{2x}\begin{pmatrix} \cos x + \sin x \\ -\sin x \end{pmatrix}$$

Notice that this is the same solution that we obtained above because $y_1(x) = c_1 e^{2x}(\sin x - \cos x) + c_2 e^{2x}(\cos x + \sin x)$ and $y_2(x) = c_1 e^{2x}\cos x - c_2 e^{2x}\sin x$ with $c_1 = \alpha$ and $c_2 = \beta$.

9. The matrix A that describes the three differential equations is $\mathsf{A} = \begin{pmatrix} 0 & 0 & 1 \\ 3 & 7 & -9 \\ 0 & 2 & -1 \end{pmatrix}$. The eigenvalues and corresponding eigenvectors are $\lambda_1 = 1$, $\lambda_2 = 2$, and $\lambda_3 = 3$ with

$\mathbf{u}_1 = (1,1,1)^\mathsf{T}$, $\mathbf{u}_2 = (1,3,2)^\mathsf{T}$, and $\mathbf{u}_3 = (1,6,3)^\mathsf{T}$. The general solution is therefore

$$\mathbf{y}(x) = \begin{pmatrix} y_1(x) \\ y_2(x) \\ y_3(x) \end{pmatrix} = c_1 \begin{pmatrix} 1 \\ 1 \\ 1 \end{pmatrix} e^x + c_2 \begin{pmatrix} 1 \\ 3 \\ 2 \end{pmatrix} e^{2x} + c_3 \begin{pmatrix} 1 \\ 6 \\ 3 \end{pmatrix} e^{3x}$$

12. The two equations below Example 2 in the text show that the statement is true.

15. The third-order equation in the problem can be written as three simultaneous first-order equations by letting $y = x_1$, $x_2 = x_1' = y'$, $x_2' = x_3 = y''$, and $x_3' = y'''$. Now, the corresponding first-order equations are

$$\begin{aligned} x_1' &= x_2 \\ x_2' &= x_3 \\ x_3' &= -a_2 x_3 - a_1 x_2 - a_0 x_1 \end{aligned}$$

The corresponding matrix is $\begin{pmatrix} 0 & 1 & 0 \\ 0 & 0 & 1 \\ -a_0 & -a_1 & -a_2 \end{pmatrix}$. The eigenvalues are given by the equation $\lambda^3 + a_2 \lambda^2 + a_1 \lambda + a_0 = 0$, which is the same as the auxiliary equation associated with the third-order differential equation. The solution to the set of three first-order equations and the solution to the third-order equation are equivalent.

CHAPTER 12

Series Solutions of Differential Equations

12.1   The Power Series Method

1. (a) $\displaystyle\sum_{n=1}^{\infty} na_n x^{n-1} = \sum_{n=0}^{\infty}(n+1)a_{n+1}x^n$ (b) $\displaystyle\sum_{n=2}^{\infty} n(n-1)a_n x^{n-2} = \sum_{n=0}^{\infty}(n+2)(n+1)a_{n+2}x^n$

(c) $\displaystyle\sum_{n=2}^{\infty}(n-2)c_{n-2}x^n = \sum_{n=0}^{\infty} nc_n x^{n+2}$

3. $11!! = 11\cdot9\cdot7\cdot5\cdot3\cdot1 = 10\ 395$ and $10!! = 10\cdot8\cdot6\cdot4\cdot2 = 3840$

6. Let $\displaystyle y(x) = \sum_{n=0}^{\infty} a_n x^n$. Then $\displaystyle y'(x) = \sum_{n=0}^{\infty} na_n x^{n-1}$ and $y'(x) + y(x) = 0$ becomes

$$\sum_{n=0}^{\infty} na_n x^{n-1} + \sum_{n=0}^{\infty} a_n x^n = \sum_{n=1}^{\infty} na_n x^{n-1} + \sum_{n=0}^{\infty} a_n x^n$$

$$= \sum_{n=0}^{\infty}\left[(n+1)a_{n+1} + a_n\right]x^n = 0$$

Therefore, $(n+1)a_{n+1} + a_n = 0$, or $a_{n+1} = -\dfrac{a_n}{n+1}$, $n \geq 0$. Therefore,

$$a_1 = -a_0, \quad a_2 = -\frac{a_1}{2} = \frac{a_0}{2}, \quad a_3 = -\frac{a_2}{3} = -\frac{a_0}{3\cdot2}, \quad a_4 = -\frac{a_3}{4} = \frac{a_0}{4\cdot3\cdot2}, \quad \text{etc.}$$

The general term is $a_n = \dfrac{(-1)^n a_0}{n!}$ and the solution is

$$y(x) = a_0 \sum_{n=0}^{\infty} \frac{(-1)^n x^n}{n!} = a_0 e^{-x}$$

as you can verify almost by inspection.

9. Substitute $\displaystyle\sum_{n=0}^{\infty} a_n x^n$ into $(1-x^2)y''(x) - 2xy'(x) + 2y(x) = 0$ to obtain

$$(1-x^2)\sum_{n=0}^{\infty} n(n-1)a_n x^{n-2} - 2x\sum_{n=0}^{\infty} na_n x^{n-1} + 2\sum_{n=0}^{\infty} a_n x^n = 0$$

$$\sum_{n=0}^{\infty} n(n-1)a_n x^{n-2} - \sum_{n=0}^{\infty} n(n-1)a_n x^n - 2\sum_{n=0}^{\infty} na_n x^n + 2\sum_{n=0}^{\infty} a_n x^n = 0$$

or

$$\sum_{n=0}^{\infty}\left[(n+2)(n+1)a_{n+2} - (n+2)(n-1)a_n \right] x^n = 0$$

or $a_{n+2} = \dfrac{n-1}{n+1}a_n$, $n \geq 0$. Thus $a_2 = -a_0$, $a_3 = 0$, $a_4 = \dfrac{a_2}{3} = -\dfrac{a_0}{3}$, $a_5 = 0$, $a_6 = \dfrac{3a_4}{5} = -\dfrac{a_0}{5}$, $a_7 = 0$, $a_8 = -\dfrac{5a_6}{7} = -\dfrac{a_0}{7}$. Therefore, $a_{2n} = \dfrac{-a_0}{2n-1}$, $n \geq 1$, and $a_{2n+1} = 0$, $n \geq 0$, and the solution is

$$\begin{aligned}
y(x) &= a_1 x + a_0\left(1 - x^2 - \frac{x^4}{3} - \frac{x^6}{5} + \cdots\right) \\
&= a_1 x + a_0\left(1 - x\sum_{n=0}^{\infty}\frac{x^{2n+1}}{2n+1}\right) = a_1 x + a_0\left(1 - \frac{x}{2}\ln\frac{1+x}{1-x}\right)
\end{aligned}$$

12.

$$\begin{aligned}
\frac{s(x)}{c(x)} &= \frac{x - \dfrac{x^3}{6} + \dfrac{x^5}{120} - \dfrac{x^7}{5040} + O(x^9)}{1 - \dfrac{x^2}{2} + \dfrac{x^4}{24} - \dfrac{x^6}{720} + O(x^8)} \\
&= \left[x - \frac{x^3}{6} + \frac{x^5}{120} - \frac{x^7}{5040} + O(x^9)\right]\left[1 + \frac{x^2}{2} - \frac{x^4}{24} + \frac{x^6}{720} + O(x^8)\right. \\
&\quad \left. + \frac{x^4}{4} - \frac{x^6}{24} + O(x^8) + \frac{x^6}{8} + O(x^8)\right] \\
&= x + \frac{x^3}{2} - \frac{x^5}{24} + \frac{x^7}{720} + O(x^9) + \frac{x^5}{4} - \frac{x^7}{24} + O(x^9) + \frac{x^7}{8} + O(x^9) \\
&\quad - \frac{x^3}{6} - \frac{x^5}{12} + \frac{x^7}{6\cdot24} + O(x^9) - \frac{x^7}{24} + O(x^9) + \cdots \\
&\quad + \frac{x^5}{120} + \frac{x^7}{240} + O(x^9) - \frac{x^7}{5040} + O(x^9) \\
&= x + \frac{x^3}{3} + \frac{2x^5}{5} + \frac{272x^7}{5040} + O(x^9) \\
&= x + \frac{x^3}{3} + \frac{2x^5}{15} + \frac{17x^7}{315} + O(x^9)
\end{aligned}$$

This is the Maclaurin expansion of $\tan x$.

15. We use the ratio test in each case.

(a) $\displaystyle\lim_{n\to\infty}\left|\frac{u_{n+1}}{u_n}\right| = \lim_{n\to\infty}\left|\frac{x^{n+1}}{2^{n+1}}\cdot\frac{2^n}{x^n}\right| = \frac{|x|}{2}$ $\qquad |x| < 2$

(b) $\displaystyle\lim_{n\to\infty}\left|\frac{u_{n+1}}{u_n}\right| = \lim_{n\to\infty}\left|\frac{x^{n+1}}{n+1}\cdot\frac{n}{x^n}\right| = |x|$ $\qquad |x| < 1$

18. Start with $\dfrac{d^2y(x)}{dx^2} + y(x) = 0$. Let $x \to -x$ to obtain $\dfrac{d^2(-x)}{dx^2} + y(-x) = 0$, which shows that $y(-x)$ is also a solution. Now $y(-x)$ must be a multiple of $y(x)$, so that we write $y(-x) = cy(x)$. Letting $x \to -x$ gives $y(x) = cy(-x) = c^2y(x)$, so $c^2 = 1$, or $c = \pm 1$.

12.2   Ordinary Points and Singular Points of Differential Equations

1. Substitute $y(x) = \displaystyle\sum_{n=0}^{\infty} a_n x^n$ into Equation 4 to get

$$\sum_{n=0}^{\infty} n(n-1)a_n x^{n-2} - \sum_{n=0}^{\infty} n(n-1)a_n x^n - 6\sum_{n=0}^{\infty} na_n x^n - 4\sum_{n=0}^{\infty} a_n x^n = 0$$

or

$$\sum_{n=0}^{\infty} \left[(n+2)(n+1)a_{n+2} - (n+4)(n+1)a_n\right]x^n = 0$$

Setting each coefficient of x^n equal to zero gives $a_{n+2} = \dfrac{n+4}{n+2}a_n$ for $n \geq 0$. Solving this result iteratively gives

$$a_2 = 2a_0 \qquad\qquad a_3 = \frac{5}{3}a_1$$

$$a_4 = \frac{3}{2}a_2 = 3a_0 \qquad\qquad a_5 = \frac{7}{5}a_3 = \frac{7}{3}a_1$$

$$a_6 = \frac{4}{3}a_4 = 4a_0 \qquad\qquad a_7 = \frac{9}{7}a_5 = \frac{9}{3}a_1$$

$$a_8 = \frac{5}{4}a_6 = 5a_0 \qquad\qquad a_9 = \frac{11}{9} = \frac{11}{3}a_1$$

$$a_{2n} = (n+1)a_0 \qquad\qquad a_{2n+1} = \frac{2n+3}{3}a_1$$

3. We use the ratio test in each case.

(a) $\displaystyle\lim_{n\to\infty}\left|\frac{u_{n+1}}{u_n}\right| = \lim_{n\to\infty}\frac{(n+2)x^{2n+2}}{(n+1)x^{2n}} = x^2$ and so $|x| < 1$.

(b) $\displaystyle\lim_{n\to\infty}\left|\frac{u_{n+1}}{u_n}\right| = \lim_{n\to\infty}\frac{(2n+5)x^{2n+3}}{(2n+3)x^{2n+1}} = x^2$ and so $|x| < 1$.

6. There is a singular point at $x = \pm 2i$. The distance of the point $x = 4$ to the points $x = \pm 2i$ is $(2^2 + 4^2)^{1/2} = (20)^{1/2} = 2\sqrt{5}$. Thus the radius of convergence is at least $2\sqrt{5}$.

9. $x^2 y''(x) = xy'(x) + y(x) = 0$. There is a regular singular point at $x = 0$ because $xp(x)$ and $x^2q(x)$ are well-behaved at $x = 0$.

12. We write the equation in the form

$$y''(x) + \frac{2}{x^3(1+x^2)}y'(x) + \frac{1}{x^3(1+x^2)}y(x) = 0$$

The point $x = 0$ is an irregular singular point because $xp(x) = \dfrac{2}{x^2(1+x^2)}$ diverges at $x = 0$. The points $x = \pm i$ are regular singular points because $(1\pm ix)p(x)$ and $(1\pm ix)^2 q(x)$ are well-behaved at $x = \pm i$.

15. Write the equation as

$$y''(x) + \frac{e^x - 1}{x^2}y'(x) + \frac{e^x + 1}{x^2}y(x) = 0$$

Now

$$xp(x) = \frac{e^x - 1}{x} = 1 + \frac{x}{2} + \cdots \qquad \text{and} \qquad x^2 q(x) = 1 + e^x = 2 + x + \cdots$$

are well-behaved at $x = 0$, so the point $x = 0$ is a regular singular point.

18. All points are ordinary points. There are no singular points.

12.3   Series Solutions Near an Ordinary Point: Legendre's Equation

1. Substitute $y(x) = \sum\limits_{n=0}^{\infty} a_n x^n$ into Equation 1 to obtain

$$\sum_{n=0}^{\infty} n(n-1)a_n x^{n-2} - \sum_{n=0}^{\infty} n(n-1)a_n x^n - 2\sum_{n=0}^{\infty} na_n x^n + \alpha(\alpha+1)\sum_{n=0}^{\infty} a_n x^n = 0$$

or

$$\sum_{n=0}^{\infty} \{(n+2)(n+1)a_{n+2} - [n^2 + n - \alpha(\alpha+1)]a_n\}x^n = 0$$

Setting the coefficient of x^n equal to zero gives

$$a_{n+2} = -\frac{(\alpha-n)(\alpha+n+1)}{(n+2)(n+1)}a_n \qquad n \geq 0$$

3. First we'll determine $f_4(x)$ and $f_5(x)$ (Problem 2). Let $\alpha = 4$ in Equation 3 and get $a_2 = -10a_0$, $a_4 = 35a_0/3$, $a_6 = a_8 = \cdots = 0$. So

$$f_4(x) = 1 - 10x^2 + \frac{35}{3}x^4 \longrightarrow 3 - 30x^2 + 35x^4$$

Similarly, letting $\alpha = 5$ in Equation 4 gives $a_3 = -14a_1/3$, $a_5 = 21a_1/5$, $a_7 = a_9 = \cdots = 0$. So

$$f_5(x) = x - \frac{14}{3}x^3 + \frac{21}{5}x^5 \longrightarrow 15x - 70x^3 + 63x^5$$

Now, substitute $f_4(x)$ into Equation 1 with $\alpha = 4$ to obtain

(a) $f_4'(x) = -60x + 140x^3$ and $f_4''(x) = -60 + 420x^2$

$$(-60 + 420x^2) + (60x^2 - 420x^4) + (120x^2 - 280x^4) + (60 - 600x^2 + 700x^4) = 0$$

(b) Similarly, we have

$$(-420x + 1260x^3) + (420x^3 - 1260x^5) - (30x - 420x^3 + 630x^5) + (450x - 2100x^3 + 1890x^5) = 0$$

6. Use $\ln(1+x) = x - \dfrac{x^2}{2} + \dfrac{x^3}{3} - \dfrac{x^4}{4} + \cdots$ and $\ln(1-x) = -x - \dfrac{x^2}{2} - \dfrac{x^3}{3} - \dfrac{x^4}{4} + \cdots$ to show that

$$\ln \frac{1+x}{1-x} = \ln(1+x) - \ln(1-x) = \frac{1}{2}\left(x + \frac{x^3}{3} + \frac{x^5}{5} + \cdots \right)$$

9. Use Equation 8 with $j = 0$.

12. $n = 0$: $P_1(x) - xP_0(x) = 0$ or

$$P_1(x) = x \qquad \text{and} \qquad P_0(x) = 1$$

$n = 1$: $2P_2(x) - 3xP_1(x) + P_0(x) = 0$ or

$$P_2(x) = \frac{3}{2}xP_1(x) - \frac{1}{2}P_0(x) = \frac{1}{2}(3x^2 - 1)$$

$n = 2$: $3P_3(x) - 5xP_2(x) + 2P_1(x) = 0$ or

$$P_3(x) = \frac{5}{3}xP_2(x) - \frac{2}{3}P_1(x) = \frac{5}{6}(3x^2 - 1)x - \frac{2}{3}x = \frac{1}{2}(5x^3 - 3x)$$

15. Start with Legendre's equation

$$(1 - x^2)\frac{d^2 P_n}{dx^2} - 2x\frac{dP_n}{dx} + n(n+1)P_n(x) = 0$$

Now let $x = \cos\theta$ and $\Theta_n(\theta) = P_n(x)$ to write

$$\frac{dP_n}{dx} = \frac{d\Theta_n}{d\theta}\frac{d\theta}{dx} = -\frac{1}{\sin\theta}\frac{d\Theta_n}{d\theta}$$

and

$$\begin{aligned}
\frac{d^2 P_n}{dx^2} &= \left[\frac{d}{d\theta}\left(-\frac{1}{\sin\theta}\frac{d\Theta_n}{d\theta} \right) \right]\left(-\frac{1}{\sin\theta} \right) \\
&= \left(\frac{\cos\theta}{\sin^2\theta}\frac{d\Theta_n}{d\theta} - \frac{1}{\sin\theta}\frac{d^2\Theta_n}{d\theta^2} \right)\left(-\frac{1}{\sin\theta} \right) \\
&= \frac{1}{\sin^2\theta}\frac{d^2\Theta_n}{d\theta^2} - \frac{\cos\theta}{\sin^3\theta}\frac{d\Theta_n}{d\theta}
\end{aligned}$$

Now

$$(1 - x^2)\frac{d^2 P_n}{dx^2} = \sin^2\theta\frac{d^2 P_n}{dx^2} = \frac{d^2\Theta_n}{d\theta^2} - \cot\theta\frac{d\Theta_n}{d\theta}$$

$$-2x\frac{dP_n}{dx} = -\cos\theta\frac{dP_n}{dx} = 2\cot\theta\frac{d\Theta_n}{d\theta}$$

and so

$$(1 - x^2)\frac{d^2 P_n}{dx^2} - 2x\frac{dP_n}{dx} + n(n+1)P_n(x) = \frac{d^2\Theta_n}{d\theta^2} + \cot\theta\frac{d\Theta_n}{d\theta} + n(n+1)\Theta_n(\theta) = 0$$

12.4   Solutions Near Regular Singular Points

1. Substitute $y(x) = \sum_{n=0}^{\infty} a_n x^{n+r}$ into Equation 2 to get

$$2\sum_{n=0}^{\infty}(n+r)(n+r-1)a_n x^{n+r-1} + 3\sum_{n=0}^{\infty}(n+r)a_n x^{n+r-1} - \sum_{n=0}^{\infty} a_n x^{n+r} = 0$$

Collecting coefficients of like powers of x gives

$$a_0 r(2r+1)x^{r-1} + \sum_{n=0}^{\infty}\{[\,2(n+1+r)(n+r) + 3(n+1+r)\,]a_{n+1} - a_n\}x^{n+r} = 0$$

or

$$a_0 r(2r+1)x^{r-1} + \sum_{n=1}^{\infty}[\,(n+r)(2n+2r+1)a_n - a_{n-1}\,]x^{n+r-1} = 0$$

3. Put $r = -1/2$ in the term in braces in Equation 4 to get $2n(n - \frac{1}{2})c_n = c_{n-1}$ or $c_n = -c_{n-1}/n(2n-1)$ for $n \geq 1$. This formula gives us

$$c_1 = \frac{c_0}{1} = c_0, \quad c_2 = \frac{c_1}{2 \cdot 3} = \frac{c_0}{3!} = \frac{2^2 c_0}{4!}, \quad c_3 = \frac{c_2}{3 \cdot 5} = \frac{2^2 c_0}{3 \cdot 5!} = \frac{2^3 c_0}{6!}, \quad \text{and so on}$$

6. Substitute $r = 1/4$ into the recursion formula to obtain $a_{n+1} = \dfrac{a_n}{(4n+5)(2n+3)-1}$ for $n \geq 0$. Thus we have successively for $n = 0,\ 1,\ 2, \ldots$

$$a_1 = \frac{a_0}{14}; \quad a_2 = \frac{a_1}{44} = \frac{a_0}{616}; \quad \cdots$$

Now substitute $r = -1/2$ into the recursion formula to obtain $a_{n+1} = \dfrac{a_n}{(4n+3)(2n+1)-1}$ for $n \geq 0$. This formula gives

$$a_1 = \frac{a_0}{2}; \quad a_2 = \frac{a_1}{20} = \frac{a_0}{40}; \quad \cdots$$

9. Substitute $y(x) = \sum_{n=0}^{\infty} a_n x^{n+r}$ into the differential equation to obtain

$$\sum_{n=0}^{\infty} 4(n+r)(n+r-1)a_n x^{n+r} + \sum_{n=0}^{\infty} a_n x^{n+r} - \sum_{n=0}^{\infty} 2a_n x^{n+r+1} = 0$$

or

$$[\,4r(r-1)+1\,]a_0 + \sum_{n=0}^{\infty}\{[\,4(n+r+1)(n+r)+1\,]a_{n+1} - 2a_n\}x^{n+r+1} = 0$$

The indicial equation is $4r(r-1) + 1 = 0$ whose two roots are $r = 1/2$ (twice.)

12. Substitute $y(x) = \sum\limits_{n=0}^{\infty} a_n x^{n+r}$ into $x^2 y''(x) + 3xy'(x) + (1 - 2x)y(x) = 0$ to obtain

$$\sum_{n=0}^{\infty}(n+r)(n+r-1)a_n x^{n+r} + 3\sum_{n=0}^{\infty}(n+r)a_n x^{n+r} + \sum_{n=0}^{\infty} a_n x^{n+r} - 2\sum_{n=0}^{\infty} a_n x^{n+r+1} = 0$$

$$(r^2 + 2r + 1)a_0 x^r + \sum_{n=0}^{\infty}\{[(n+1+r)(n+r) + 3(n+1+r) + 1]a_{n+1} - 2a_n\}x^{n+r+1} = 0$$

Let $r = -1$, then $a_{n+1} = \dfrac{2a_n}{(n+1)^2}$ for $n \geq 0$. Thus,

$$a_1 = 2a_0, \quad a_2 = \frac{2a_1}{2^2} = \frac{2^2 a_0}{(2!)^2}; \quad a_3 = \frac{2a_2}{3^2} = \frac{2^3 a_0}{(3!)^2}; \quad \cdots$$

15. From Equation 20, we find that

$$y_1(x) \;=\; x^{1/2}\left[1 - \frac{3}{4}x + \frac{9}{64}x^2 - \frac{3}{256}x^3 + O(x^4)\right]$$

$$y_1^2(x) \;=\; x\left[1 - \frac{3}{2}x + \left(\frac{9}{16} + \frac{9}{32}\right)x^2 - \left(\frac{6}{256} + \frac{27}{128}\right)x^3 + O(x^4)\right]$$

$$\;=\; x\left[1 - \frac{3}{2}x + \frac{27}{32}x^2 - \frac{15}{64}x^3 + O(x^4)\right]$$

and

$$\frac{1}{y_1^2(x)} \;=\; \frac{1}{x}\left[1 + \frac{3}{2}x - \frac{27}{32}x^2 + \frac{15}{64}x^3 + \cdots + \frac{9}{4}x^2 - \frac{81}{32} + \cdots + \frac{27}{8}x^3 + O(x^4)\right]$$

$$\;=\; \frac{1}{x}\left[1 + \frac{3}{2}x + \frac{45}{32}x^2 + \frac{69}{64}x^3 + O(x^4)\right]$$

Now $\int p\,dx = 0$, so

$$y_2(x) \;=\; y_1(x)\int \frac{dx}{y_1^2(x)} = y_1(x)\left[\ln x + \frac{3}{2}x + \frac{45}{64}x^2 + \frac{23}{64}x^3 + O(x^4)\right]$$

$$\;=\; y_1(x)\ln x + x^{3/2}\left[\frac{3}{2} + \left(\frac{45}{64} - \frac{9}{8}\right)x + \left(\frac{23}{64} - \frac{135}{256} + \frac{27}{128}\right)x^2 + O(x^3)\right]$$

$$\;=\; y_1(x)\ln x + x^{3/2}\left[\frac{3}{2} - \frac{27}{64}x + \frac{11}{256}x^2 + O(x^3)\right]$$

18. Substitute $y(x) = \sum\limits_{n=0}^{\infty} a_n x^n$ into Equation 25 to obtain

$$\sum_{n=0}^{\infty} n(n-1)a_n x^{n-1} - \sum_{n=0}^{\infty} n(n-1)a_n x^n + 2\sum_{n=0}^{\infty} n a_n x^{n-1} - 2\sum_{n=0}^{\infty} n a_n x^n + 2\sum_{n=0}^{\infty} a_n x^n = 0$$

or

$$\sum_{n=0}^{\infty}[(n+2)(n+1)a_{n+1}-(n+2)(n-1)a_n]x^n=0$$

Setting the coefficients of x^n to zero gives $a_{n+1}=\dfrac{n-1}{n+1}a_n$ for $n\ge0$. Solving successively for $n=0,\ 1,\ 2,\ldots$ gives $a_1=-a_0;\ a_2=0;\ a_3=0,\ \cdots$ or $y_1(x)=a_0(1-x)$.

12.5   Bessel's Equation

1. The zero-order Bessel's equation is $x^2y''(x)+xy'(x)+x^2y(x)=0$. Substitute $y(x)=\sum_{n=0}^{\infty}a_nx^n$ into this equation to obtain

$$\sum_{n=0}^{\infty}n(n-1)a_nx^n+\sum_{n=0}^{\infty}na_nx^n+\sum_{n=0}^{\infty}a_nx^{n+2}=0$$

or $\displaystyle\sum_{n=0}^{\infty}n^2a_nx^n+\sum_{n=0}^{\infty}a_nx^{n+2}=0$ or $a_1x+\displaystyle\sum_{n=0}^{\infty}[(n+2)^2a_{n+2}+a_n]x^{n+2}=0$

Set each coefficient of x^n equal to zero to find $a_1=0$ and $a_{n+2}=-\dfrac{a_n}{(n+2)^2}$ for $n\ge0$. Setting $n=0,\ 1,\ 2,\ \ldots$ yields

$$a_2=-\frac{a_0}{2^2},\qquad a_4=-\frac{a_2}{4^2}=\frac{a_0}{2^2\cdot4^2}=\frac{a)}{2^4(2!)^2},\qquad a_6=-\frac{a_4}{6^2}=-\frac{a_0}{2^2\cdot4^2\cdot6^2}=-\frac{a_0}{2^6(3!)^2},$$

and so on. The general term is

$$a_{2n}=\frac{(-1)^na_0}{2^{2n}(n!)^2}\qquad n\ge1$$

3. Start with $y_2(x)=J_0(x)\ln x+\displaystyle\sum_{n=0}^{\infty}b_nx^{n+1}$ where

$$J_0(x)=1-\frac{x^2}{4}+\frac{x^4}{64}-\frac{x^6}{36\cdot64}+\cdots$$

Now

$$y_2'(x)=J_0'(x)\ln x+\left(\frac{1}{x}-\frac{x}{4}+\frac{x^3}{64}+\cdots\right)+\sum_{n=0}^{\infty}(n+1)b_nx^n$$

and

$$y_2''(x)=J_0''(x)\ln x+\left(-\frac{1}{2}+\frac{x^2}{16}-\frac{x^4}{384}+\cdots\right)+\left(-\frac{1}{x^2}-\frac{1}{4}+\frac{3x^2}{64}+\cdots\right)+\sum_{n=0}^{\infty}n(n+1)b_nx^{n-1}$$

Substitute these results into Equation 1 to obtain

$$[x^2 J_0''(x) + x J_0'(x) + x^2 J_0(x)]\ln x + \left[-\frac{x^2}{2} + \frac{x^4}{16} + O(x^6)\right] + \left[-1 - \frac{x^2}{4} + \frac{3x^4}{64} + O(x^6)\right]$$

$$+ \sum_{n=0}^{\infty} n(n+1)b_n x^{n+1} + \left[1 - \frac{x^2}{4} + \frac{x^4}{64} + O(x^6)\right] + \sum_{n=0}^{\infty}(n+1)b_n x^{n+1} + \sum_{n=0}^{\infty} b_n x^{n+3} = 0$$

The factor multiplying $\ln x$ is equal to zero because $J_0(x)$ satisfies Equation 1 with $\nu = 0$. Setting the coefficients of x^n equal to zero gives

$$b_0 = 0, \qquad 4b_1 - 1 = 0, \qquad 9b_2 + b_0 = 0, \qquad \frac{1}{8} + 16b_3 + b_1 = 0, \qquad \text{and so on}$$

These results gives us $b_0 = 0$, $b_1 = 1/4$, $b_2 = 0$, $b_3 = -3/128$, which yields Equation 8.

6. Start with Equation 19,

$$J_n(x) = \sum_{j=0}^{\infty} \frac{(-1)^j}{\Gamma(j+1)\Gamma(j+1+n)} \left(\frac{x}{2}\right)^{2j+n}$$

Now replace n by $-n$ to obtain

$$J_{-n}(x) = \sum_{j=0}^{\infty} \frac{(-1)^j}{\Gamma(j+1)\Gamma(j+1-n)} \left(\frac{x}{2}\right)^{2j-n} = \sum_{j=n}^{\infty} \frac{(-1)^j}{\Gamma(j+1)\Gamma(j+1-n)} \left(\frac{x}{2}\right)^{2j-n}$$

where we have used the fact that $1/\Gamma(z) = 0$ when z is 0 or a negative integer. Now change the summation index from j to $k = j - n$ to write

$$J_{-n}(x) = \sum_{k=0}^{\infty} \frac{(-1)^{k+n}}{\Gamma(k+n+1)\Gamma(k+1)} \left(\frac{x}{2}\right)^{2k+n} = (-1)^n J_n(x)$$

9. Start with $y(x) = J_\nu(ix) = I_\nu(x)$. Let $z = ix$ and differentiate two times to obtain

$$\frac{dI_\nu}{dx} = \frac{dJ_\nu(ix)}{dx} = \frac{dJ_\nu(z)}{dz}\frac{dz}{dx} = i\frac{dJ_\nu(z)}{dz}$$

and

$$\frac{d^2 I_\nu}{dx^2} = \frac{d^2 J_\nu(ix)}{dx^2} = i^2\frac{d^2 J_\nu(z)}{dz^2} = -\frac{d^2 J_\nu(z)}{dz^2}$$

Now substitute these results into Equation 35 to obtain

$$-x^2\frac{d^2 J_\nu(z)}{dz^2} + ix\frac{dJ_\nu(z)}{dz} - (x^2 + \nu^2)J_\nu = 0$$

Let $z = ix$ to write

$$z^2\frac{d^2 J_\nu(z)}{dz^2} + z\frac{dJ_\nu(z)}{dz} - (\nu^2 - z^2)J_\nu = 0$$

which is Equation 1.

12. Start with Equation 35, $x^2 y''(x) + xy'(x) - (x^2 + \nu^2)y(x) = 0$. Let $x = iz$. Then

$$\frac{dy}{dx} = \frac{dy}{dz}\frac{dz}{dx} = -iy'(z) \qquad \text{and} \qquad \frac{d^2y}{dx^2} = -\frac{d^2y}{dz^2} = -y''(z)$$

Substituting these into the above equation gives

$$-z^2[-y''(z)] + iz[-iy'(z)] - (-z^2 + \nu^2)y(z) = 0$$

or $z^2 y''(z) + zy'(z) + (z^2 - \nu^2)y(z) = 0$.

15. Comparing the equation $x^2 y''(x) + 5xy'(x) + x^2 y(x) = 0$ to Equation 42, we see that $1 - 2\alpha = 5$, $\beta^2\gamma^2 = 1$, $\gamma = 1$, and $\alpha^2 - \nu^2\gamma^2 = 0$, or $\alpha = -2$, $\beta = 1$, $\gamma = 1$, and $\nu = 2$. So the general solution is

$$y(x) = c_1 x^{-2} J_2(x) + c_2 x^{-2} Y_2(x)$$

18. From Equations 30 and 31,

$$J_\nu(x) = \sum_{j=0}^{\infty} \frac{(-1)^j}{\Gamma(j+1)\Gamma(j+1+\nu)} \left(\frac{x}{2}\right)^{2j+\nu} \qquad \nu > 0$$

and

$$J_{-\nu}(x) = \sum_{j=0}^{\infty} \frac{(-1)^j}{\Gamma(j+1)\Gamma(j+1-\nu)} \left(\frac{x}{2}\right)^{2j-\nu} \qquad \nu > 0$$

The first term of each expansion gives

$$J_\nu(x) = \frac{1}{\Gamma(1+\nu)} \left(\frac{x}{2}\right)^\nu \qquad \text{and} \qquad J_{-\nu}(x) = \frac{1}{\Gamma(1-\nu)} \left(\frac{x}{2}\right)^{-\nu}$$

Using these terms, we have

$$J_\nu'(x) = \frac{\nu}{2\Gamma(1+\nu)} \left(\frac{x}{2}\right)^{\nu-1} \qquad \text{and} \qquad J_{-\nu}'(x) = -\frac{\nu}{2\Gamma(1-\nu)} \left(\frac{x}{2}\right)^{-\nu-1}$$

Therefore,

$$
\begin{aligned}
J_\nu(x)J_{-\nu}'(x) - J_\nu'(x)J_{-\nu}(x) &= \frac{1}{2^\nu\Gamma(1+\nu)x} \cdot \frac{(-\nu)2^\nu}{\Gamma(1-\nu)} - \frac{\nu}{2^\nu\Gamma(1+\nu)x} \cdot \frac{2^\nu}{\Gamma(1-\nu)} \\
&= -\frac{2\nu}{[\Gamma(1+\nu)\Gamma(1-\nu)]\,x} = -\frac{2\nu}{\nu\Gamma(\nu)\Gamma(1-\nu)\,x} = -\frac{2\sin\pi\nu}{\pi x}
\end{aligned}
$$

21. The derivative of $u(x) = u_1(x)u_2(x)\ldots u_n(x)$ is

$$u'(x) = u_1'(x)u_2(x)\cdots u_n(x) + u_1(x)u_2'(x)\ldots u_n(x) + \cdots + u_1(x)u_2(x)\ldots u_n'(x)$$

Now multiply and divide the right side by $u(x) = u_1(x)u_2(x)\ldots u_n(x)$ to get

$$u'(x) = u(x)\left[\frac{u_1'(x)}{u_1(x)} + \frac{u_2'(x)}{u_2(x)} + \cdots + \frac{u_n'(x)}{u_n(x)}\right]$$

12.6   Bessel Functions

1. Multiply through by x^2 and let $z = \alpha x$ to get

$$x^2 \frac{d^2y}{dx^2} + x\frac{dy}{dx} + \alpha^2 x^2 y(x) = 0$$

$$z^2 \frac{d^2y}{dx^2} + z\frac{dy}{dz} + z^2 y(z) = 0$$

3. Integrate by parts, letting "u" $= \ln x$ and "dv" $= x\, J_0(x)\, dx$.

$$\int x \ln x \, J_0(x)\, dx = x \ln x \, J_1(x) - \int J_1(x)\, dx = x \ln x \, J_1(x) + J_0(x)$$

6.

$$\frac{d}{dx}[\, x^{-\nu} J_\nu(x)\,] = \frac{d}{dx} \sum_{n=0}^{\infty} \frac{(-1)^n x^{2n}}{2^{2n+\nu} n! \, \Gamma(n+1+\nu)} = \sum_{n=0}^{\infty} \frac{(-1)^n x^{2n-1}}{2^{2n+\nu-1}(n-1)!\,\Gamma(n+1+\nu)}$$

$$= \sum_{n=0}^{\infty} \frac{(-1)^{n+1} x^{2n+1}}{2^{2n+1+\nu} n! \, \Gamma(n+2+\nu)}$$

$$= -x^{-\nu} \sum_{n=0}^{\infty} \frac{(-1)^n}{n! \, \Gamma(n+2+\nu)} \left(\frac{x}{2}\right)^{2n+\nu+1} = -x^{-\nu} J_{\nu+1}(x)$$

9. First note that

$$\int_0^{2\pi} \sin^{2n+1}\theta \, d\theta = \int_0^{2\pi} \cos^{2n+1}\theta \, d\theta = 0 \qquad n = 0,\ 1,\ 2,\ \ldots$$

and that

$$\int_0^{2\pi} \sin^{2n}\theta \, d\theta = \int_0^{2\pi} \cos^{2n}\theta \, d\theta = 4\int_0^{\pi/2} \sin^{2n}\theta \, d\theta = 4\int_0^{\pi/2} \cos^{2n}\theta \, d\theta$$

$$= 4\frac{1\cdot 3\cdot 5 \ldots (2n-1)}{2\cdot 4\cdot 6 \ldots (2n)} \frac{\pi}{2} = \frac{(2n-1)!\,\pi}{2^{2n-2}(n-1)!\, n!} \qquad n = 1,\ 2,\ \ldots$$

Therefore,

$$J_0(x) = \frac{1}{2\pi} \int_0^{2\pi} e^{ix\sin\theta}\, d\theta = \frac{1}{2\pi} \sum_{n=0}^{\infty} \frac{(ix)^n}{n!} \int_0^{2\pi} \sin^n\theta \, d\theta$$

$$= \frac{2}{\pi} \sum_{n=0}^{\infty} \frac{(ix)^{2n}}{(2n)!} \int_0^{\pi/2} \sin^{2n}\theta \, d\theta$$

$$= \sum_{n=0}^{\infty} \frac{(-1)^n x^{2n}}{(2n)!} \cdot \frac{(2n-1)!}{2^{2n-1}(n-1)!\, n!}$$

$$= \sum_{n=0}^{\infty} \frac{(-1)^n}{(n!)^2} \left(\frac{x}{2}\right)^{2n}$$

The proof that $J_0(x) = \dfrac{1}{2\pi} \displaystyle\int_0^{2\pi} e^{ix\cos\theta}\, d\theta$ is the same.

12. Simply use the fact that $\displaystyle\int_0^{\pi} \sin n\theta \sin m\theta\, d\theta = \delta_{nm}\pi$.

15. For $a > 0$,

$$\int_0^{\infty} e^{-ax} e^{ibx}\, dx = \int_0^{\infty} e^{(ib-a)x}\, dx = \frac{1}{a - ib} = \frac{a + ib}{a^2 + b^2}$$

The real part is $a/(a^2 + b^2)$.

18. Integrate $J_0''(x) + \dfrac{1}{x} J_0'(x) + J_0(x) = 0$ once to obtain

$$\left[J_0'(x) \right]_0^{\infty} + \int_0^{\infty} \frac{J_0'(x)}{x}\, dx + \int_0^{\infty} J_0(x)\, dx = 0$$

Now use $J_0'(x) = -J_1(x)$ to write

$$\int_0^{\infty} \frac{J_1(x)}{x}\, dx = \int_0^{\infty} J_0(x)\, dx = 1$$

21.

$$
\begin{aligned}
\int_0^{\infty} e^{-ax^2} J_n(bx) x^{n+1}\, dx &= \sum_{m=0}^{\infty} \frac{(-1)^m b^{2m+n}}{2^{2m+n} m!(m+n)!} \int_0^{\infty} e^{-ax^2} x^{2m+2n+1}\, dx \\
&= \sum_{m=0}^{\infty} \frac{(-1)^m b^{2m+n}}{2^{2m+n} m!(n+m)!} \cdot \frac{(n+m)!}{2a^{n+m+1}} \\
&= \frac{b^n}{(2a)^{n+1}} \sum_{m=0}^{\infty} \frac{(-1)^m}{m!} \left(\frac{b^2}{4a} \right)^m = \frac{b^n}{(2a)^{n+1}} e^{-b^2/4a}
\end{aligned}
$$

24. Start with the two equations

$$x^2 u''(x) + x u'(x) + (\alpha_i^2 x^2 - n^2) u(x) = 0$$

$$x^2 v''(x) + x v'(x) + (\alpha_j^2 x^2 - n^2) v(x) = 0$$

where $u(x) = J_n(\alpha_i x)$ and $v(x) = J_n(\alpha_j x)$. Multiply the first equation by $v(x)$ and the second by $u(x)$ and subtract to get

$$x[u''(x)v(x) - u(x)v''(x)] + [u'(x)v(x) - u(x)v'(x)] + (\alpha_i^2 - \alpha_j^2) x u(x) v(x) = 0$$

which we can write as

$$\frac{d}{dx} \{ x[u'(x)v(x) - u(x)v'(x)] \} + (\alpha_i^2 - \alpha_j^2) x u(x) v(x) = 0$$

Now integrate from 0 to 1 to obtain

$$\left[x\{u'(x)v(x) - u(x)v'(x)\} \right]_0^1 = \int_0^1 (\alpha_i^2 - \alpha_j^2)xu(x)v(x)\,dx$$

or

$$\int_0^1 (\alpha_i^2 - \alpha_j^2)xJ_n(\alpha_i x)J_n(\alpha_j x) = x\left[\alpha_j J_n'(\alpha_i)J_n(\alpha_j) - \alpha_i J_n(\alpha_i)J_n'(\alpha_j) \right]$$

But α_i and α_j are distinct zeros of $J_n(x)$, and so we have

$$\int_0^1 xJ_n(\alpha_i x)J_n(\alpha_j x)\,dx = 0$$

27. Start with

$$J_0(x) = \frac{1}{2\pi}\int_0^{2\pi} e^{ix\sin\theta}\,d\theta = \frac{1}{2\pi}\int_0^{2\pi} e^{ix\cos\theta}\,d\theta$$

and let $x = iz$ to write

$$I_0(z) = J_0(iz) = \frac{1}{2\pi}\int_0^{2\pi} e^{-z\sin\theta}\,d\theta = \frac{1}{2\pi}\int_0^{2\pi} e^{-z\cos\theta}\,d\theta$$

Now let $\psi = \theta - \pi$ to write $\sin\theta = \sin(\psi + \pi) = -\sin\psi$ and

$$I_0(x) = \frac{1}{2\pi}\int_{-\pi}^{\pi} e^{x\sin\psi}\,d\psi = \frac{1}{2\pi}\int_{-\pi}^{\pi} e^{x\sin\theta}\,d\theta$$

And now let $\cos\theta = \cos(\psi + \pi) = -\cos\psi$ and

$$I_0(x) = \frac{1}{2\pi}\int_{-\pi}^{\pi} e^{x\cos\psi}\,d\psi = \frac{1}{2\pi}\int_{-\pi}^{\pi} e^{x\cos\theta}\,d\theta$$

CHAPTER 13

Qualitative Methods for Nonlinear Differential Equations

13.1 The Phase Plane

1. Use the identity $a\sin(\omega t + \phi) = A\cos\phi\sin\omega t + A\sin\phi\cos\omega t$ with $x_0 = A\sin\phi$ and $v_0/\omega = A\cos\phi$ and

$$A^2\sin^2\phi + A^2\cos^2\phi = A^2 = x_0^2 + \frac{v_0^2}{\omega^2}$$

3. The time that it takes to go from point A to point B along a trajectory is given by $\tau_{AB} = \int_{t(A)}^{t(B)} dt = \int_A^B \frac{dx}{\dot{x}}$, where we have used $\dot{x} = dx/dt$, or $dt = dx/\dot{x}$. We use Equation 3 to write $\dot{x} = \left(\frac{2E}{m} - \omega^2 x^2\right)^{1/2}$. The amplitude is the value of x when $\dot{x} = 0$, or $A = (2E/m\omega^2)^{1/2} = (2E/k)^{1/2}$. The period of a simple harmonic oscillator is

$$\tau = 4\int_0^{(2E/k)^{1/2}} \frac{dx}{\left(\dfrac{2E}{m} - \omega^2 x^2\right)^{1/2}} = \frac{4}{\omega}\int_0^{(2E/k)^{1/2}} \frac{dx}{\left(\dfrac{2E}{k} - x^2\right)^{1/2}}$$

$$= \frac{4}{\omega}\left[\sin^{-1}\frac{x}{(2E/k)^{1/2}}\right]_0^{(2E/k)^{1/2}} = \frac{4\pi}{2\omega} = \frac{2\pi}{\omega}$$

6. (a) Let $y = \dot{x}$ and write $\dot{x} = y$ and $\dot{y} = -\gamma y + \beta x^3 - kx^2$.

(b) Let $y = \dot{x}$ and write $\dot{x} = y$ and $\dot{y} = \epsilon y\left(1 - \frac{1}{3}y^2\right) - x$.

9. Substitute $y(1) = 2$ into $y = mx$.

12. The eigenvalues and corresponding eigenvectors are $\lambda_\pm = -1 \pm 2i$ and $\mathbf{v}_\pm = (1 \mp i, 1)^\mathsf{T}$. Therefore, the general solution is $\mathbf{x}(t) = c_1\mathbf{v}_+ e^{-t}e^{2it} + c_2\mathbf{v}_- e^{-t}e^{-2it}$ or

$$\begin{pmatrix} x(t) \\ y(t) \end{pmatrix} = c_1\begin{pmatrix} 1-i \\ 1 \end{pmatrix}e^{-t}e^{2it} + c_2\begin{pmatrix} 1+i \\ 1 \end{pmatrix}e^{-t}e^{-2it}$$

or

$$
\begin{aligned}
x(t) &= e^{-t}\left[\,c_1\cos 2t + ic_1\sin 2t - ic_1\cos 2t + c_1\sin 2t + c_2\cos 2t - ic_2\sin 2t\right.\\
&\qquad\left. + ic_2\cos 2t + c_2\sin 2t\,\right]\\
&= e^{-t}\left[\,(c_1 + c_2)\cos 2t - i(c_2 - c_1)\sin 2t + i(c_2 - c_1)\cos 2t + (c_1 + c_2)\sin 2t\,\right]\\
&= e^{-t}\left[\,\alpha\cos 2t - \beta\sin 2t + \beta\cos 2t + \alpha\sin 2t\,\right]
\end{aligned}
$$

and

$$
\begin{aligned}
y(t) &= e^{-t}\left[\,c_1\cos 2t + ic_1\sin 2t + c_2\cos 2t - i\sin 2t\,\right]\\
&= e^{-t}\left[\,(c_1 + c_2)\cos 2t + i(c_1 - c_2)\sin 2t\,\right]\\
&= e^{-t}\left[\,\alpha\cos 2t - \beta\sin 2t\,\right]
\end{aligned}
$$

where we have set $\alpha = c_1 + c_2$ and $\beta = i(c_2 - c_1)$. We write these equations in matrix notation as

$$
\mathbf{x}(t) = \begin{pmatrix} \alpha + \beta \\ \alpha \end{pmatrix} e^{-t}\cos 2t + \begin{pmatrix} \alpha - \beta \\ -\beta \end{pmatrix} e^{-t}\sin 2t
$$

We could also take real and imaginary parts of

$$
\begin{aligned}
\begin{pmatrix} 1 - i \\ 1 \end{pmatrix} e^{-t}e^{2it} &= \begin{pmatrix} 1 - i \\ 1 \end{pmatrix} e^{-t}(\cos 2t + i\sin 2t)\\
&= e^{-t}\begin{pmatrix} \cos 2t + i\sin 2t - i\cos 2t + \sin 2t \\ \cos 2t + i\sin 2t \end{pmatrix}\\
&= e^{-t}\begin{pmatrix} \cos 2t + \sin 2t \\ \cos 2t \end{pmatrix} + ie^{-t}\begin{pmatrix} \sin 2t - \cos 2t \\ \sin 2t \end{pmatrix}
\end{aligned}
$$

So

$$
\begin{aligned}
\mathbf{x}(t) &= c_1 e^{-t}\begin{pmatrix} \cos 2t + \sin 2t \\ \cos 2t \end{pmatrix} + c_2 e^{-t}\begin{pmatrix} \sin 2t - \cos 2t \\ \sin 2t \end{pmatrix}\\
&= e^{-t}\left[\begin{pmatrix} c_1 - c_2 \\ c_1 \end{pmatrix}\cos 2t + \begin{pmatrix} c_1 + c_2 \\ c_2 \end{pmatrix}\sin 2t\right]
\end{aligned}
$$

This is the same result as $\mathbf{x}(t)$ above if we let $c_1 = \alpha$ and $c_2 = -\beta$.

15. In this case, $\dfrac{\dot{y}}{\dot{x}} = \dfrac{dy}{dx} = -\dfrac{\omega^2 x}{y}$ or $\omega^2 x\, dx + y\, dy = 0$. Integration gives

$$
\frac{1}{2}\omega^2 x^2 + \frac{1}{2}y^2 = \text{constant}
$$

18. Expand $\sin\theta$ about the points $(0, \pm n\pi)$ where n is even:

$$
\sin\theta = \sin n\pi + \cos n\pi(\theta - n\pi) + \cdots = 0 + (\theta - n\pi) + \cdots \qquad (n \text{ even})
$$

Let $\theta' = \theta - n\pi$ to get Equation 16 (with $\theta' \to \theta$). For odd values of n

$$
\sin\theta = 0 - (\theta - n\pi) + \cdots \qquad (n \text{ odd})
$$

Let $\theta' = \theta - n\pi$ to get Equation 17 (with $\theta' \to \theta$).

21. Differentiate Equation 18 with respect to θ to obtain

$$\frac{d\Omega}{d\omega} = -\frac{\omega^2 \sin\theta}{\Omega}$$

For $0 \le \theta \le \pi$, $0 \le \sin\theta$ and so $d\Omega/d\omega < 0$ for $\Omega > 0$. For $\pi \le \theta \le 2\pi$, $\sin\theta \le 0$, and so $d\Omega/d\omega > 0$ for $\Omega > 0$. Opposite signs occur for $\Omega < 0$.

13.2   Critical Points in the Phase Plane

1. Because $\tau = t - t_0$, $\dfrac{dx}{dt} = \dfrac{dx}{d\tau}$ and $\dfrac{dy}{dt} = \dfrac{dy}{d\tau}$. The initial condition is $\tau = 0$ rather than $t = 0_0$.

3. If the trajectories cross, then there would exist two possible trajectories with the same initial conditions (the point at which the trajectories cross).

6. We first must do Problems 4 and 5.

(a) The critical point occurs at $x = y = 1$. Therefore, we substitute $x = u+1$ and $y = v+1$ into the two differential equations to obtain

$$\dot{u} = 1 - (1+u)(1+v) = -u - v - uv \qquad \text{and} \qquad \dot{v} = (1+u)(1+v) - (v+1) = u + uv$$

(b) The critical points occur at $x = \pm 1$ and $y = 1$. Therefore, we substitute $x = u+1$ and $y = v+1$ into the two differential equations to obtain

$$\dot{u} = 1 - (1+v) = -v \qquad \text{and} \qquad \dot{v} = 1 + 2u + u^2 - 1 - 2v - v^2 = 2u - 2v + u^2 - v^2$$

At the other critical point, we substitute $x = u - 1$ and $y = v + 1$ into the two differential equations to obtain

$$\dot{u} = -v \qquad \text{and} \qquad \dot{v} = u^2 - 2u + 1 - 1 - 2v - v^2 = -2u - 2v + u^2 + v^2$$

(c) There are four critical points: $(0,0)$, $(0,4)$, $(1/3,0)$, and $(1,2)$.
At $(0,0)$, we substitute $x = u$ and $y = v$ into the two differential equations to obtain

$$\dot{u} = u - 3u^2 + uv \qquad \text{and} \qquad \dot{v} = 4v - v^2 - 2uv$$

At $(0,4)$, we substitute $x = u$ and $y = v + 4$ into the two differential equations to obtain

$$\begin{aligned} \dot{u} &= u - 3u^2 + u(v+4) = 5u - 3u^2 + uv \\ \dot{v} &= 4(v+4) - (v+4)^2 - 2u(v+4) = -8u - 4v - v^2 - 2uv \end{aligned}$$

At $(1/3,0)$, we substitute $x = u + 1/3$ and $y = v$ into the two differential equations to obtain

$$\begin{aligned} \dot{u} &= u + \frac{1}{3} - 3\left(u + \frac{1}{3}\right)^2 + \left(u + \frac{1}{3}\right)v = -u + \frac{1}{3}v - 3u^2 + uv \\ \dot{v} &= 4v - v^2 - 2\left(u + \frac{1}{3}\right)v = \frac{10}{3}v - v^2 - 2uv \end{aligned}$$

At $(1,2)$, we substitute $x = u+1$ and $y = v+2$ into the two differential equations to obtain

$$
\begin{aligned}
\dot{u} &= (1+u) - 3(1+u)^2 + (1+u)(2+v) = -3u + v + uv - 3u^2 \\
\dot{v} &= 4(2+v) - (2+v)^2 - 2(1+u)(2+v) = -4u - 2v - 2uv - v^2
\end{aligned}
$$

(d)   There are two critical points. One occurs at $x = -1$ and $y = -1$. Therefore, we substitute $x = u - 1$ and $y = v - 1$ into the two differential equations to obtain

$$
\dot{u} = (u - 1) - (v - 1) = u - v \qquad \dot{v} = u^2 - 2u + 1 - 1 = -2u + u^2
$$

The other occurs at $x = 1$ and $y = 1$, Therefore, we substitute $x = u + 1$ and $y = v + 1$ into the two differential equations to obtain

$$
\dot{u} = (u + 1) - (v + 1) = u - v \qquad \dot{v} = 1 + 2u + u^2 - 1 = 2u + u^2
$$

Finally, then, the coefficient matrix in each case is

(a) $\begin{pmatrix} -1 & -1 \\ 1 & 0 \end{pmatrix}$ (b) $\begin{pmatrix} 0 & -1 \\ 2 & -2 \end{pmatrix}$ and $\begin{pmatrix} 0 & -1 \\ -2 & -2 \end{pmatrix}$

(c) $\begin{pmatrix} 1 & 0 \\ 0 & 4 \end{pmatrix}, \begin{pmatrix} 5 & 0 \\ -8 & -4 \end{pmatrix}, \begin{pmatrix} -1 & 1/3 \\ 0 & 10/3 \end{pmatrix}$, and $\begin{pmatrix} -3 & 1 \\ -4 & -2 \end{pmatrix}$

(d) $\begin{pmatrix} 1 & -1 \\ -2 & 0 \end{pmatrix}$ and $\begin{pmatrix} 1 & -1 \\ 2 & 0 \end{pmatrix}$

9. Let $v = mu$ and use the condition $v(-1 + \sqrt{2}) = 1$ to obtain $m = \dfrac{1}{\sqrt{2} - 1}$, or $v = \dfrac{u}{\sqrt{2} - 1}$.

12. Substituting Equations 11 (with $\lambda = 1$) into 13 gives

$$
c_2 e^t + (c_1 + c_2 t)e^t = (c_1 + c_2 t)e^t \qquad \text{and} \qquad c_4 e^t + (c_3 + c_4)e^t = (c_1 + c_2 t)e^t + (c_3 + c_4 t)e^t
$$

or $c_2 = 0$ and $c_4 = c_1 + c_2$. The solutions to these equations gives $c_2 = 0$, $c_1 = c_4$, and c_3 arbitrary, which gives Equation 14.

15.   Start with $u = Ae^{-t}\cos 2t - Be^{-t}\sin 2t = e^{-t}(A\cos 2t - B\sin 2t) = Ce^{-t}\cos(2t + \alpha)$ where $A = C\cos\alpha$ and $B = C\sin\alpha$.
   Similarly, $v = e^{-t}(B\cos 2t + A\sin 2t) = Ce^{-t}\sin(2t + \alpha)$. Now write

$$
x = r\cos\theta = Ce^{-t}\cos(2t + \alpha) \qquad \text{and} \qquad y = r\sin\theta = Ce^{-t}\sin(2t + \alpha)
$$

Thus, $r^2 = C^2 e^{-2t}$, or $r = Ce^{-t}$ and $\theta = 2t + \alpha$.

18. We'll use the real and imaginary parts of $\lambda = i$ and $\mathbf{u} = (2 + i, 1)^\mathsf{T}$, which gives

$$
\begin{aligned}
\begin{pmatrix} 2 + i \\ 1 \end{pmatrix} e^{it} &= \begin{pmatrix} 2 + i \\ 1 \end{pmatrix}(\cos t + i\sin t) \\
&= \begin{pmatrix} 2\cos t + i\cos t + 2i\sin t - \sin t \\ \cos t + i\sin t \end{pmatrix} \\
&= \begin{pmatrix} 2\cos t - \sin t \\ \cos t \end{pmatrix} + i\begin{pmatrix} \cos t + 2\sin t \\ \sin t \end{pmatrix}
\end{aligned}
$$

Thus, we can write

$$\mathbf{v}(t) = c_1 \begin{pmatrix} 2\cos t - \sin t \\ \cos t \end{pmatrix} + c_2 \begin{pmatrix} \cos t + 2\sin t \\ \sin t \end{pmatrix}$$

$$= \begin{pmatrix} 2c_1 + c_2 \\ c_1 \end{pmatrix} \cos t + \begin{pmatrix} 2c_2 - c_1 \\ c_2 \end{pmatrix} \sin t$$

This result is the same as that in Example 5 if we let $u_0 = 2c_1 + c_2$ and $v_0 = c_1$, in which case we get $c_1 = v_0$ and $c_2 = u_0 - 2v_0$.

13.3   Stability of Critical Points

1. The eigenvalues, $\lambda_1 = -1$ and $\lambda_2 = -1$, are equal and negative with $x(t) = x_0 e^{-t}$ and $y(t) = y_0 e^{-t}$. The critical point is an asymptotically stable proper node.

3. The eigenvalues, $\lambda = -1 \pm \sqrt{5}$, are real, unequal, and of opposite sign. The critical point is a saddle point (unstable.)

6. The eigenvalues, $\lambda = -2 \pm i$, are complex conjugates with the real part less than zero, and so the critical point is an asymptotically stable spiral point.

9. The critical point is given by $x + y - 1 = 0$ and $x - 3y - 5 = 0$, or $x = 2$, $y = -1$. The linearized equations are $\dot{u} = u + v$ and $\dot{v} = u - 3v$. The eigenvalues of this system, $-1 \pm \sqrt{5}$, are real, unequal, and of opposite sign. The critical point is a saddle point (unstable.)

12. The eigenvalues are $\lambda = -1 \pm ia^{1/2}$. If $a = 0$, then the critical point is an asymptotically stable node; if $a > 0$, then the critical point is an asymptotically stable spiral point; if $-1 < a < 0$, then the critical point is an asymptotically stable node; and if $a < -1$, then the critical point is a saddle point.

15. There are two critical points: $(2, 1)$ and $(-2, -1)$.

(a) The equations linearized about $(2, 1)$ are $\dot{u} = u + 2v$ and $\dot{v} = u - 2v$ with eigenvalues $\lambda_\pm = \dfrac{-1 \pm \sqrt{17}}{2}$. Therefore, the critical point is a saddle point (unstable.)

(b) The equations linearized about $(-2, -1)$ are $\dot{u} = -u - 2v$ and $\dot{v} = u - 2v$ with eigenvalues $\lambda_\pm = \dfrac{-3 \pm i\sqrt{7}}{2}$. Therefore, the critical point is an asymptotically stable spiral point.

18. Differentiate $x = r\cos\theta$ and $y = r\sin\theta$ to get

$$\dot{x} = \dot{r}\cos\theta - r\sin\theta\,\dot{\theta} \qquad \text{and} \qquad \dot{(y)} = \dot{r}\sin\theta + r\cos\theta\,\dot{\theta}$$

Substitute these into Equation 4 to get

$$\dot{r}\cos\theta - r\sin\theta\,\dot{\theta} = r\sin\theta + r\cos\theta\,(1 - r^2) \qquad \dot{r}\sin\theta + r\cos\theta\,\dot{\theta} = -r\cos\theta + r\sin\theta\,(1 - r^2)$$

Multiply the first equation by $\cos\theta$ and the second by $\sin\theta$, and add to get $\dot{r} = r(1 - r^2)$. Multiply the first equation by $\sin\theta$ and the second by $\sin\theta$, and subtract to get $r\dot{\theta} = -r$, or $\dot{\theta} = 1$.

21. There are three critical points: $(0,0)$, $(-0.851\ 171, 0.616\ 666)$, and $(0.851\ 171, -0.616\ 666)$.

(a) $(0,0)$ The eigenvalues are $\dfrac{1 \pm \sqrt{5}}{2}$ and so the critical point is a saddle point.

(b) $(-0.851\ 171, 0.616\ 666)$ We let $u = x + 0.851\ 171$ and $v = y - 0.616\ 666$, which gives the linearized equations $\dot{u} = 2.17348u + v$ and $\dot{v} = u + 2.14083v$. The eigenvalues (3.15724 and 1.15702) are real, unequal, and both positive, so the critical point is an unstable node.

(c) $(0.851\ 171, -0.616\ 666)$ We let $u = x - 0.851\ 171$ and $v = y + 0.616\ 666$, which gives the linearized equations $\dot{u} = 2.17348u + v$ and $\dot{v} = u + 2.14083v$. The eigenvalues, 3.15729 and 1.15702, are real, unequal, and both positive, so the critical point is an unstable node.

24. Start with

$$\oint_C (-Q\,dx + P\,dy) = \int_S \left(\frac{\partial P}{\partial x} + \frac{\partial Q}{dy} \right) dx dy$$

Let $\dot{x} = P$ and $\dot{y} = Q$ to write

$$\oint_C (-Q\,dx + P\,dy) = \oint_C (-\dot{y}\,dx + \dot{x}\,dy) = \oint_C (-\dot{y}\dot{x} + \dot{x}\dot{y})\,dt = 0$$

You can see that it is necessary that $\partial P/\partial x + \partial Q/\partial y$ change sign in S. Thus, if $\partial P/\partial x + \partial Q/\partial y$ does not change sign in S, there can be no closed trajectory.

13.4 Nonlinear Oscillators

1. The force is $-x + x^3$, and so the potential is $\dfrac{x^2}{2} - \dfrac{x^4}{4}$. The accompanying figure shows that $V(x)$ goes through a maximum at $x = \pm 1$.

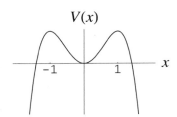

3. The slope corresponding to $(1, \pm\sqrt{2})^{\mathsf{T}}$ is $m = \pm\sqrt{2}$. Use the formula $y - y_0 = m(x - x_0)$ with $x_0 = 1$, $y_0 = 0$ at the point $(1, 0)$ and $x_0 = -1$, $y_0 = 0$ at the point $(-1, 0)$.

6. $x = 4$, $y = 0$ is a critical point.

9. First we must solve Problem 7. Start with $\dot{x} = y$ and $\dot{y} = -x + \dfrac{x^2}{4}$. There are critical points at $(0,0)$ and $(4,0)$. The eigenvalues about the critical point $(0,0)$ are $\pm i$, so according to Poincaré's theorem, it is *either* a center or a spiral point. Physically, however, we know that the equation represents an oscillator for small values of x, so we expect the origin to be a center. The eigenvalues about $(4,0)$ are ± 1, so the critical point is a saddle point.

The trajectories are obtained by multiplying $\ddot{x} + x - x^2/4 = 0$ by $\dot{x}$ and integrating to obtain

$$\frac{y^2}{2} + \frac{x^2}{2} - \frac{x^3}{12} = \text{constant} = C$$

These curves are plotted in the accompanying figure for various values of C.

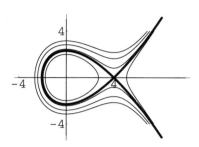

The critical points are at $(0,0)$ and $(4,0)$. The separatrix is the trajectory that passes through the point $(4,0)$, which gives us that $C = 8/3$. (See the figure above.) Therefore, the time that it takes for the system to reach the point $(4,0)$ is given by

$$\tau = \int_0^4 \frac{dx}{\dot{x}} = \int_0^4 \frac{dy}{y} = \int_0^4 \frac{dx}{\left(\dfrac{16}{3} - x^2 + \dfrac{x^3}{6}\right)^{1/2}}$$

Numerical integration shows that this integral diverges because of the upper limit of $x = 4$.

12. The force is given by $f(x) = -9x + x^3$, and so the potential energy is given by $V(x) = \dfrac{9x^2}{2} - \dfrac{x^4}{4}$. The accompanying figure shows that $V(x)$ has maxima at $x = \pm 3$.

$$V(x)$$

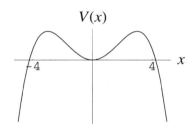

15. The following plot of $V(x)$ against x shows that the critical points are at $x = 0$ (saddle) and $x = \pm 1$ (center).

$$V(x)$$

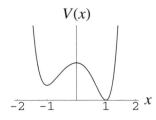

The phase portraits can be obtained by multiplying

$$\ddot{x} = -\frac{dV}{dx} = \frac{3}{5}x + \frac{3}{8}x^2 - \frac{3}{10}x^4 - \frac{3}{5}x^5$$

by $\dot{x}$ and integrating to obtain

$$y^2 + \frac{1}{2} - \frac{3x^2}{5} - \frac{x^3}{4} + \frac{3x^5}{20} + \frac{x^6}{5} = \text{constant}$$

where $y = \dot{x}$. The accompanying figure shows a family of trajectories.

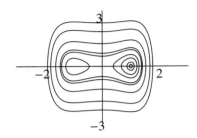

18. (a) Equation 12 gives $F(x) = \displaystyle\int_0^x \epsilon(u^2 - 1)\, du = \epsilon\left(\frac{x^3}{3} - x\right)$ and $G(x) = \displaystyle\int_0^x u^3\, du = \frac{x^4}{4}$.
Now

1. $G(x) \to \infty$ as $x \to \infty$

2. $F(x) < 0$ when $x < \sqrt{3}$, $F(x) > 0$ when $x > \sqrt{3}$, and $F(x) \to \infty$ monotonically as $x \to \infty$ when $x > \sqrt{3}$. Therefore, there is a limit cycle.

(b) Equation 12 gives $F(x) = \displaystyle\int_0^x \epsilon(2u^2 - 1)\, du = \epsilon\left(\frac{2}{3}x^2 - x\right)$ and $G(x) = \displaystyle\int_0^x (u + u^3)\, du = \frac{x^2}{2} +$
Now

1. $G(x) \to \infty$ as $x \to \infty$

2. $F(x) < 0$ when $x < 1$, $F(x) > 0$ when $x > 1$, and $F(x) \to \infty$ monotonically as $x \to \infty$. Therefore, there is a limit cycle.

13.5   Population Dynamics

1. Write Equation 2 as $\dfrac{dP}{P(a - bP)} = dt$ and integrate to get $-\dfrac{1}{a}\ln\dfrac{a - bP}{P} = t + \text{constant}$,

or $\dfrac{a - bP}{P} = Ae^{-at}$. Using the fact that $P = P_0$ when $t = 0$, we get $A = \dfrac{a - bP_0}{P_0}$, which
gives

$$P(t) = \frac{aP_0}{bP_0 + (a - bP_0)e^{-at}}$$

3. Substitute $x = u + c/d$ and $y = v + a/b$ into Equation 6 to get

$$\begin{aligned}
\dot{u} &= a\left(u + \frac{c}{d}\right) - b\left(u + \frac{c}{d}\right)\left(v + \frac{a}{b}\right) \\
&= au + \frac{ac}{d} - buv - \frac{bc}{d}v - au - \frac{ac}{d} = -\frac{bc}{d}v + \cdots
\end{aligned}$$

and

$$\begin{aligned}
\dot{v} &= -c\left(v + \frac{a}{b}\right) + d\left(u + \frac{c}{d}\right)\left(v + \frac{a}{b}\right) \\
&= -cv - \frac{ac}{b} + duv + cv + \frac{ad}{b}u + \frac{ac}{b} = \frac{ad}{b}u + \cdots
\end{aligned}$$

These two equations are equivalent to Equation 7.

6. Divide $\dot{x}$ by $\dot{y}$ to get

$$\frac{\dot{x}}{\dot{y}} = \frac{dx}{dy} = \frac{-2x + 0.0050xy}{4y - 0.015xy} = \frac{x(2 - 0.0050y)}{y(0.015x - 4)}$$

Integration gives (see Equation 8 with $a = 2$, $b = 0.0050$, $c = 4$, and $d = 0.0150$)

$$2\ln y - 0.0050y + 4\ln x - 0.015x = \text{constant}$$

The phase portrait

$$2\ln y - 0.0050y + 4\ln x - 0.015x = 2\ln(400) - 0.0050(400) + 4\ln(60) - 0.015(60) = 25.46$$

is shown in the accompanying figure.

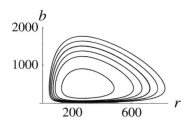

9. Both species go almost extinct.

12. Equations 10 linearized about the critical point $(3/2, 0)$ are $\dot{u} = -3u - 3v/2$ and $\dot{v} = 2v$. The corresponding matrix is $A = \begin{pmatrix} -3 & -3/2 \\ 0 & 2 \end{pmatrix}$ and its eigenvalues are -3 and 2, and the corresponding eigenvectors are $(1, 0)^\mathsf{T}$ and $(-3/10, 1)^\mathsf{T}$. The general solution to the equations is

$$\begin{pmatrix} u \\ v \end{pmatrix} = c_1 \begin{pmatrix} 1 \\ 0 \end{pmatrix} e^{-3t} + c_2 \begin{pmatrix} -3/10 \\ 1 \end{pmatrix} e^{2t}$$

Therefore, $y = 0$ $(t \to -\infty)$ and $y = (-10/3)(x - 3/2)$ $(t \to \infty)$ are the directions of the asymptotes.

15. There are three critical points in the first quadrant: $(0, 0)$, $(0, 4)$, and $(3, 0)$. The origin (with $\lambda = 12$ and 16) is an unstable node and the critical point $(3, 0)$ (with $\lambda = 4$ and -12) is a saddle point. The critical point $(0, 4)$ (with $\lambda = -12$ and -16) is a stable node. Thus, the y species survives and the x species becomes extinct.

18. The concentration of species X increases according to $k_1 AX$ and decreases according to $k_1 XY$, and so $\dot{X} = k_1 XA - k_2 XY$. Similarly, the concentration of species Y increases according to $k_2\mathrm{XY}$ and decreases according to $k_3\mathrm{Y}$, and so $\dot{\mathrm{Y}} = k_2\mathrm{XY} - k_3\mathrm{Y}$. Dividing $\dot{\mathrm{X}}$ by $\dot{\mathrm{Y}}$ gives

$$\frac{\dot{\mathrm{X}}}{\dot{\mathrm{Y}}} = \frac{d\mathrm{X}}{d\mathrm{Y}} = \frac{\mathrm{X}(k_1\mathrm{A}_0 - k_2\mathrm{Y})}{\mathrm{Y}(k_2\mathrm{X} - k_3)}$$

or using Equation 8 with $a = k_1\mathrm{A}_0 = 1$, $b = k_2 = 0.500$, $c = k_3 = 0.100$, and $d = k_2 = 0.500$ gives

$$-0.500(\mathrm{X} + \mathrm{Y}) + 0.100 \ln \mathrm{X} + \ln \mathrm{Y} = \mathrm{constant} = -1.00$$

This is a closed curve in the phase plane (see the accompanying figure), so $\mathrm{X}(t)$ and $\mathrm{Y}(t)$ are periodic functions.

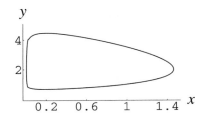

CHAPTER 14

Orthogonal Polynomials and Sturm-Liouville Problems

14.1 Legendre Polynomials

1.

$$P_0(x) = 1$$

$$P_1(x) = \frac{1}{2}\left(\frac{2!x}{0!1!1!}\right) = x$$

$$P_2(x) = \frac{1}{4}\left(\frac{4!}{0!2!2!}x^2 - \frac{2!}{1!1!0!}\right) = \frac{3}{2}x^2 - \frac{1}{2} = \frac{1}{2}(3x^2 - 1)$$

$$P_3(x) = \frac{1}{8}\left(\frac{6!}{0!3!3!}x^3 - \frac{4!}{1!2!1!}x\right) = \frac{1}{8}(20x^3 - 12x) = \frac{1}{2}(5x^3 - 3x)$$

3. The change $x \to -x$ introduces a factor $(-1)^{n-2j}$ in the summation. Because $(-1)^{2j} = 1$ for any integer value of j, the factor is actually $(-1)^n$.

6. Multiply Equation 9 by t^n and sum from 0 to ∞ to obtain

$$\sum_{n=0}^{\infty}(n + 1)t^n P_{n+1}(x) - x\sum_{n=0}^{\infty}(2n + 1)t^n P_n(x) + \sum_{n=0}^{\infty} nt^n P_{n-1}(x) = 0$$

Let's look at each term separately, using $G(x, t) = 1 + tP_1(x) + t^2 P_2(x) + \cdots$

$$\sum_{n=0}^{\infty}(n + 1)t^n P_{n+1}(x) = P_1 + 2tP_2 + 3t^2 P_3 + \cdots = G_t(x, t)$$

$$\sum_{n=0}^{\infty}(2n + 1)t^n P_n(x) = P_0(x) + 3tP_1(x) + 5t^2 P_2(x) + \cdots = G(x, t) + 2tG_t(x, t)$$

and

$$\sum_{n=0}^{\infty} nt^n P_{n-1}(x) = tP_0 + 2t^2 P_1 + 3t^3 P_2 + \cdots = t(P_0 + 2tP_1 + 3t^2 P_2 + \cdots)$$

$$= t\left[G(x, t) + tG_t(x, t)\right]$$

Putting all this together gives

$$G_t - 2xtG_t - xG + t^2G_t + tG = 0$$

Collect like terms and integrate to get

$$\int \frac{x-t}{1-2xt+t^2}dt = \frac{dG}{G}$$

or $\ln G(x,t) = -\dfrac{1}{2}\ln(1-2xt+t^2) + c$ or

$$G(x,t) = \frac{\alpha}{(1-2xt+t^2)^{1/2}}$$

Using the fact that $G(x,0) = P_0(x) = 1$, we find that $\alpha = 1$.

9. $G(1,t) = (1-t)^{-1} = \displaystyle\sum_{n=0}^{\infty} P_n(1)t^n = \sum_{n=0}^{\infty} t^n \qquad |t| < 1$

12. Starting with $G(x,t) = 1/(1-2xt+t^2)^{1/2}$ from Problem 6, we have

$$G_x = \sum_{n=0}^{\infty} t^n P_n'(x) = \frac{t}{(1-2xt+t^2)^{3/2}}$$

and

$$G_t = \sum_{n=0}^{\infty} nt^{n-1} P_n(x) = \frac{x-t}{(1-2xt+t^2)^{3/2}}$$

Multiply G_x by $(1-t^2)/t$ and G_t by $2t$ and subtract to obtain

$$\frac{1-t^2}{t} \cdot \frac{t}{(1-2xt+t^2)^{3/2}} - \frac{2t(x-t)}{(1-2xt+t^2)^{3/2}} = \frac{1-2xt+t^2}{(1-2xt+t^2)^{3/2}}$$

$$= \frac{1}{(1-2xt+t^2)^{1/2}} = \sum_{n=0}^{\infty} t^n P_n(x)$$

We also have

$$\frac{1-t^2}{t} \sum_{n=0}^{\infty} t^n P_n'(x) - 2\sum_{n=0}^{\infty} nt^n P_n(x) = \sum_{n=0}^{\infty} t^n P_n(x)$$

or

$$\sum_{n=0}^{\infty} t^{n-1} P_n'(x) - \sum_{n=0}^{\infty} t^{n+1} P_n'(x) - \sum_{n=0}^{\infty} 2nt^n P_n(x) = \sum_{n=0}^{\infty} t^n P_n(x)$$

which is the equation given in the statement of the problem. Now, equate like powers of t to obtain $P_{n+1}'(x) - P_{n-1}'(x) = (2n+1)P_n(x)$.

15. Use Equation 9 to write

$$xP_m(x) = \frac{m+1}{2m+1}P_{m+1}(x) + \frac{m}{2m+1}P_{m-1}(x)$$

Now, substitute this result into $\int_{-1}^{1} x P_n(x) P_m(x)\, dx$ to obtain

$$\int_{-1}^{1} x P_m(x) P_n(x)\, dx = \frac{m+1}{2m+1}\int_{-1}^{1} P_n(x)P_{m+1}(x)\, dx + \frac{m}{2m+1}\int_{-1}^{1} P_n(x)P_{m-1}(x)\, dx$$

$$= \frac{2n}{(2n+1)(n+1)}\delta_{n,m+1} + \frac{2(n+1)}{(2n+1)(2n+3)}\delta_{n,m-1}$$

18. The sequence of partial sums in Equation 25 is monotonically increasing *and* bounded by the left side. Thus, the series converges, and so $h_n a_n^2$ must go to zero as n increases.

21. Using $u = \dfrac{2x-\alpha-\beta}{\beta-\alpha}$, we see that $u = 1$ when $x = \beta$ and $u = -1$ when $x = \alpha$.

14.2   Orthogonal Polynomials

1. Start with $\phi_0(x) = c_0$. We require that

$$I = \int_{-\infty}^{\infty} \phi_0^2(x)e^{-x^2}\, dx = c_0^2 \int_{-\infty}^{\infty} e^{-x^2}\, dx = \pi^{1/2}c_0^2$$

which gives $c_0 = \pi^{-1/4}$. Now let $\phi_1(x) = c_1 x + c_2$ and require that it be orthogonal to $\phi_0(x)$, or that

$$\int_{-\infty}^{\infty} \phi_0(x)\phi_1(x)\, dx = 0$$

which gives $c_2 = 0$. The normalization condition gives

$$I = \int_{-\infty}^{\infty} \phi_1^2(x)e^{-x^2}\, dx = c_1^2 \int_{-\infty}^{\infty} x^2 e^{-x^2}\, dx = c_1^2 \frac{\pi^{1/2}}{2}$$

which gives $c_1 = 2^{1/2}/\pi^{1/4}$ and $\phi_1(x) = \dfrac{2^{1/2}}{\pi^{1/4}}x$. Now let $\phi_2(x) = c_3 x^2 + c_4 x + c_5$. The two orthogonality conditions give

$$\int_{-\infty}^{\infty} \phi_2(x)\phi_0(x)\, dx = 0 = \frac{1}{\pi^{1/4}}\left(c_3\frac{\pi^{1/2}}{2} + c_5\pi^{1/2}\right)$$

and

$$\int_{-\infty}^{\infty} \phi_2(x)\phi_1(x)\, dx = 0 = \frac{2^{1/2}}{\pi^{1/4}}\left(c_4\frac{\pi^{1/2}}{2}\right)$$

from which we find that $c_4 = 0$ and $c_3 = -2c_5$, or that $\phi_2(x) = c_3(2x^2-1)$. Now we require that

$$\int_{-\infty}^{\infty} e^{-x^2}\phi_2^2(x)\, dx = c_3^2\int_{-\infty}^{\infty}(2x^2-1)^2 e^{-x^2}\, dx = c_3^2 2\pi^{1/2} = 1$$

or $c_3 = 1/2^{1/2}\pi^{1/4}$ so that $\phi_2(x) = \dfrac{1}{2^{1/2}\pi^{1/4}}(2x^2-1)$

3.

$$
\begin{aligned}
H_4(x) &= 2xH_3(x) - 6H_2(x) = 2x(8x^3 - 12x) - 6(4x^2 - 2) \\
&= 16x^4 - 48x^2 + 12
\end{aligned}
$$

6. The generating function has the form

$$
G(x,t) = H_0' + H_1 t + \frac{1}{2!}H_3 t^2 + \frac{1}{3!}H_4 t^3 + \cdots
$$

Multiply each term of the recursion formula by $t^n/n!$ and sum to get

$$
\sum_{n=0}^{\infty} H_{n+1}(x)\frac{t^n}{n!} = H_1 + H_2 t + \frac{1}{2!}H_3 t^2 + \frac{1}{3!}H_4 t^3 + \cdots = G_t(x,t)
$$

$$
\sum_{n=0}^{\infty} H_n(x)\frac{t^n}{n!} = G(x,t)
$$

$$
\sum_{n=0}^{\infty} nH_{n-1}(x)\frac{t^n}{n!} = H_0 t + H_1 t^2 + H_2\frac{t^3}{2!} + H_3\frac{t^4}{3!} + \cdots = tG(x,t)
$$

Putting everything together, we get

$$
G_t(x,t) - 2xG(x,t) + 2tG(x,t) = 0
$$

or $\ln G(x,t) = 2xt - t^2 + \text{constant}$ or $G(x,t) = Ae^{2x(t-t^2)}$. Using $G(x,0) = 1$ gives $A = 1$.

9. Multiply a generating function expressed in terms of t by one expressed in terms of u:

$$
\frac{e^{-xt/(1-t)}e^{-xu/(1-u)}}{(1-t)(1-u)} = \sum_{n=0}^{\infty}\sum_{m=0}^{\infty} L_n(x)L_m(x)t^n u^m
$$

Now multiply both sides by e^{-x} and integrate from 0 to ∞ to get

$$
\frac{1}{(1-t)(1-u)}\int_0^{\infty} e^{-x}e^{-xt/(1-t)}e^{-xu/(1-u)}dx
$$

$$
= \frac{1}{(1-t)(1-u)}\int_0^{\infty} e^{-x\alpha}dx \quad \left(\alpha = \frac{1+tu}{(1-t)(1-u)}\right)
$$

$$
= \frac{1}{(1-t)(1-u)}\frac{(1-t)(1-u)}{1+tu} = \frac{1}{1+tu}\sum_{n=0}^{\infty}(tu)^n
$$

The fact that this result is a function of only tu shows that $\int_0^{\infty} e^{-x}L_n(x)L_m(x)\,dx = \delta_{nm}$.

12. Differentiate D_N^2 with respect to α_N to get

$$
\frac{dD_N^2}{d\alpha_N} = \frac{d}{d\alpha_N}\left[\int_a^b r(x)f^2(x)\,dx - 2\alpha_N\int_a^b r(x)f(x)\phi_n(x)\,dx + \alpha_N^2\int_a^b r(x)\phi^2(x)\,dx\right]
$$

$$
= -2\int_a^b r(x)f(x)\phi_N(x)\,dx + 2\alpha_N\int_a^b r(x)\phi_N^2(x)\,dx = 0
$$

or $\alpha_N = \dfrac{\displaystyle\int_a^b r(x)f(x)\phi_N(x)\,dx}{\displaystyle\int_a^b r(x)\phi_N^2(x)\,dx}$

15. Write $\sin 2\pi x = \displaystyle\sum_{n=0}^{\infty} a_n T_m(x)$ for $-1 \le x \le 1$, multiply both sides by $(1-x^2)^{-1/2}T_m(x)$, and integrate to obtain

$$\int_{-1}^{1}(1-x^2)^{-1/2}T_m(x)\sin 2\pi x\,dx = \sum_{n=0}^{\infty} a_n\delta_{nm}h_n = \begin{cases} a_m\pi/2 & m \text{ odd} \\ 0 & m \text{ even} \end{cases}$$

The accompanying figure shows $\sin 2\pi x$ as $f(x) = \displaystyle\sum_{n=0}^{m} a_{2n+1}T_{2n+1}(x)$ for $m = 2, 3, 4,$ and 5.

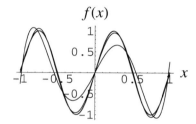

18. Write $e^{-x^2} = \displaystyle\sum_{n=0}^{\infty} a_n e^{-x} L_n(x)$ for $x \ge 0$. Multiply both sides by $L_m(x)$ and integrate to obtain

$$\int_0^{\infty} e^{-x^2} L_m(x)\,dx = \sum_{n=0}^{\infty} a_n \int_0^{\infty} e^{-x} L_n(x)L_m(x)\,dx$$

$$= \sum_{n=0}^{\infty} a_n h_n \delta_{nm} = a_m$$

The figure on the next page compares e^{-x} and $f(x) = \displaystyle\sum_{n=0}^{\infty} a_n e^{-x} L_n(x)$ for $N = 10$ and 20. The curves are essentially indistinguishable.

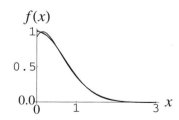

To verify Bessel's inequality in this case, minimize

$$D^2 = \left\{ \int_0^{\infty} e^x \left[e^{-x^2} - \sum_{n=0}^{\infty} \alpha_n e^{-x} L_n(x) \right] dx \right\}^2$$

with respect to the α_n to obtain

$$\int_0^\infty e^{-x^2+x}\,dx = 0.982\ 009 \overset{?}{\leq} \sum_{n=0}^\infty a_n^2 h_n = \sum_{n=0}^N a_n^2 = S_N$$

The following figure shows S_N plotted against N.

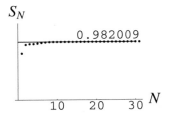

14.3   Sturm-Liouville Theory

1. We divide $a(x)y''(x) + b(x)y'(x) + c(x)y(x) + \lambda d(x)y(x) = 0$ by $a(x)$ and multiply by $p(x) = \exp\left(\int \dfrac{b(x)}{a(x)}\,dx\right)$ to obtain

$$p(x)y''(x) + \frac{p(x)b(x)}{a(x)}y'(x) + \frac{p(x)c(x)}{a(x)}y(x) + \lambda\frac{p(x)d(x)}{a(x)}y(x) = 0 \tag{1}$$

Now note that

$$[\,p(x)y'(x)\,]' = p(x)y''(x) + p'(x)y'(x) = p(x)y''(x) + \frac{b(x)}{a(x)}p(x)y'(x)$$

Therefore, Equation 1 becomes

$$[\,p(x)y'(x)\,]' + \frac{p(x)c(x)}{a(x)}y(x) + \lambda\frac{p(x)d(x)}{a(x)}y(x) = 0$$

or

$$[\,p(x)y'(x)\,]' + q(x)y(x) + \lambda r(x)y(x0 = 0$$

where $q(x) = p(x)c(x)/a(x)$ and $r(x) = p(x)d(x)/a(x)$.

3. The right side of Equation 7 is

$$p(b)y_m^{*\,'}(b)y_n(b) - p(b)y_m^*(b)y_n'(b) - p(a)y_m^{*\,'}(a)y_n(a) + p(a)y_m^*(a)y_n'(a)$$

Now use Equations 3 to eliminate the factors which are not primes:

$$p(b)\left[-\frac{\beta_2}{\beta_1}y_m^{*\,'}(b)y_n'(b) + y_m^{*\,'}(b)\frac{\beta_2}{\beta_1}y_n'(b)\right] - p(a)\left[-\frac{\alpha_2}{\alpha_1}y_m^{*\,'}(a)y_n'(a) + y_m^{*\,'}(a)\frac{\alpha_2}{\alpha_1}y_n'(a)\right] = 0$$

6.

$$\int_0^1 \sin \beta_n x \sin \beta_m x \, dx = \left[\frac{\sin(\beta_n - \beta_m)x}{2(\beta_n - \beta_m)} - \frac{\sin(\beta_n + \beta_m)x}{2(\beta_n + \beta_m)} \right]_0^1$$

$$= \left[\frac{\sin \beta_n x \cos \beta_m x - \cos \beta_n \sin \beta_m x}{2(\beta_n - \beta_m)} \right]_0^1$$

$$- \left[\frac{\sin \beta_n x \cos \beta_m x + \cos \beta_n \sin \beta_m x}{2(\beta_n + \beta_m)} \right]_0^1$$

$$= \left[\frac{\beta_m \sin \beta_n x \cos \beta_m x - \beta_n \cos \beta_n x \sin \beta_m x}{\beta_n^2 - \beta_m^2} \right]_0^1$$

$$= \frac{\beta_m \sin \beta_n \cos \beta_m - \beta_n \cos \beta_n \sin \beta_m}{\beta_n^2 - \beta_m^2}$$

Now, $\tan \beta_n = -\beta_n/5$ is equivalent to $\beta_n \cos \beta_n = -5 \sin \beta_n$, and so we have

$$\int_0^1 \sin \beta_n x \sin \beta_m x \, dx = \frac{-5 \sin \beta_n \sin \beta_m + 5 \sin \beta_n \sin \beta_m}{\beta_n^2 - \beta_m^2} = 0$$

9. The general solution is $y(x) = c_1 \cos \lambda x + c_1 \sin \lambda x$, which gives $y'(x) = -c_1 \lambda \sin \lambda x + c_2 \lambda \cos \lambda x$. Now $y'(0) = 0$ gives $c_2 = 0$ and $y'(1) = 0$ gives $c_1 \lambda \sin \lambda = 0$, or $\lambda = n\pi$ with $n = 1, \, 2, \, \ldots$. Therefore, $y(x) = \cos n\pi x$ with $n = 1, \, 2, \, \ldots$.

12. Chebyshev's equation is $(1 - x^2)y''(x) - xy'(x) + n^2 y(x) = 0$ and so

$$p(x) = \exp \left\{ - \int \frac{x \, dx}{1 - x^2} \right\} = \exp \left\{ \frac{1}{2} \ln(1 - x^2) \right\} = (1 - x^2)^{1/2}$$

Therefore,

$$[(1 - x^2)^{1/2} y'(x)]' + n^2 (1 - x^2)^{-1/2} y(x) = 0$$

is Chebyshev's equation in Sturm-Liouville form.

15. (a) $p(x) = x$, $q(x) = 0$, and $r(x) = x$. Both $p(x) \geq 0$ and $r(x) \geq 0$ on $[0, \infty)$, and so the equation represents a singular Sturm-Liouville problem.

(b) $p(x) = 1 - x^2$, $q(x) = 0$, and $r(x) = 1$. The factor $p(x) = 1 - x^2 = 0$ when $x = \pm 1$, and so the equation represents a singular Sturm-Liouville problem.

(c) We must first write the equation in the form of Equation 2, using $p(x) = \exp \left\{ \int -\frac{x}{1 - x^2} \, dx \right\} = \frac{1}{2} e^{\ln(1 - x^2)} = (1 - x^2)^{1/2}$, $q(x) = 0$, and $r(x) = (1 - x^2)^{-1/2}$. Because $p(x) = 0$ at $x = \pm 1$, the equation represents a singular Sturm-Liouville problem.

18. Start with

$$y(x) = c_1 \cos \lambda^{1/2} x + c_2 \sin \lambda^{1/2} x \qquad \text{and} \qquad y'(x) = -c_1 \lambda^{1/2} \sin \lambda^{1/2} x + c_2 \lambda^{1/2} \cos \lambda^{1/2} x$$

Consider the case in which $\lambda \neq 0$ first. Then

$$y(0) + y'(0) = 0 = c_1 + c_2 \lambda^{1/2} \qquad \text{and} \qquad y(1) = 0 = c_1 \cos \lambda^{1/2} + c_2 \sin \lambda^{1/2} = 0$$

These two equations lead to $c_2(\lambda_n^{1/2}\cos\lambda_n^{1/2} - \sin\lambda_n^{1/2}) = 0$, or to $\tan\lambda_n^{1/2} = \lambda_n^{1/2}$. Thus, we have

$$y_n(x) = c_2(\sin\lambda_n^{1/2}x - \lambda_n^{1/2}\cos\lambda_n^{1/2}x) \quad \text{or} \quad y_n(x) = \lambda_n^{1/2}\cos\lambda_n^{1/2}x - \sin\lambda_n^{1/2}x$$

Now consider the case in which $\lambda = 0$. Then integration of the differential equation gives $y_0(x) = \alpha + \beta x$. The boundary condition says that

$$y(0) + y'(0) = 0 = \alpha + \beta = 0 \quad \text{and} \quad y(1) = 0 = \alpha + \beta = 0$$

Therefore, $\alpha = -\beta$ and $y_0(x) = x - 1$.

14.4   Eigenfunction Expansions

1. The normalized eigenfunctions are $\phi_n(x) = 2^{1/2}\sin n\pi x$ for $n = 1, 2, \ldots$. We have $f(x) = \sum_{n=1}^{\infty} a_n \phi_n(x)$.

$$a_n = 2^{1/2}\int_0^1 x^2(1-x)\sin n\pi x \, dx = -\frac{2^{3/2}}{n^3\pi^3}[1 + (-1)^n 2]$$

$$f(x) = -\frac{4}{\pi^3}\sum_{n=1}^{\infty}\frac{1 + (-1)^n 2}{n^3}\sin n\pi x$$

The accompanying figure shows $f(x)$ plotted against x using three and four terms along with $x^2(1-x)$.

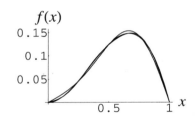

3. The normalized eigenfunctions are $\phi_n(x) = 2^{1/2}\sin\dfrac{(2n-1)\pi x}{2}$ for $n = 1, 2, \ldots$.

$$f(x) = \sum_{n=1}^{\infty} a_n \phi_n(x) = 2^{1/2}\sum_{n=1}^{\infty} a_n \sin\frac{(2n-1)\pi x}{2}$$

$$\begin{aligned}
a_n &= 2^{1/2}\int_0^1 x(1-x)^2\sin\frac{(2n-1)\pi x}{2}\, dx = 2^{1/2}\left\{\frac{32}{\pi^3}\left[\frac{1}{(2n-1)^3} - \frac{3/\pi}{(2n-1)^4}(-1)^{n+1}\right]\right\}\\
&= 2^{1/2}\frac{32}{\pi^3}\frac{1}{(2n-1)^3}\left[1 + \frac{(-1)^n(3/\pi)}{2n-1}\right]
\end{aligned}$$

$$f(x) = \frac{64}{\pi^3}\sum_{n=1}^{\infty}\frac{1}{(2n-1)^3}\left[1 + \frac{(-1)^n(3/\pi)}{2n-1}\right]\sin\frac{(2n-1)\pi x}{2}$$

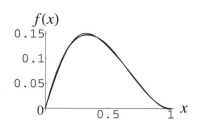

The accompanying figure shows $f(x)$ plotted against x using four and six terms along with $x(1-x)^2$.

6. $f(x) = \displaystyle\sum_{n=0}^{\infty} a_n P_n(x)$ and $\displaystyle\int_{-1}^{1} f(x) P_n(x)\, dx = a_n \cdot \frac{2}{2n+1}$.

$$a_0 = \frac{1}{2} \int_{-1}^{1} \sin \pi x\, dx = 0$$

$$a_1 = \frac{3}{2} \int_{-1}^{1} x \sin \pi x\, dx = 3 \left[\frac{\sin \pi x}{\pi^2} - \frac{x \cos \pi x}{\pi} \right]_0^1 = \frac{3}{\pi}$$

$$a_2 = a_4 = a_6 = \cdots = a_{2n} = 0$$

$$a_3 = \frac{7}{2} \int_{-1}^{1} \sin \pi x\, P_3(x)\, dx$$

$$= \frac{7}{2} \left[\frac{5(3\pi^2 x^2 - 6)}{\pi^4} \sin \pi x - \frac{5(\pi^2 x^3 - 6x)}{\pi^3} \cos \pi x - 3\frac{\sin \pi x}{\pi^2} + \frac{3x}{\pi} \cos \pi x \right]_0^1$$

$$= \frac{7}{2} \left(\frac{2}{\pi} - \frac{30}{\pi^3} \right) = \frac{7}{\pi^3}(\pi^2 - 15)$$

$$f(x) = \frac{3x}{\pi} + \frac{7}{\pi^3}(\pi^2 - 15)\frac{5x^3 - 3x}{2} + \cdots + \frac{11(945 - 105\pi^2 + \pi^4)}{\pi^5} P_5(x)$$

The accompanying figure shows $f(x)$ plotted against x along with $\sin \pi x$.

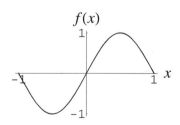

9. Using $a_n = -\dfrac{2^{3/2}}{n^3 \pi^3}\left[1 - (-1)^n 2\right]$ from Problem 1 and $h_n = 1$, we have

$$\sum_{n=1}^{\infty} h_n a_n^2 = \sum_{n=1}^{\infty} \frac{8}{\pi^6} \frac{5 + (-1)^n 4}{n^6} = \frac{40}{\pi^6} \sum_{n=1}^{\infty} \frac{1}{n^6} - \frac{32}{\pi^6} \sum_{n=1}^{\infty} \frac{(-1)^{n+1}}{n^6}$$

$$= \frac{40}{945} - 32 \cdot \frac{31}{32 \cdot 945} = \frac{9}{945} = \frac{1}{105}$$

To evaluate the sums, refer to Section 3.7 or Chapter 23 of Abramowitz and Stegun. Now,

$$\int_0^1 f^2(x)\, dx = \int_0^1 x^4(1-x)^2\, dx = \frac{1}{105}$$

12. The normalized eigenfunctions are $y_n(x) = 2^{1/2} \sin n\pi x$ and the a_n are given by

$$a_n = \int_0^{1/2} (-1)2^{1/2} \sin n\pi x \, dx + \int_{1/2}^1 2^{1/2} \sin n\pi x \, dx = \frac{2^{5/2} \cos(n\pi/2) \sin^2(n\pi/4)}{n\pi}$$

and $f(x) = 2^{1/2} \sum_{n=1}^{\infty} a_n \sin n\pi x \, dx$. The following figures show $f(x)$ calculated to 100 terms and 500 terms and they also show that $f(1/2) = 0$.

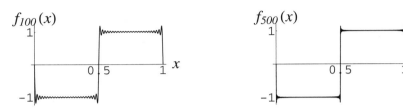

15. The left side of Equation 7 is finite, and so the right side is bounded, in which case, $a_n \to 0$ as $n \to \infty$.

14.5   Green's Functions

1. Use Equation 5 with $\mathcal{L} = -\dfrac{d^2}{dx^2}$, $\mu = 1$, $r(x) = 1$, and $g(x) = -f(x)$. Then Equation 4 gives $\phi_n''(x) + \lambda_n \phi_n(x) = 0$. The boundary conditions, $\phi_n(0) = 0$ and $\phi_n'(1) = 0$, yield

$$\phi_n(x) = 2^{1/2} \sin \frac{(2n-1)\pi x}{2} \qquad \text{and} \qquad \lambda_n = \frac{(2n-1)^2 \pi^2}{4} \qquad n = 1,\, 2\, \dots$$

Therefore, Equation 13 gives

$$G(x, z) = 2\sum_{n=1}^{\infty} \frac{\sin \dfrac{(2n-1)\pi x}{2} \sin \dfrac{(2n-1)\pi z}{2}}{\dfrac{(2n-1)^2 \pi^2}{4} - 1} = 8\sum_{n=1}^{\infty} \frac{\sin \dfrac{(2n-1)\pi x}{2} \sin \dfrac{(2n-1)\pi z}{2}}{(2n-1)^2 \pi^2 - 4}$$

3. Use Equation 5 with $\mathcal{L} = -d^2/dx^2$, $\mu = 1$, $r(x) = 1$, and $g(x) = -f(x)$. Then Equation 4 gives $\phi_n''(x) + \lambda_n \phi_n(x) = 0$. The boundary conditions, $\phi_n'(0) = 0$ and $\phi_n'(1) = 0$, give $\phi_n(x) = 2^{1/2} \cos n\pi x$ and $\lambda_n = n^2 \pi^2$ with $n = 1,\, 2,\, \dots$. Therefore, Equation 13 gives

$$G(x, z) = 2\sum_{n=1}^{\infty} \frac{\cos n\pi x \cos n\pi z}{n^2 \pi^2 - 1}$$

6. The general solution is $G(x, z) = c_1 \cos x + c_2 \sin x$. The boundary conditions, $G'(0, z) = 0$ gives $c_1 = 0$ and $G'(1, z) = 0$ gives $c_3 \cos 1 - c_4 \sin 1 = 0$ or $c_3 = c_4 \tan 1$. Therefore,

$$G(x, z) = \begin{cases} c_2 \cos x & x < z \\ c_4 \left[\dfrac{\cos(x-1)}{\cos 1} \right] & x > z \end{cases}$$

The continuity condition and the jump condition give

$$c_2 \cos z = \frac{c_4}{\cos 1} \cos(z-1) \qquad \text{and} \qquad -\frac{c_4}{\cos 1} \sin(z-1) + c_2 \sin z = -1$$

Therefore, we have

$$\frac{c_4}{\cos 1} [\sin z \cos(z-1) - \cos z \sin(z-1)] = -\cos z$$

or $c_4 \dfrac{\sin 1}{\cos 1} = -\cos z$. The continuity condition gives $c_2 = -\dfrac{\cos(z-1)}{\sin 1}$, and so we finally have

$$G(x, z) = \begin{cases} -\dfrac{\cos(z-1)\cos x}{\sin 1} & 0 \le x < z \\[3mm] -\dfrac{\cos(x-1)\cos z}{\sin 1} & z < x \le 1 \end{cases}$$

9. Start with $y''(x) + y(x) = x$ and assume that the particular solution is given by $y_p(x) = \alpha x + \beta$. Substitute this into the differential equation to get

$$y_p''(x) + y_p(x) = \alpha x + \beta = x$$

from which we see that $\alpha = 1$ and $\beta = 0$. Therefore, $y_p(x) = x$ and

$$y(x) = c_1 \sin x + c_2 \cos x + x$$

The boundary conditions give $y(0) = 0 = c_2 = 0$ and $y(1) = c_1 \sin 1 + 1 = 0$, or $c_1 = -\dfrac{1}{\sin 1}$. Therefore,

$$y(x) = x - \frac{\sin x}{\sin 1}$$

which is the same result as in Example 1.

12. We assume that the particular solution is given by

$$y_p(x) = u_1(x) \sin x + u_2(x) \cos x$$

The method of variation of parameters leads to the equations

$$u_1(x) = \int x \cos x \, dx = \cos x + x \sin x$$

$$u_2(x) = -\int x \sin x \, dx = -\sin x + x \cos x$$

Substituting these results into $y_p(x)$ above gives $y_p(x) = x$ and so

$$y(x) = c_1 \sin x + c_2 \cos x + x$$

The boundary conditions give $c_2 = 0$ and $c_1 \cos 1 + 1 = 0$. Therefore,

$$y(x) = x - \frac{\sin x}{\cos 1}$$

This is the same result as in Problem 11.

15. Start with $y(x) = \sum_{n=1}^{\infty} a_n \sin \dfrac{(2n-1)\pi x}{2}$ with

$$
\begin{aligned}
a_n &= 2 \int_0^1 y(x) \sin \frac{(2n-1)\pi x}{2} \, dx \\
&= 2 \int_0^1 x \sin \frac{(2n-1)\pi x}{2} \, dx - \frac{2}{\cos 1} \int_0^1 \sin x \sin \frac{(2n-1)\pi x}{2} \, dx \\
&= 2 \left[\frac{\sin \dfrac{(2n-1)\pi x}{2}}{(2n-1)^2 \pi^2/4} - \frac{x \cos \dfrac{(2n-1)\pi x}{2}}{(2n-1)\pi/2} \right]_0^1 \\
&\quad - \frac{1}{\cos 1} \left[\frac{\sin \left[x - \dfrac{(2n-1)\pi x}{2} \right]}{x - \dfrac{(2n-1)\pi x}{2}} - \frac{\sin \left[x + \dfrac{(2n-1)\pi x}{2} \right]}{x + \dfrac{(2n-1)\pi x}{2}} \right]_0^1 \\
&= 8 \left[\frac{(-1)^{n+1}}{(2n-1)^2 \pi^2} \right] - \frac{1}{\cos 1} \left[\frac{-2 \cos 1 (-1)^{n+1}}{1 - (2n-1)^2 \pi^2/4} \right] \\
&= \frac{8}{\pi^2} \frac{(-1)^{n+1}}{(2n-1)^2} - \frac{8(-1)^{n+1}}{(2n-1)^2 \pi^2 - 4} = \frac{32(-1)^n}{\pi^2 (2n-1)^2 \left[(2n-1)^2 \pi^2 - 4 \right]}
\end{aligned}
$$

Therefore,

$$
y(x) = \frac{32}{\pi^2} \sum_{n=1}^{\infty} \frac{(-1)^n \sin \dfrac{(2n-1)\pi x}{2}}{(2n-1)^2 \left[(2n-1)^2 \pi^2 - 4 \right]}
$$

which is the same result as Problem 2. The figure on the following page shows $y(x)$ plotted against x for three terms along with $f(x) = x - \dfrac{\sin x}{\cos 1}$.

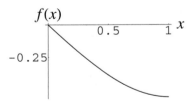

18. Start with the equation $\dfrac{d^2 G}{dx^2} = -\delta(x - z)$ whose solution is

$$
G(x) = \begin{cases} c_1 x + c_2 & x < z \\ c_3 x + c_4 & x > z \end{cases}
$$

The boundary conditions, $G(0) = G(1) = 0$, give us $c_2 = 0$ and $c_3 = -c_4$. Therefore,

$$
G(x) = \begin{cases} c_1 x & x < z \\ c_3 (1 - x) & x > z \end{cases}
$$

Continuity of $G(x, z)$) at $x = z$ gives $c_1 z = c_3(1 - z)$ and the discontinuity gives $c_3 + c_1 = 1$. So $c_1 = 1 - z$ and $c_3 = z$ and

$$G(x) = \begin{cases} x(1 - z) & 0 \leq x < z \\ z(1 - x) & z \leq x \leq 1 \end{cases}$$

21. If the eigenfunctions are complex, Equation 8 reads

$$f_n = \int_a^b r(x) f(x) \phi_n^*(x)\, dx = \int_a^b r(z) f(z) \phi_n^*(z)\, dz$$

Substituting this result into Equation 11 gives

$$y(x) = \int_a^b \left[\sum_{n=1}^{\infty} \frac{\phi_n(x)\phi_n^*(z)}{\lambda_n - \mu} \right] r(z) f(z)\, dz = \int_a^b G(x, z) g(z)\, dz$$

where

$$G(x, z) = \sum_{n=1}^{\infty} \frac{\phi_n(x)\phi_n^*(z)}{\lambda_n - \mu}$$

Notice now that $G^*(x, z) = G(z, x)$.

24. Start with the general solution to Equation 18

$$G(x) = \begin{cases} c_1 \sin x + c_2 \cos x & a \leq x < z \\ c_3 \sin x + c_4 \cos x & z < x \leq b \end{cases}$$

Using the boundary conditions $G(a, z) = G'(b, z) = 0$ gives

$$c_1 \sin a + c_2 \cos a = 0 \quad \text{and} \quad c_3 \cos b - c_4 \sin b = 0$$

This gives us

$$G(x) = \begin{cases} c_1 \sin x - c_1 \tan a \cos x & a \leq x < z \\ c_4 \tan b \sin x + c_4 \cos x & z < x \leq b \end{cases}$$

The continuity condition and the jump discontinuity condition give

$$c_1(\sin z - \tan a \cos z) = c_4(\tan b \sin z + \cos z)$$

and

$$c_4 \tan b \cos z - c_4 \sin z - c_1 \cos z - c_1 \tan a \sin z = -1$$

These equation can be written as

$$c_1 \frac{\sin(z - a)}{\cos a} = c_4 \frac{\cos(z - b)}{\cos b} \quad \text{and} \quad c_4 \frac{\sin(z - b)}{\cos b} + c_1 \frac{\cos(z - a)}{\cos a} = 1$$

Solving for c_1 and c_4 yields

$$c_4 \left[\cos(a - b) \right] = \sin(z - a) \cos b \quad \text{and} \quad c_1 = \frac{\cos a \cos(z - b)}{\cos(a - b)}$$

Therefore,

$$G(x, z) = \begin{cases} \dfrac{\cos(b - z) \sin(x - a)}{\cos(a - b)} & a \leq x < z \\[4mm] \dfrac{\sin(z - a) \cos(b - x)}{\cos(a - b)} & z < x \leq b \end{cases}$$

CHAPTER 15

Fourier Series

15.1 Fourier Series as Eigenfunction Expansions

1. Using $\sin ax = \dfrac{e^{iax} - e^{-iax}}{2i}$ and $\sin bx = \dfrac{e^{ibx} - e^{-ibx}}{2i}$ we get

$$
\begin{aligned}
\sin ax \sin bx &= \left(\frac{e^{iax} - e^{-iax}}{2i}\right)\left(\frac{e^{ibx} - e^{-ibx}}{2i}\right) \\
&= \frac{e^{i(a+b)x} + e^{-i(a+b)x}}{-4} + \frac{e^{i(b-a)x} + e^{-i(b-a)x}}{4} \\
&= \frac{1}{2}\cos(b-a)x - \frac{1}{2}\cos(a+b)x
\end{aligned}
$$

Now use $\cos ax = \dfrac{e^{iax} + e^{-iax}}{2}$ and $\cos bx = \dfrac{e^{ibx} + e^{-ibx}}{2}$ to write

$$
\begin{aligned}
\cos ax \cos bx &= \left(\frac{e^{iax} + e^{-iax}}{2}\right)\left(\frac{e^{ibx} + e^{-ibx}}{2}\right) \\
&= \frac{e^{i(a+b)x} + e^{-i(b-a)x} + e^{i(b-a)x} + e^{-i(a+b)x}}{4} \\
&= \frac{1}{2}\cos(a+b)x + \frac{1}{2}\cos(a-b)x
\end{aligned}
$$

3. Using the result of Problem 1, we write

$$
\int \sin ax \sin bx = \frac{1}{2}\int \cos(b-a)x\,dx - \frac{1}{2}\int \cos(a+b)x\,dx = \frac{\sin(a-b)x}{2(a-b)} - \frac{\sin(a+b)x}{2(a+b)}
$$

and

$$
\int \cos ax \cos bx = \frac{1}{2}\int \cos(b-a)x\,dx + \frac{1}{2}\int \cos(a+b)x\,dx = \frac{\sin(a+b)x}{2(a+b)} + \frac{\sin(a-b)x}{2(a-b)}
$$

6. Using the results of Problem 3, we have

$$
\begin{aligned}
\int_{-\pi}^{\pi} \sin nx \sin mx\,dx &= \left[\frac{\sin(n-m)x}{2(n-m)} - \frac{\sin(n+m)x}{2(n+m)}\right]_{-\pi}^{\pi} \\
&= 0 \qquad \text{when } n \text{ and } m \text{ are integers} \quad (\text{but } n \neq m)
\end{aligned}
$$

and

$$\int_{-\pi}^{\pi} \cos nx \cos mx \, dx = \left[\frac{\sin(n-m)x}{2(n-m)} + \frac{\sin(n+m)x}{2(n+m)} \right]_{-\pi}^{\pi}$$
$$= 0 \quad \text{when } n \text{ and } m \text{ are integers} \quad (\text{but } n \neq m)$$

If $n = m$, $\lim\limits_{n \to m} \dfrac{\sin(n-m)x}{2(n-m)} = \dfrac{x}{2}$, and so the result is π in each case.

Using the result of Problem 5, we have

$$\int_{-\pi}^{\pi} \cos nx \sin mx \, dx = \left[-\frac{\cos(n-m)x}{2(n-m)} - \frac{\cos(n+m)x}{2(n+m)} \right]_{-\pi}^{\pi}$$
$$= 0 \quad \text{when } n \text{ and } m \text{ are integers (but } n \neq m)$$

and

$$\int_{-\pi}^{\pi} \cos nx \sin nx \, dx = \left[\frac{1}{2n} \sin^2 nx \right]_{-\pi}^{\pi} = 0$$

Also notice that $\cos nx \sin mx$ is an odd function of x.

9. Start with $\phi_n''(x) + \lambda_n \phi_n(x) = 0$, whose general solution is

$$\phi_n(x) = c_1 \cos \lambda_n^{1/2} x + c_2 \sin \lambda_n^{1/2} x$$

The periodic boundary conditions, $\phi_n(l) = \phi_n(-l)$, give

$$c_1 \cos \lambda_n^{1/2} l + c_2 \sin \lambda_n^{1/2} l = c_1 \cos \lambda_n^{1/2} l - c_2 \sin \lambda_n^{1/2} l$$

or $c_2 \sin \lambda_n^{1/2} l = 0$ or $\lambda_n^{1/2} l = n\pi$. Both c_1 and c_2 are arbitrary.

12. Using Equations 11 and 12, we get

$$a_n = \frac{1}{\pi} \int_{-\pi}^{\pi} x^2 \cos nx \, dx = \frac{4(-1)^n}{n^2} \qquad n = 1, 2, \ldots$$

$$a_0 = \frac{2\pi^2}{3}$$

$$b_n = \frac{1}{\pi^2} \int_{-\pi}^{\pi} x^2 \sin nx \, dx = 0$$

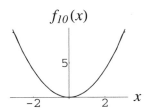

$f_{10}(x)$

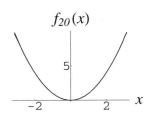

$f_{20}(x)$

Therefore, Equation 9 gives

$$f(x) = \frac{\pi^2}{3} - 4 \left(\cos x - \frac{\cos 2x}{4} + \frac{\cos 3x}{9} - \frac{\cos 4x}{16} + \cdots \right)$$

The above figure shows $f(x)$ for 10 and 20 terms along with $f(x) = x^2$.

15. Using Equations 11 and 12, we get

$$a_n = \frac{1}{l}\left[\int_0^l \frac{x}{l}\cos\frac{n\pi x}{l}\,dx + \int_l^{2l}\left(\frac{2l-x}{l}\right)\cos\frac{n\pi x}{l}\,dx\right] = \begin{cases} -\dfrac{4}{n^2\pi^2} & n \text{ odd} \\ 0 & n \text{ even} \end{cases}$$

$$a_0 = 1$$

$$b_n = 0$$

Therefore,

$$f(x) = \frac{1}{2} - \frac{4}{\pi^2}\sum_{n=1}^{\infty}\frac{1}{(2n-1)^2}\cos\frac{(2n-1)\pi x}{l}$$

The following figure shows $f(x)$ for $l = 3$ using 10 terms along with

$$f(x) = \begin{cases} x/l & 0 \le x < l \\ (2l-x)/l & l \le x < 2l \end{cases}.$$

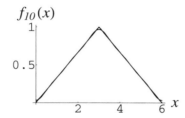

18. Using Equations 11 and 12, we have

$$a_n = \frac{1}{\pi}\int_{-\pi}^{\pi} |\cos x|\cos nx\,dx = \begin{cases} \dfrac{4(-1)^{n/2+1}}{\pi}\dfrac{1}{n^2-1} & n \text{ even} \\ 0 & n \text{ odd} \end{cases}$$

$$a_0 = \frac{4}{\pi}$$

$$b_n = 0 \qquad n = 1,\,2,\,\ldots$$

Therefore,

$$f(x) = \frac{2}{\pi} + \frac{4}{\pi}\sum_{n=1}^{\infty}\frac{(-1)^{n+1}}{4n^2-1}\cos 2nx$$

The following figure shows $f(x)$ plotted against x using 10 terms along with $f(x) = |\cos x|$.

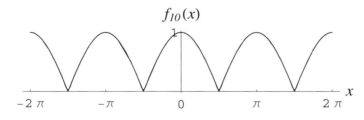

21. Consider $x = \dfrac{b+a}{2} + \dfrac{b-a}{2\pi}u$. If $x = b$, then we get $\dfrac{b-a}{2} = \dfrac{b-a}{2\pi}u$, or $u = \pi$. If $x = a$, then we get $\dfrac{a-b}{2} = \dfrac{b-a}{2\pi}u$, or $u = -\pi$. Also

$$F(u + 2\pi n) = f\left[\frac{b+a}{2} + \frac{b-a}{2\pi}(u + 2\pi n)\right] = f\left[\frac{b+a}{2} + \frac{b-a}{2\pi}u + n(b-a)\right]$$

where $n = 0,\ \pm 1,\ \pm 2,\ \ldots$. Thus, if $F(u)$ is periodic with period 2π, then $f(x)$ is periodic with period $b - a$.

24. $\displaystyle\int_{-l}^{l} e^{i(n-m)\pi x/l}dx = \int_{-l}^{l}\left[\cos\frac{(n-m)\pi x}{l} + i\sin\frac{(n-m)\pi x}{l}\right]dx = 0$ if $m \neq n$ because each integral is over one cycle.

27. See the solution to Problem 12.

15.2   Sine and Cosine Series

1. Let $f_e(x)$ be an even function of x and $f_o(x)$ be an odd function.

(a) Let $g(x) = f_e(x)f_e(x)$. Then $g(-x) = f_e(-x)f_e(-x) = f_e(x)f_e(x) = g(x)$.

(b) Let $g(x) = f_o(x)f_o(x)$. Then $g(-x) = f_o(-x)f_o(-x) = (-1)^2 f_o(x)f_o(x) = g(x)$.

(c) Let $g(x) = f_e(x)f_o(x)$. Then $g(-x) = f_e(-x)f_o(-x) = -f_e(x)f_o(x) = -g(x)$

3. (a) $f(x) = \tan 3x = \dfrac{\sin 3x}{\cos 3x}$ and so $\tan 3x$ is an odd function of x because $\sin 3x$ is an odd function of x and $\cos 3x$ is an even function.

(b) $f(x) = x - \sinh x$ is an odd function of x because both x and $\sinh x$ are odd functions.

(c) $f(x) = x + \cos x$ is neither even nor odd because x is an odd function but $\cos x$ is an even function.

(d) $f(x) = xe^{-x}$ is neither because $f(-x) = -xe^x$.

(e) $f(x) = \dfrac{e^x}{(1 \pm e^x)^2}$ is an even function of x because

$$f(-x) = \frac{e^{-x}}{(1 \pm e^{-x})^2} = \frac{e^{2x}e^{-x}}{e^{2x}(1 \pm e^{-x})^2} = \frac{e^x}{(e^x \pm 1)^2} = \frac{e^x}{(1 \pm e^x)^2} = f(x)$$

(f) $f(x) = \begin{cases} -1 & x < 0 \\ 1 & x > 0 \end{cases}$ is an odd function of x because $f(-x) = f(x)$.

6. (a)

$$f_{\mathrm{e}}(x) = \begin{cases} \dfrac{\sin x}{x} - x\cos x & x < 0 \\[2mm] \dfrac{\sin x}{x} + x\cos x & x > 0 \end{cases}$$

$$f_{\mathrm{o}}(x) = \begin{cases} -\dfrac{\sin x}{x} + x\cos x & x < 0 \\[2mm] \dfrac{\sin x}{x} + x\cos x & x > 0 \end{cases}$$

(b)

$$f_{\mathrm{e}}(x) = \begin{cases} -\sinh x - x & x < 0 \\[2mm] \sinh x + x & x > 0 \end{cases}$$

$$f_{\mathrm{o}}(x) = \begin{cases} \sinh x + x & x < 0 \\[2mm] \sinh x + x & x > 0 \end{cases}$$

sinx/x + x cosx

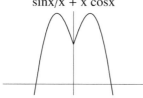

odd extension of x cosx + sinx/x

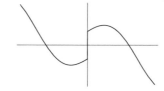

x + sinhx

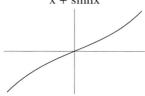

even extension of x+sinhx

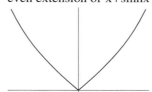

All four functions are plotted above.

9. We use the odd extension $f(x) = x^2$

$$f_{\mathrm{o}}(x) = \begin{cases} -x^2 & -l < x < 0 \\ x^2 & 0 < x < l \end{cases}$$

and then

$$b_n = \frac{1}{l}\int_{-l}^{l} f_{\mathrm{o}}(x)\sin\frac{n\pi x}{l}\,dx = -\frac{4l^2 + (-1)^n 2(n^2\pi^2 - 2)l^2}{n^3\pi^3} \qquad n = 1,\,2,\,\ldots$$

Thus

$$f_{\mathrm{o}}(x) = \sum_{n=1}^{\infty} \frac{-2l^2\left[\,2 + (-1)^n(n^2\pi^2 - 2)\,\right]}{n^3\pi^3}\sin\frac{n\pi x}{l}$$

The figure on the following page shows $f(x)$ calculated to 100 terms for $l = 2$.

12. The function to be expanded is $\sin x$, and so its Fourier sine series is simply $\sin x$. In other words, $b_1 = 1$ and $b_n = 0$ for $n \neq 1$.

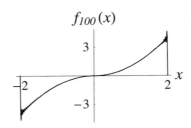

$$f_{100}(x)$$

15. The even extension of $f(x) = \pi - x$, $0 < x < \pi$ is

$$f_e(x) = \begin{cases} \pi + x & -\pi < x < 0 \\ \pi - x & 0 < x < \pi \end{cases}$$

The Fourier coefficients are given by

$$\begin{aligned} a_n &= \frac{1}{\pi}\int_{-\pi}^{\pi} f_e(x)\cos nx\, dx = \frac{2}{\pi n^2}(1 - \cos n\pi) \\ &= \begin{cases} \dfrac{4}{\pi n^2} & n \text{ odd} \\[2mm] 0 & n \text{ even} \end{cases} \qquad n = 1,\ 2,\ \ldots \\ a_0 &= \pi \end{aligned}$$

Therefore,

$$f_e(x) = \frac{\pi}{2} + \frac{4}{\pi}\sum_{n=1}^{\infty} \frac{\cos(2n-1)x}{(2n-1)^2}$$

The following figure shows $f_e(x)$ calculated to 10 terms.

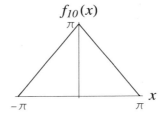

$$f_{10}(x)$$

15.3   Convergence of Fourier Series

1. Start with

$$\begin{aligned} S_N^2(x) &= \left[\frac{a_0}{2} + \sum_{n=1}^{\infty}\left(a_n\cos\frac{n\pi x}{l} + b_n\sin\frac{n\pi x}{l}\right)\right]^2 \\ &= \frac{a_0^2}{4} + a_0\sum_{n=1}^{\infty}\left(a_n\cos\frac{n\pi x}{l} + b_n\sin\frac{n\pi x}{l}\right) + \sum_{n=1}^{\infty}\sum_{m=1}^{\infty} a_n a_m \cos\frac{n\pi x}{l}\cos\frac{m\pi x}{l} \\ &\quad + \sum_{n=1}^{\infty}\sum_{m=1}^{\infty} a_n b_m \cos\frac{n\pi x}{l}\sin\frac{m\pi x}{l} + \sum_{n=1}^{\infty}\sum_{m=1}^{\infty} b_n b_m \sin\frac{n\pi x}{l}\sin\frac{m\pi x}{l} \end{aligned}$$

Now integrate from $-l$ to l using the orthogonality relations of $\left\{\cos\dfrac{n\pi x}{l}\right\}$ and $\left\{\sin\dfrac{n\pi x}{l}\right\}$ to obtain

$$\int_{-l}^{l} S_N^2(x)\, dx = \frac{l a_0^2}{2} + l \sum_{n=1}^{\infty} (a_n^2 + b_n^2)$$

3. Using Equation 9, we have

$$
\begin{aligned}
\frac{1}{l} \int_{-l}^{l} f^2(x)\, dx &= \frac{1}{l} \int_0^l x^2\, dx = \frac{l^2}{3} \\
&= \frac{l^2}{8} + \frac{4l^2}{\pi^4} \sum_{n=1}^{\infty} \frac{1}{(2n-1)^4} + \frac{l^2}{\pi^2} \sum_{n=1}^{\infty} \frac{1}{n^2}
\end{aligned}
$$

or

$$\frac{5}{24} = \frac{4}{\pi^4} \sum_{n=1}^{\infty} \frac{1}{(2n-1)^4} + \frac{1}{\pi^2} \frac{\pi^2}{6}$$

or

$$\sum_{n=1}^{\infty} \frac{1}{(2n-1)^4} = \frac{\pi^4}{96}$$

6. Use Equation 9:

$$\int_0^2 (x^3 - 4x)^2\, dx = \frac{1024}{105} = \left(\frac{96}{\pi^3}\right)^2 \sum_{n=1}^{\infty} \frac{1}{n^6}$$

and so $\displaystyle\sum_{n=1}^{\infty} \frac{1}{n^6} = \frac{\pi^6}{945}$.

9.

$$f(0) = \frac{2l^2}{3} + \frac{4l^2}{\pi^2} \sum_{n=1}^{\infty} \frac{(-1)^{n+1}}{n^2} = \frac{2l^2}{3} + \frac{l^2}{3} = l^2$$

$$f(l) = \frac{2l^2}{3} + \frac{4l^2}{\pi^2} \sum_{n=1}^{\infty} \frac{(-1)}{n^2} = \frac{2l^2}{3} - \frac{l^2}{3} = 0$$

The answers make sense because there are no discontinuities in the periodic extension of $f(x)$. (See Figure 15.2.)

12. Integrate

$$x^2 = \frac{4l^2}{\pi^2} \left[\frac{\pi^2}{12} - \sum_{n=1}^{\infty} \frac{(-1)^{n+1}}{n^2} \cos \frac{n\pi x}{l} \right]$$

from 0 to x to obtain

$$\frac{x^3}{3} = \frac{l^2 x}{3} + \frac{4l^3}{\pi^3} \sum_{n=1}^{\infty} \frac{(-1)^n}{n^3} \sin \frac{n\pi x}{l}$$

Now substitute the Fourier series for x given in Example 3 to write

$$x^3 = \frac{2l^3}{\pi}\sum_{n=1}^{\infty}\frac{(-1)^{n+1}}{n}\sin\frac{n\pi x}{l} + \frac{12l^3}{\pi^3}\sum_{n=1}^{\infty}\frac{(-1)^n}{n^3}\sin\frac{n\pi x}{l}$$

$$= \frac{2l^3}{\pi}\sum_{n=1}^{\infty}\left[\frac{(-1)^{n+1}}{n} + \frac{6}{\pi^2}\frac{(-1)^n}{n^3}\right]\sin\frac{n\pi x}{l}$$

This series with 500 terms is plotted below for $l = 2$. As the figure shows, the above Fourier series represents the periodic extensions of x^3,

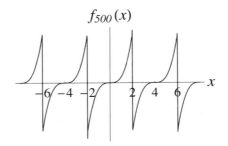

$f_{500}(x)$

15. (a) $f(x) = |\cos x|$ is continuous, but has discontinuous derivatives at $x = n\pi/2$, with $n = \pm 1, \pm 2, \ldots$, and so its Fourier coefficients go as $1/n^2$.

(b) $f(x) = \begin{cases} -1 & -l \leq x < 0 \\ 1 & 0 \leq x < l \end{cases}$ is discontinuous at $x = nl$, with $n = 0, \pm 1, \ldots$, and so its Fourier coefficients go as $1/n$.

(c) $f(x) = x^3 - x$ is continuous and $f'(x) = 3x^2 - 1$ is continuous over $[-1, 1)$, and so its Fourier coefficients go as $1/n^3$.

(d) $f(x) = (x-1)^2$ is discontinuous over $[-1, 1)$, and so its Fourier coefficients go as $1/n$.

15.4   Fourier Series and Ordinary Differential Equations

1. The functions $\cos\dfrac{n\pi x}{l}$ and $\sin\dfrac{n\pi x}{l}$ are linearly independent.

3. The coefficients of the Fourier sine expansion of $x - l\sin x/\sin l$ are given by

$$b_n = \frac{1}{l}\int_{-l}^{l}\left(x - \frac{l\sin x}{\sin l}\right)\sin\frac{n\pi x}{l} = \frac{2}{l}\int_{0}^{l}\left(x - \frac{l\sin x}{\sin l}\right)\sin\frac{n\pi x}{l}$$

$$= -\frac{2l(-1)^n}{n\pi} + \frac{2ln\pi(-1)^n}{n^2\pi^2 - l^2} = \frac{2l^3(-1)^n}{n\pi(n^2\pi^2 - l^2)}$$

Therefore,

$$f(x) = \frac{2l^3}{\pi} \sum_{n=1}^{\infty} \frac{(-1)^n}{n(n^2\pi^2 - l^2)} \sin \frac{n\pi x}{l}$$

Both $f(x)$ (for 10 terms) and $x - l \sin x / \sin l$ are plotted below.

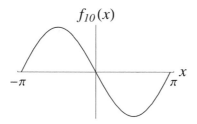

$f_{10}(x)$

6. The a_n are given by

$$
\begin{aligned}
a_n &= \frac{1 - \cos l}{\sin l} \cdot \frac{2}{l} \int_0^l \cos x \cos \frac{n\pi x}{l}\, dx - \frac{2}{l} \int_0^l \sin x \cos \frac{n\pi x}{l}\, dx + \frac{2}{l} \int_0^l x \cos \frac{n\pi x}{l}\, dx \\[1mm]
&= \frac{2(1 - \cos l)}{l \sin l} \left[\frac{\sin(1 - n\pi/l)x}{2(1 - n\pi/l)} + \frac{\sin(1 + n\pi/l)x}{2(1 + n\pi/l)} \right]_0^l \\[1mm]
&\quad - \frac{2}{l} \left[-\frac{\cos(1 - n\pi/l)x}{2(1 - n\pi/l)} - \frac{\cos(1 + n\pi/l)x}{2(1 + n\pi/l)} \right]_0^l + \frac{2}{l} \left[\frac{\cos n\pi x/l}{n^2\pi^2/l^2} + \frac{x \sin(n\pi x/l)}{n\pi/l} \right]_0^l \\[1mm]
&= \frac{(1 - \cos l)}{l \sin l} \left[\frac{\sin(l - n\pi)}{1 - n\pi/l} + \frac{\sin(l + n\pi)}{1 + n\pi/l} \right] \\[1mm]
&\quad + \frac{1}{l} \left[\frac{\cos(l - n\pi)}{1 - n\pi/l} + \frac{\cos(l + n\pi)}{1 + n\pi/l} - \frac{1}{1 - \pi n/l} - \frac{1}{1 + n\pi/l} \right] + \frac{2l}{n^2\pi^2}[(-1)^n - 1)] \\[1mm]
&= \frac{(1 - \cos l)(-1)^n}{l} \left[\frac{2}{1 - n^2\pi^2/l^2} \right] + \frac{1}{l}[(-1)^n \cos l] \left[\frac{2}{1 - n^2\pi^2/l^2} \right] \\[1mm]
&\quad - \frac{1}{l} \left[\frac{2}{1 - n^2\pi^2/l^2} \right] + \frac{2l}{n^2\pi^2}[(-1)^n - 1] \\[1mm]
&= \frac{2}{l(1 - n^2\pi^2/l^2)}[(-1)^n - 1] + \frac{2l}{n^2\pi^2}[(-1)^n - 1] \\[1mm]
&= 2l[(-1)^n - 1]\frac{l^2}{n^2\pi^2(l^2 - n^2\pi^2)} \qquad n \neq 0
\end{aligned}
$$

and

$$a_0 = \frac{2(1 - \cos l)}{l \sin l} \sin l + \frac{2}{l}(\cos l - 1) + l = l \qquad n \neq 0$$

Therefore,

$$f(x) = \frac{l}{2} + \frac{2l^3}{\pi^2} \sum_{n=1}^{\infty} \frac{(-1)^n - 1}{n^2(l^2 - n^2n^2)} \cos \frac{n\pi x}{l}$$

which is the same result as in Problem 4. The figure on the following page shows $f(x)$ (for 10 terms) plotted against x.

9. The complementary solution is $y_c(x) = c_1 \cos \beta x + c_2 \sin \beta x$ and the particular solution is $y_p(x) = \alpha + \gamma x$. Substitute $y(x) = y_c(x) + y_p(x)$ into $y''(x) + \beta^2 y(x) = x$ to obtain

$$y''(x) + \beta^2 y(x) = \alpha \beta^2 + \gamma \beta^2 x = x$$

or $\alpha = 0$ and $\gamma = 1/\beta^2$. Therefore,

$$y(x) = c_1 \cos \beta x + c_2 \sin \beta x + \frac{x}{\beta^2}$$

$f_{10}(x)$

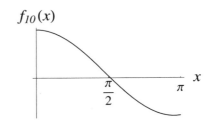

The boundary conditions, $y(0) = y(1) = 0$ give $c_1 = 0$ and $c_2 = -\dfrac{1}{\beta^2 \sin \beta}$. The complete solution is

$$y(x) = \frac{x}{\beta^2} - \frac{\sin \beta x}{\beta^2 \sin \beta}$$

12. Substitute $x(t) = \displaystyle\sum_{n=1}^{\infty} x_n(t)$ and $F(t) = \displaystyle\sum_{n=1}^{\infty} b_n \sin \omega_n t$ into Equation 7 to obtain

$$\sum_{n=1}^{\infty} (m\ddot{x}_n + \gamma \dot{x}_n + k x_n) = \sum_{n=1}^{\infty} b_n \sin \omega_n t$$

or

$$\sum_{n=1}^{\infty} (m\ddot{x}_n + \gamma \dot{x}_n + k x_n - b_n \sin \omega_n t) = 0$$

15. If $m = 4$, $\gamma = 1/10$, $k = 98$, and $\omega = 1$, then Equation 17 becomes

$$x^{\mathrm{sp}}(t) \quad = \quad -\frac{4F_0}{\pi}[0.0106 \sin(t - 0.00106) + 0.00538 \sin(3t - 0.0048)$$
$$+ \ 0.0970 \sin(5t - 0.250) + 0.00146 \sin(7t - 0.0076) + \cdots]$$

The coefficient of the third term is much larger than any of the others. The following figure shows $x^{\mathrm{sp}}(t)$ given above compared to just the third term alone (for $-4F_0/\pi = 1$.)

x^{sp}

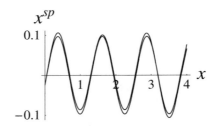

CHAPTER 16

Partial Differential Equations

16.1 Some Examples of Partial Differential Equations

1. Substitute Fick's law, $\mathbf{J} = -D \operatorname{grad} c$, into the continuity equation $\dfrac{\partial c}{\partial t} + \operatorname{div} \mathbf{J} = 0$ to get $\dfrac{\partial c}{\partial t} = -\operatorname{div} \mathbf{J} = D \operatorname{div} \operatorname{grad} c = D \nabla^2 c$.

3. Let $x \pm vt = z$ and write

$$\frac{\partial u}{\partial x} = \frac{\partial f}{\partial (x \pm vt)} \frac{d(x \pm vt)}{dx} = \frac{\partial f}{\partial z}$$

we also have

$$\frac{\partial^2 u}{\partial x^2} = \frac{\partial f^2}{\partial z^2}$$

$$\frac{\partial u}{\partial t} = \frac{\partial f}{\partial (x \pm vt)} \frac{d(x \pm vt)}{dt} = \pm v \frac{\partial f}{\partial z}$$

$$\frac{\partial^2 u}{\partial t^2} = v^2 \frac{\partial^2 f}{\partial z^2}$$

to get $v^2 \dfrac{\partial^2 u}{\partial x^2} = \dfrac{\partial^2 u}{\partial t^2}$.

6. Fick's law with a drift in the x direction is given by $\mathbf{J} = -D \operatorname{grad} c - \beta c \mathbf{i}$. Substitute this into the continuity equation, $\dfrac{\partial c}{\partial t} + \operatorname{div} \mathbf{J} = 0$, to get

$$\frac{\partial c}{\partial t} = -\operatorname{div} \mathbf{J} = D \frac{\partial^2 c}{\partial x^2} + \beta \frac{\partial u}{\partial x}$$

9. Start with $\dfrac{\partial T}{\partial t} = \alpha^2 \dfrac{\partial^2 T}{\partial x^2} - hT$. Using $T(x,t) = \phi(x,t) e^{-ht}$, we have

$$\frac{\partial^2 T}{\partial x^2} = \frac{\partial^2 \phi}{\partial x^2} e^{-ht} \qquad \text{and} \qquad \frac{\partial T}{\partial t} = \frac{\partial \phi}{\partial t} e^{-ht} - h\phi e^{-ht}$$

Substitute these two results into $\dfrac{\partial T}{\partial t} = \alpha^2 \dfrac{\partial^2 T}{\partial x^2} - hT$ to get

$$\frac{\partial \phi}{\partial t} e^{-ht} - h\phi e^{-ht} = \alpha^2 \frac{\partial \phi}{\partial t} e^{-ht} - h\phi e^{-ht}$$

or $\dfrac{\partial \phi}{\partial t} = \alpha^2 \dfrac{\partial^2 \phi}{\partial t^2}$.

12. Substitute $u(x,t) = f(x)\sin vt$ into $\dfrac{\partial^2 u}{\partial t^2} = v^2 \dfrac{\partial^2 u}{\partial x^2}$ to get

$$\frac{\partial^2 u}{\partial x^2} = \frac{d^2 f}{dx^2}\sin vt \qquad \text{and} \qquad \frac{\partial^2 u}{\partial t^2} = -v^2 f(x)\sin vt$$

and we have $\dfrac{d^2 f}{dx^2} + f(x) = 0$. The substitution makes sense physically because it is based upon the physical fact that the temporal motion is harmonic.

15. Referring to Figure 16.1, we see that for small displacements, $\tau_2 \sin\beta \approx \tau_2 \tan\beta = \tau(x + \Delta x)u_x(x + \Delta x)$ and $\tau_1 \sin\alpha \approx \tau_1 \tan\alpha = \tau(x)u_x(x)$. Equation 8 becomes

$$\frac{\tau(x + \Delta x)u_x(x + \Delta x) - \tau(x)u_x(x)}{\Delta x} = \rho(x)\frac{\partial^2 u}{\partial t^2}$$

and so we get $\dfrac{\partial}{\partial x}\left[\tau(x)\dfrac{\partial u}{\partial x}\right] = \rho(x)\dfrac{\partial^2 u}{\partial t^2}$.

16.2   Laplace's Equation

1. The general solutions to Equations 10 are

$$X(x) = c_1 e^{k^{1/2}x} + c_2 e^{-k^{1/2}x} \qquad \text{and} \qquad Y(y) = c_3 \cos k^{1/2}x + c_4 \sin k^{1/2}x$$

The boundary conditions $X(0) = X(a) = 0$ give

$$\begin{aligned} X(0) &= 0 = c_1 + c_2 = 0 \\ X(a) &= 0 = c_1 e^{k^{1/2}a} + c_2 e^{-k^{1/2}a} = 0 \end{aligned}$$

These two equations can be written as $2c_1 \sinh k^{1/2}a = 0$. If $k > 0$, then $c_1 = 0 = c_2$.

3. Start with Equation 10. The boundary conditions, $u(0,y) = u(a,y) = 0$ (i.e., $X(0) = X(a) = 0$), give $X(x) = \sin n\pi x/a$ for $n = 1,\ 2,\ \dots$. Using this result ($k^2 = n^2\pi^2/a^2$) in the second of Equations 10 gives

$$Y_n(y) = c_1 \cosh \frac{n\pi y}{a} + c_2 \sinh \frac{n\pi y}{a}$$

The boundary condition $u_y(x,0) = 0$ ($Y'(a) = 0$) gives $c_2 = 0$. So far we have

$$u_n(x,y) = c_n \sin \frac{n\pi x}{a} \cosh \frac{n\pi y}{a} \qquad \text{and} \qquad u(x,y) = \sum_{n=1}^{\infty} c_n \sin \frac{n\pi x}{a} \cosh \frac{n\pi y}{a}$$

We can determine the c_n from the boundary condition, $u_y(x, b) = u_0 x(a - x)$, which gives us

$$u_y(x, b) = u_0 x(a - x) = \sum_{n=1}^{\infty} c_n \frac{n\pi}{a} \sin \frac{n\pi x}{a} \sinh \frac{n\pi b}{a}$$

and

$$\frac{a}{2} c_n \frac{n\pi}{a} \sinh \frac{n\pi b}{a} = u_0 \int_0^a x(a - x) \sin \frac{n\pi x}{a} \, dx = u_0 \frac{2a^3}{n^3 \pi^3} [1 - (-1)^n]$$

Therefore, $c_n = \dfrac{4u_0 a^3}{n^4 \pi^4} \dfrac{[1 - (-1)^n]}{\sinh n\pi b/a}$ and

$$u(x, y) = \frac{4a^3 u_0}{\pi^4} \sum_{n=1}^{\infty} \frac{1 - (-1)^n}{n^4 \sinh n\pi b/a} \sin \frac{n\pi x}{a} \cosh \frac{n\pi y}{a}$$

The following figure shows $u(x, y)$ plotted against x and y for $a = 2$ and $b = 3$ using five terms. We also show $u_y(x, b)$ compared to $x(a - x)$.

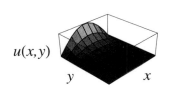

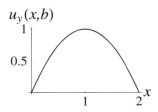

6. First let $u(x, y) = v(x, y) + w(x, y)$ to get

$$\nabla^2 u = \nabla^2 v + \nabla^2 w = 0 + 0 = 0$$

Now the boundary conditions are

$$u(0, y) = v(0, y) + w(0, y) = f_1(y) + 0 = f_1(y)$$
$$u(a, y) = v(a, y) + w(a, y) = f_2(y) + 0 = f_2(y)$$

which lead to the boundary value problem

$$\nabla^2 v = 0 \qquad \begin{array}{l} 0 < x < a \\ 0 < y < b \end{array}$$

with $v(0, y) = f_1(y)$, $v(a, y) = f_2(y)$, and $v(x, 0) = v(x, b) = 0$. Similarly, we have

$$\nabla^2 w = 0 \qquad \begin{array}{l} 0 < x < a \\ 0 < y < b \end{array}$$

with $w(0, y) = w(a, y) = 0$, $w(x, 0) = g_1(x)$, and $w(x, b) = g_2(x)$. Pictorially, we have

You can now take this procedure one step further and use, for example,

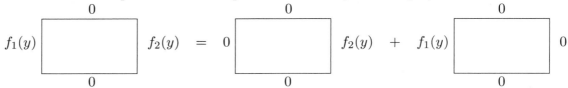

9. We express the problem as the sum of three separate problems:

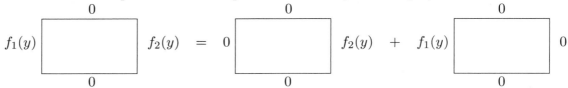

1. Equations 10 (with $k = \lambda^2$) give

$$X(x) = c_1 \cos \lambda x + c_2 \sin \lambda x \qquad \text{and} \qquad Y(y) = c_3 \cosh \lambda x + c_4 \sinh \lambda x$$

The boundary conditions, $X(0) = X(a) = 0$, give $c_1 = 0$ and $\lambda = n\pi/a$, or

$$X_n(x) = \sin \frac{n\pi x}{a} \qquad n = 1, \ 2, \ \ldots$$

The boundary conditions $Y(0) = 0$ and $Y(a) = u_0$ give $c_3 = 0$ and $c_4 = u_0 / \sinh n\pi$, or

$$Y(y) = \frac{u_0}{\sinh n\pi} \sinh \frac{n\pi y}{a}$$

The general solution is

$$u_1(x, y) = \sum_{n=1}^{\infty} c_n \frac{u_0}{\sinh n\pi} \sin \frac{n\pi x}{a} \sinh \frac{n\pi y}{a}$$

The c_n are given by the condition $u(x, a) = u_0$, or by

$$c_n \frac{a}{2} = u_0 \int_0^a \sin \frac{n\pi x}{a} \, dx = \frac{a u_0}{n\pi} [\, 1 - (-1)^n \,]$$

and so

$$u_1(x, y) = \frac{2u_0}{\pi} \sum_{n=1}^{\infty} \frac{1 - (-1)^n}{n \sinh n\pi} \sin \frac{n\pi x}{a} \sinh \frac{n\pi y}{a}$$

2. The solution to the second part can be obtained simply by interchanging x and y in the previous part.

$$u_2(x, y) = \frac{2u_0}{\pi} \sum_{n=1}^{\infty} \frac{1 - (-1)^n}{n \sinh n\pi} \sinh \frac{n\pi x}{a} \sin \frac{n\pi y}{a}$$

3. For the third part, Equations 10 give

$$Y_n(y) = \sin \frac{n\pi y}{a} \quad n = 1, \ 2, \ \ldots \qquad \text{and} \qquad X_n(x) = c_1 \cosh \frac{n\pi x}{a} + c_2 \sinh \frac{n\pi x}{a}$$

The boundary condition $X_n(a) = 0$ gives $c_1 = -c_2 \tanh n\pi$ and so we can write

$$X_n(x) = \sinh \frac{n\pi x}{a} - \tanh n\pi \cosh \frac{n\pi x}{a}$$

The general solution is

$$u_3(x, y) = \sum_{n=1}^{\infty} c_n X_n(x) Y_n(y)$$

$$= \sum_{n=1}^{\infty} c_n \left(\sinh \frac{n\pi x}{a} - \tanh n\pi \cosh \frac{n\pi x}{a} \right) \sin \frac{n\pi y}{a}$$

The c_n are given by the condition $u(0, y) = u_0$, or by

$$\frac{a}{2} c_n \tanh n\pi = -u_0 \int_0^a \sin \frac{n\pi y}{a} \, dy = -\frac{u_0 a}{n\pi} [1 - (-1)^n]$$

and so

$$u_3(x, y) = -\frac{2u_0}{\pi} \sum_{n=1}^{\infty} \frac{1 - (-1)^n}{n \tanh n\pi} \left(\sinh \frac{n\pi x}{a} - \tanh n\pi \cosh \frac{n\pi x}{a} \right) \sin \frac{n\pi y}{a}$$

Finally, then

$$
\begin{aligned}
u(x, y) &= u_1(x, y) + u_2(x, y) + u_3(x, y) \\
&= \frac{2u_0}{\pi} \sum_{n=1}^{\infty} \frac{1 - (-1)^n}{n \sinh n\pi} \left(\sin \frac{n\pi x}{a} \sinh \frac{n\pi y}{a} + \sinh \frac{n\pi x}{a} \sin \frac{n\pi y}{a} \right. \\
&\qquad \left. - \cosh n\pi \sinh \frac{n\pi x}{a} \sin \frac{n\pi y}{a} + \sinh n\pi \cosh \frac{n\pi x}{a} \sin \frac{n\pi y}{a} \right) \\
&= \frac{2u_0}{\pi} \sum_{n=1}^{\infty} \frac{1 - (-1)^n}{n \sinh n\pi} \left[\sin \frac{n\pi x}{a} \sinh \frac{n\pi y}{a} + \sinh \frac{n\pi x}{a} \sin \frac{n\pi y}{a} - \sinh \frac{n\pi (x - a)}{a} \sin \frac{n\pi y}{a} \right]
\end{aligned}
$$

The following figure shows $u(x, y)$ to 100 terms for $a = 2$.

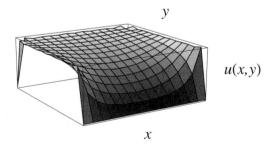

and the three figures on the following page show $f(x, y)$ along its boundaries.

12. Express $\cos n(\theta - x)$ as $(e^{in(\theta - x)} + e^{-in(\theta - x)})/2$ to write

$$\sum_{n=1}^{\infty} \left(\frac{r}{a} \right)^n \cos n(\theta - x) = \frac{1}{2} \sum_{n=1}^{\infty} \left(\frac{r}{a} \right)^n \left[e^{in(\theta - x)} + e^{-in(\theta - x)} \right]$$

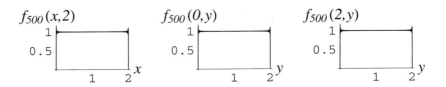

Now use the geometric series to put this summation into closed form

$$\sum_{n=1}^{\infty} \left(\frac{r}{a}\right)^n \cos n(\theta - x) = \frac{1}{2}\left[\frac{1}{1 - \frac{r}{a}e^{i(\theta-x)}} + \frac{1}{1 - \frac{r}{a}e^{-i(\theta-x)}} - 2\right]$$

$$= \frac{1}{2}\left[\frac{1 - \frac{r}{a}e^{-i(\theta-x)} + 1 - \frac{r}{a}e^{i(\theta-x)}}{1 - \frac{r}{a}(e^{i(\theta-x)} + e^{-i(\theta-x)}) + \frac{r^2}{a^2}} - 2\right]$$

$$= \frac{1}{2}\left[\frac{2 - \frac{2r}{a}\cos(\theta - x)}{1 - \frac{2r}{a}\cos(\theta - x) + \frac{r^2}{a^2}}\right]$$

$$= \frac{a^2 - ar\cos(\theta - x)}{a^2 - 2ar\cos(\theta - x) + r^2} - 1$$

15. Start with $u(r, \theta) = \dfrac{a^2 - r^2}{2\pi}\displaystyle\int_{-\pi}^{\pi} \dfrac{f(x)\,dx}{a^2 - 2ar\cos(x - \theta) + r^2}$ from Problem 13. Using the integral

$$\int_{-\pi}^{\pi} \frac{u_0\,dx}{a^2 - 2ar\cos(x - \theta) + r^2} = \frac{2\pi u_0}{(a - r)(a + r)}$$

gives $u(r, \theta) = \dfrac{a^2 - r^2}{2\pi} \cdot \dfrac{2\pi u_0}{a^2 - r^2} = u_0$ if $f(x) = u_0$.

18. This problem is known as Dirichlet's problem in an annulus. Start with

$$\frac{\partial^2 u}{\partial r^2} + \frac{1}{r}\frac{\partial u}{\partial r} + \frac{1}{r^2}\frac{\partial^2 u}{\partial \theta^2} = 0 \qquad a < r < b$$

writing $u(r, \theta) = R(r)\Theta(\theta)$ leads to two ordinary differential equations whose solutions are

$$R(r) = \begin{cases} c_1 + c_2 \ln r & n = 0 \\ c_3 r^n + c_4 r^{-n} & n = 1,\ 2,\ \dots \end{cases}$$

$$\Theta(\theta) = \begin{cases} \dfrac{a_0}{2} & n = 0 \\ a_n \cos n\theta + b_n \sin n\theta & n = 1,\ 2,\ \dots \end{cases}$$

We don't reject the r^{-n} solution in this case because the point $r = 0$ does not lie within the annulus. The complete solution is

$$u(r, \theta) = \frac{1}{2}(a_0 + b_0 \ln r) + \sum_{n=1}^{\infty}(a_n r^n + b_n r^{-n})\cos n\theta + \sum_{n=1}^{\infty}(c_n r^n + d_n r^{-n})\sin n\theta$$

Applying the boundary conditions gives

$$u(1,\theta) = \sin^2\theta = \frac{a_0}{2} + \sum_{n=1}^{\infty}(a_n + b_n)\cos n\theta + \sum_{n=1}^{\infty}(c_n + d_n)\sin n\theta$$

and

$$u(2,\theta) = 0 = \frac{1}{2}(a_0 + b_0\ln 2) + \sum_{n=1}^{\infty}(2^n a_n + 2^{-n}b_n)\cos n\theta + \sum_{n=1}^{\infty}(2^n c_n + 2^{-n}d_n)\sin n\theta$$

Using the fact that $\sin^2\theta = (1 - \cos 2\theta)/2$, these equations are satisfied by

$$a_0 = 1 \qquad b_0 = -\frac{1}{\ln 2} \qquad a_2 = \frac{1}{30} \qquad b_2 = -\frac{16}{30}$$

with all the other constants equal to zero. Therefore, our solution is

$$u(r,\theta) = \frac{1}{2}\left(1 - \frac{\ln r}{\ln 2}\right) + \frac{1}{30}\left(r^2 - \frac{16}{r^2}\right)\cos 2\theta$$

You can readily show that $u(r,\theta)$ satisfies Laplace's equation and the boundary conditions.

21. Start with Equation 29. The $b_n = 0$ because the potential must be finite at $r = 0$. So far, we have

$$u(r,\theta) = \sum_{n=0}^{\infty} a_n r^n P_n(\cos\theta)$$

The boundary condition gives

$$u(a,\theta) = u_0\cos 2\theta = \sum_{n=0}^{\infty} a_n a^n P_n(\cos\theta)$$

Multiply both sides by $P_n(\cos\theta)$ and integrate from 0 to π to obtain

$$a_n a^n \frac{2}{2n+1} = u_0\int_0^{\pi}(\cos 2\theta)P_n(\cos\theta)\sin\theta\, d\theta = u_0\int_{-1}^{1}(2x^2 - 1)P_n(x)\, dx$$

$$= \begin{cases} -\dfrac{2u_0}{3} & n = 0 \\[2mm] \dfrac{8u_0}{15} & n = 2 \\[2mm] 0 & \text{otherwise} \end{cases}$$

Therefore, $a_0 = -\dfrac{u_0}{3}$, $a_2 = \dfrac{4u_0}{3a^2}$ and so

$$u(r,\theta) = -\frac{u_0}{3} + \frac{4u_0}{3}\left(\frac{r^2}{a^2}\right)P_2(\cos\theta)$$

You can readily show that $u(r,\theta)$ satisfies Laplace's equation and the boundary conditions.

24. Let $f = g$ in Green's first identity:

$$\iint_R (f\nabla^2 f + \nabla f \cdot \nabla f)dS = \oint_B f\frac{\partial f}{\partial n}\, ds = 0$$

Now $\nabla^2 f = 0$ and $f = 0$ on B, so we have

$$\iint_R \nabla f \cdot \nabla f\, dS = 0$$

which implies that $f = $ constant, which we can take to be zero.

27. Let $f = g$ in Green's first identity (Problem 24) and then let $f = u_1 - u_2$ to see that $\nabla(u_1 - u_2) = 0$, or that u_1 and u_2 differ by a constant.

16.3   The One-Dimensional Wave Equation

1. The general solution to the first of Equations 3 is $X(x) = c_1 \sin \beta x + c_2 \cos \beta x$. The boundary conditions, $X(0) = X(l) = 0$, require that $c_2 = 0$ and $\sin \beta l = 0$, or that $\beta l = n\pi$. Therefore,

$$X_n(x) = \sin\frac{n\pi x}{l} \qquad \text{and} \qquad T_n(t) = a_n \cos\frac{n\pi vt}{l} + b_n \sin\frac{n\pi vt}{l}$$

and

$$u_n(x,t) = (a_n \cos\omega_n t + b_n \sin\omega_n t)\sin\frac{n\pi x}{l}$$

3. $\sin\dfrac{n\pi x}{l} = 0$ when $x = \dfrac{l}{n}, \dfrac{2l}{n}, \dfrac{3l}{n}, \ldots, \dfrac{(n-1)l}{n}$.

6. Equation 10 gives

$$a_n = \frac{2}{l}\int_0^l \sin\frac{3\pi x}{l}\sin\frac{n\pi x}{l}\, dx = \delta_{03}$$

and using Equation 12 gives $b_n = 0$. Therefore,

$$u(x,t) = \cos\frac{3\pi vt}{l}\sin\frac{3\pi x}{l}$$

You can readily show that $u(x,t)$ satisfies the one-dimensional wave equation, the boundary conditions, and the initial conditions.

9. Equation 10 gives

$$
\begin{aligned}
a_n &= \frac{2}{l}\int_0^l u_0 x(l-x)\sin\frac{n\pi x}{l}\, dx \\
&= \frac{2u_0}{l}\cdot\frac{2l^3}{n^3\pi^3}[1-(-1)^n] = \begin{cases} \dfrac{8l^2 u_0}{n^3\pi^3} & n \text{ odd} \\ 0 & n \text{ even} \end{cases}
\end{aligned}
$$

and Equation 12 gives us

$$b_n = \frac{2}{\omega_n l} \int_0^l u_{t0} \sin \frac{n\pi x}{l}\, dx = \begin{cases} \dfrac{4u_0}{\omega_n n\pi} & n \text{ odd} \\[2mm] 0 & n \text{ even} \end{cases}$$

Therefore, using $\omega_n = n\pi v/l$, we get

$$\begin{aligned} u(x,t) &= \frac{8l^2 u_0}{\pi^3} \sum_{n=1}^{\infty} \frac{1}{(2n-1)^3} \cos \frac{(2n-1)\pi vt}{l} \sin \frac{(2n-1)\pi x}{l} \\ &+ \frac{4u_{t0}l}{v\pi^2} \sum_{n=1}^{\infty} \frac{1}{(2n-1)^2} \sin \frac{(2n-1)\pi vt}{l} \sin \frac{(2n-1)\pi x}{l} \end{aligned}$$

12. Substitute $u(x,t) = X(x)T(t)$ into $\tau u_{xx} = \rho(x)u_{tt}$ to obtain

$$\frac{\tau X''(x)}{\rho(x)X(x)} = -\lambda \qquad \text{or} \qquad \tau X''(x) = \lambda \rho(x)X(x) = 0$$

18. $u(x,t) = \dfrac{1}{2}\left[\sin \dfrac{\pi(x+vt)}{l} + \sin \dfrac{\pi(x-vt)}{l}\right] = \sin \dfrac{\pi x}{l} \cos \dfrac{v\pi t}{l}$

21. Express $\nabla^2 u = k^2 u$ in spherical coordinates: $\dfrac{1}{r^2}\dfrac{d}{dr}r^2\dfrac{du}{dr} = k^2 u$. Now let $u = v/r$ and write

$$\begin{aligned} \frac{du}{dr} &= \frac{v'}{r} - \frac{v}{r^2} \\ r^2 \frac{du}{dr} &= rv' - v \\ \frac{d}{dr}r^2\frac{du}{dr} &= v' + rv'' - v' = rv'' \end{aligned}$$

and

$$\frac{1}{r^2}\frac{d}{dr}r^2\frac{du}{dr} = \frac{v''}{r} = \frac{1}{r}\frac{d^2 v}{dr^2} = k^2 u = k^2 \frac{v}{r}$$

Thus, we have $\dfrac{d^2 v}{dr^2} = k^2 v$.   To show that the general solution to the equation is $u(r,t) = \dfrac{1}{r}f(r+vt) + \dfrac{1}{r}g(r-vt)$, apply Equation 22 with ru replacing u.

16.4   The Two-Dimensional Wave Equation

1. According to Equation 14, $\omega_{mn} = \omega_{nm}$ if $a = b$.

3. We'll do Problem 2 first. The sum of the kl and lk modes of a square membrane is given by

$$u_{kl}(x,y,t) + u_{lk}(x,y,t) = A_{kl}\sin\frac{k\pi x}{a}\sin\frac{l\pi y}{a}\cos(\omega_{kl} + \phi_{kl}) + A_{lk}\sin\frac{l\pi x}{a}\sin\frac{k\pi y}{a}\cos(\omega_{lk} + \phi_{lk})$$

Because $\omega_{kl} = \omega_{lk}$, the frequency of the sum is ω_{kl}. Now $c_1 u_{kl} + c_2 u_{lk}$ also has frequency ω_{kl}.

6. Use Equation 16 with $f(x, y) = 1$.

$$
\begin{aligned}
a_{nm} &= \frac{4}{ab} \int_0^a \sin\frac{n\pi x}{a}\, dx \int_0^b \sin\frac{m\pi y}{b}\, dy \\
&= \frac{4}{ab} \left\{ \frac{a[1-(-1)^n]}{n\pi} \right\} \left\{ \frac{b[1-(-1)^n]}{m\pi} \right\} \quad \begin{array}{l} n = 1,\ 2,\ \ldots \\ m = 1,\ 2,\ \ldots \end{array} \\
&= \left\{ \begin{array}{cc} \dfrac{16}{\pi^2}\dfrac{1}{nm} & \text{if both } n \text{ and } m \text{ are odd} \\ 0 & \text{otherwise} \end{array} \right.
\end{aligned}
$$

Therefore,

$$
f(x, y) = \frac{16}{\pi^2} \sum_{n=1}^{\infty} \sum_{m=1}^{\infty} \frac{\sin(2n-1)\pi x/a}{2n-1} \cdot \frac{\sin(2m-1)\pi y/b}{2m-1}
$$

The values of $f(x, y)$ at the corners for $a = b = 2$ are $f(0.0005, 0.0005) = f(0.0005, 1.9995) = f(1.9995, 0.0005) = f(1.9995, 1.9995) = 0.981$.

9. First write $\cos^3\dfrac{2\pi y}{b}$ as

$$
\cos^3\frac{2\pi y}{b} = \frac{1}{4}\left(3\cos\frac{2\pi y}{b} + \cos\frac{6\pi y}{b}\right)
$$

so that

$$
4\cos\frac{2\pi x}{a}\cos^3\frac{2\pi y}{b} = 3\cos\frac{2\pi x}{a}\cos\frac{2\pi y}{b} + \cos\frac{2\pi x}{a}\cos\frac{6\pi y}{b}
$$

12. Separation of variables, $u(x, y, t) = X(x)Y(y)T(t)$, gives

$$
X''(x) + p^2 X(x) = 0 \quad \text{and} \quad Y''(y) + q^2 Y(y) = 0 \quad \text{and} \quad T''(t) + \beta^2 v^2 T(t) = 0
$$

where $p^2 + q^2 = \beta^2$. The boundary conditions, $u(0, y, t) = u(a, y, t) = 0$ require that $X_n(x) = \sin\dfrac{n\pi x}{a}$ for $n = 1,\ 2,\ \ldots$. The boundary conditions, $u_t(x, 0, t) = u_t(x, b, t) = 0$ require that $Y_m(y) = \cos\dfrac{m\pi y}{b}$ for $m = 0,\ 1,\ 2,\ \ldots$. The general solution is given by

$$
\begin{aligned}
u(x, y, t) &= \frac{1}{2} \sum_{n=1}^{\infty} (a_{n0}\cos\omega_{n0}t + b_{n0}\cos\omega_{n0}t)\sin\frac{n\pi x}{a} \\
&\quad + \sum_{n=1}^{\infty} \sum_{m=1}^{\infty} (a_{nm}\cos\omega_{nm}t + b_{nm}\sin\omega_{nm}t)\sin\frac{n\pi x}{a}\cos\frac{m\pi y}{b}
\end{aligned}
$$

where $\omega_{nm} = v\pi\left(\dfrac{n^2}{a^2} + \dfrac{m^2}{b^2}\right)^{1/2}$. For $u(x, y, 0) = f(x, y)$ and $u_t(x, y, 0) = 0$, we have

$$
a_{nm} = \frac{4}{ab} \int_0^a dx \int_0^b dy\, f(x, y)\sin\frac{n\pi x}{a}\cos\frac{m\pi y}{b}
$$

and $b_{nm} = 0$. For $f(x,y) = \sin^3 \dfrac{\pi x}{a} \cos^2 \dfrac{2\pi y}{b}$, we write

$$f(x,y) = \frac{1}{4}\left(3\sin\frac{\pi x}{a} - \sin\frac{3\pi x}{a}\right)\frac{1}{2}\left(1 + \cos\frac{4\pi y}{b}\right)$$

and find that

$$
\begin{aligned}
a_{nm} &= \frac{1}{2ab}\int_0^a dx \int_0^b dy \left[3\sin\frac{\pi x}{a} + 3\sin\frac{\pi x}{a}\cos\frac{4\pi y}{b} - \sin\frac{3\pi x}{a}\right.\\
&\qquad\left. - \sin\frac{3\pi x}{a}\cos\frac{4\pi y}{b}\right]\sin\frac{n\pi x}{a}\cos\frac{m\pi y}{b}\\
&= \frac{3}{4}\delta_{n1}\delta_{m0} + \frac{3}{8}\delta_{n1}\delta_{m4} - \frac{1}{4}\delta_{n3}\delta_{m0} - \frac{1}{8}\delta_{n3}\delta_{n4}
\end{aligned}
$$

Remembering that the δ_{m0} terms are to be divided by 2 (as in Equation 15.1.9), we write

$$
\begin{aligned}
u(x,y,t) &= \frac{3}{8}\sin\frac{\pi x}{a}\cos\omega_{10}t + \frac{3}{8}\sin\frac{\pi x}{a}\cos\frac{4\pi y}{b}\cos\omega_{14}t - \frac{1}{8}\sin\frac{3\pi x}{a}\cos\omega_{30}t\\
&\quad - \frac{1}{8}\sin\frac{3\pi x}{a}\cos\frac{4\pi y}{b}\cos\omega_{34}t
\end{aligned}
$$

15. From the definition

$$Y_0(x) = \frac{2}{\pi}\left(\gamma + \ln\frac{x}{2}\right)J_0(x) + \sum_{n=1}^{\infty}\frac{(-1)^{n+1}H_n}{(n!)^2}\left(\frac{x}{2}\right)^{2n}$$

we see that $Y_0(x) \to -\infty$ as $x \to 0$ because $\ln x \to -\infty$ as $x \to 0$.

18. Refer to Equation 36.

16.5   The Heat Equation

1. The solution to Equation 4 can satisfy the boundary conditions only if $X(x) = c_1\cos\lambda x + c_2\sin\lambda x$.

3. The solution to the equation in x is $X_n(x) = \sin\dfrac{n\pi x}{l}$ for $n = 1,\ 2,\ \ldots$. The c_n in Equation 9 are given by

$$
\begin{aligned}
c_n &= \frac{2}{l}\int_0^l \frac{2T_0 x}{l}\sin\frac{n\pi x}{l}\,dx + \frac{2}{l}\int_{l/2}^l (l-x)\sin\frac{n\pi x}{l}\,dx\\
&= \frac{4T_0}{l^2}\left(\frac{2l^2\sin n\pi/2}{n^2\pi^2}\right) = \frac{8T_0}{\pi^2}\frac{\sin n\pi/2}{n^2}
\end{aligned}
$$

Some specific values are

$$c_1 = \frac{8T_0}{\pi^2} \qquad c_2 = 0 \qquad c_3 = -\frac{8T_0}{9\pi^2} \qquad c_4 = 0 \qquad c_5 = \frac{8T_0}{25\pi^2}$$

and

$$T(x,t) = \frac{8T_0}{\pi^2}\sum_{n=1}^{\infty}\frac{(-1)^{n+1}\sin(2n-1)\pi x/a}{(2n-1)^2}e^{-\alpha^2(2n-1)^2\pi^2 t/l^2}$$

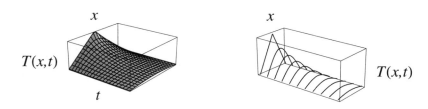

The result is plotted above for $0 \le t \le 1$ in steps of 0.10.

6. The solutions to the equation in x are $\sin \dfrac{n\pi x}{a}$, $n = 1,\ 2,\ \ldots$ and

$$c_n = \frac{2T_0}{l} \int_0^l \sin \frac{n\pi x}{l}\, dx = \frac{2T_0}{\pi} \left[\frac{1 - (-1)^n}{n} \right]$$

and

$$T(x,t) = \frac{4T_0}{\pi} \sum_{n=1}^{\infty} \frac{\sin(2n-1)\pi x/l}{2n-1} e^{-(2n-1)^2 \alpha^2 \pi^2 t/l^2}$$

The result is plotted below for $0 \le t \le 1$ in steps of 0.10.

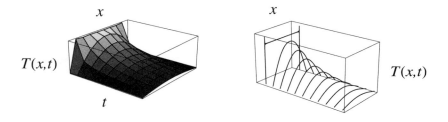

9. Let $v(x,t) = T(x,t) - T_0$ and the problem is the same as Problem 4. Here is the solution to Problem 4:

The solutions to the equations in x and t are

$$X_n(x) = \sin \frac{n\pi x}{l} \quad n = 1,\ 2,\ \ldots \qquad \text{and} \qquad T_n(t) = e^{-\alpha^2 n^2 \pi^2 t/l^2}$$

and c_n in Equation 9 are given by

$$c_n = \frac{2}{l} \int_0^l dx\, x(l-x) \sin \frac{n\pi x}{l} = \frac{4l^2}{\pi^3} \frac{\left[1 - (-1)^n \right]}{n^3}$$

Therefore, the solution to Problem 4 is

$$\begin{aligned}
T(x,t) &= \frac{4l^2}{\pi^3} \sum_{n=1}^{\infty} \frac{1-(-1)^n}{n^3} \sin \frac{n\pi x}{l} e^{-l^2 n^2 \pi^2 t/l^2} \\
&= \frac{8l^2}{\pi^3} \sum_{n=1}^{\infty} \frac{\sin(2n-1)\pi x/l}{(2n-1)^2} e^{-\alpha^2 (2n-1)^2 \pi^2 t/l^2}
\end{aligned}$$

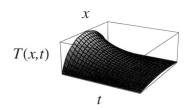
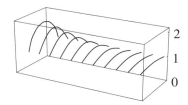

and so the solution to Problem 9 is

$$T(x,t) = T_0 + \frac{8l^2}{\pi^3} \sum_{n=1}^{\infty} \frac{\sin(2n-1)\pi x/l}{(2n-1)^2} e^{-\alpha^2(2n-1)^2\pi^2 t/l^2}$$

This result is plotted above for 10 terms and $l = 2$ and $T_0 = 1$.

12. The general solution to the differential equation in x is

$$X(x) = c_1 \cos \lambda x + c_2 \sin \lambda x$$

The boundary conditions, $X'(-l) = X'(l) = 0$, give

$$-c_1 \sin \lambda l + c_2 \cos \lambda l = 0$$
$$c_1 \sin \lambda l + c_2 \cos \lambda l = 0$$

There are two possible solutions to these equations:

(1) $c_2 = 0$ and $\sin \lambda l = 0$ ($\lambda = n\pi/l$, $n = 0$, 1, 2, ...)

(2) $c_1 = 0$ and $\cos \lambda l = 0$ ($\lambda = n\pi/2l$, n odd)

Only the first of these can later satisfy the initial condition, so we write

$$c(x,t) = \frac{a_0}{2} + \sum_{n=1}^{\infty} a_n \cos\left(\frac{n\pi x}{l}\right) e^{-n^2\pi^2 Dt/l^2}$$

The initial condition, $c(x,0) = c_0\delta(x)$, gives $a_n = c_0/l$ for $n = 0$, 1, 2, ... and so

$$c(x,t) = \frac{c_0}{l}\left[\frac{1}{2} + \sum_{n=1}^{\infty} \cos\left(\frac{n\pi x}{l}\right) e^{-n^2\pi^2 Dt/l^2}\right]$$

The following figure shows $c(x,t)$ for various values of t.

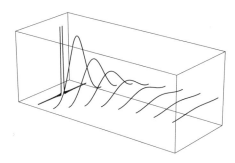

15. Simply use the orthogonality of $\{\sin n\pi x/l.\}$

18. Start with the steady-state expression $s(x) = \alpha + \beta x$. The boundary conditions, $T_x(0,t) = 0$ and $T(l,t) = T_1$, become $\dfrac{\partial s}{\partial x} = 0$ at $x = 0$, which gives $\beta = 0$ and $s(l) = T_1 = \alpha$. Thus, $s(x) = T_1$ and

$$T(x,t) = T_1 + v(x,t)$$

Now, the boundary conditions on $v(x,t)$ are $v_x(0,t) = 0$ and $v(l,t) = 0$. The general solution to the differential equation in x is

$$v(x,t) = c_1 \cos \lambda x + c_2 \sin \lambda x$$

The boundary conditions are such that $v_x(0,t) = 0$ gives $c_2 = 0$ and $v(l,t) = 0$ gives $\cos \lambda l = 0$, Therefore, $\lambda_n l = n\pi/2$ for n odd and

$$T(x,t) = T_1 + \sum_{\substack{n=1 \\ \text{odd}}}^{\infty} a_n \cos\left(\frac{n\pi x}{2l}\right) e^{-n^2\pi^2\alpha^2 t/4l^2}$$

The a_n are given by the initial conditions, $T(x,0) = T_0$ and $v(x,0) = T(x,0) - T_1 = T_0 - T_1$:

$$a_n = \frac{2}{l}\int_0^l (T_0 - T_1)\cos\frac{n\pi x}{2l}\,dx = \frac{2}{l}\cdot\frac{2l}{\pi}(T_0 - T_1)\frac{\sin n\pi/2}{n}$$

Therefore,

$$T(x,t) = T_1 + \frac{4}{\pi}(T_0 - T_1)\sum_{n=1}^{\infty}\frac{(-1)^{n+1}}{2n-1}\cos\frac{(2n-1)\pi x}{2l}e^{-(2n-1)^2\pi^2\alpha^2 t/4l^2}$$

The following figure shows $T(x,t)$ plotted against x for several values of t for $T_0 = 2$, $T_1 = 5$, and $l = 3$.

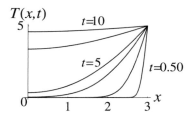

(Note that the conditions $T(x,0) = T_0$ and $T(l,t) = T_1$ are contradictory, but only at the one point $x = l$.)

21. Using $T(x,t) = v(x,t)e^{-ht}$, we have

$$T_{xx} = v_{xx}e^{-ht} \qquad \text{and} \qquad T_t = v_t e^{-ht} - hve^{-ht}$$

Substituting these results into $T_{xx} = \dfrac{1}{\alpha^2}T_t + aT$ gives

$$v_{xx}e^{-ht} = \frac{1}{\alpha^2}v_t e^{-ht} - \frac{h}{\alpha^2}ve^{-ht} + ave^{-ht}$$

which shows that we should take $h = a\alpha^2$.

16.6   The Schrödinger Equation

1. Start with

$$-\frac{\hbar^2}{2m}\left(\frac{\partial^2\psi}{\partial x^2}+\frac{\partial^2\psi}{\partial y^2}+\frac{\partial^2\psi}{\partial z^2}\right)=E\psi(x,y,z) \qquad \begin{aligned} 0&\le x\le a \\ 0&\le y\le b \\ 0&\le z\le c \end{aligned}$$

The boundary conditons are

$$\begin{aligned} \psi(0,y,z) &= \psi(a,y,z)=0 & \text{for all } y \text{ and } z \\ \psi(x,0,z) &= \psi(x,b,z)=0 & \text{for all } x \text{ and } z \\ \psi(x,y,0) &= \psi(x,y,c)=0 & \text{for all } x \text{ and } y \end{aligned}$$

We shall use the method of separation of variables. so we write $\psi(x,y,z)=X(x)Y(y)Z(z)$. Substitute this into the above Schrödinger equation and then divide through by $X(x)Y(y)Z(z)$ to obtain

$$-\frac{\hbar^2}{2m}\frac{X''}{X}-\frac{\hbar^2}{2m}\frac{Y''}{Y}-\frac{\hbar^2}{2m}\frac{Z''}{Z}=E$$

Each of the three terms on the left-hand side of this equation is a function of only x, y, or z. Because x, y, and z are independent variables, each term must equal a constant for the above equation to be valid for all values of x, y, or z. Thus we have $E_x+E_y+E_z=E$, where E_x, E_y, and E_z are constants and

$$-\frac{\hbar^2}{2m}\frac{X''}{X}=E_x \qquad -\frac{\hbar^2}{2m}\frac{Y''}{Y}=E_y \qquad -\frac{\hbar^2}{2m}\frac{Z''}{Z}=E_z$$

Now, the boundary conditions are

$$X(0)=X(a)=0 \qquad Y(0)=Y(b)=0 \qquad Z(0)=Z(c)=0$$

Thus, each of the three equations is the same as for the one-dimensional case of a particle in a box. Thus

$$\begin{aligned} X(x) &= A_x\sin\frac{n_x\pi x}{a} & n_x=1,\ 2,\ 3,\ \ldots \\ Y(y) &= A_y\sin\frac{n_y\pi y}{b} & n_y=1,\ 2,\ 3,\ \ldots \\ Z(z) &= A_z\sin\frac{n_z\pi z}{c} & n_z=1,\ 2,\ 3,\ \ldots \end{aligned}$$

Then the solution is

$$\psi_{n_x,n_y,n_z}(x,y,z)=A_xA_yA_z\sin\frac{n_x\pi x}{a}\sin\frac{n_y\pi y}{b}\sin\frac{n_z\pi z}{c}$$

with n_x, n_y, and n_y independently assuming the values 1, 2, ...

3. Equation 1 becomes

$$-\frac{\hbar^2}{2m}\left(\frac{\partial^2\psi}{\partial x^2}+\frac{\partial^2\psi}{\partial y^2}\right)=E\psi(x,y) \qquad \begin{aligned} 0&\le x\le a \\ 0&\le y\le b \end{aligned}$$

Equations 2 become

$$\psi(0,y) = \psi(a,y) = 0 \qquad \text{for all } y$$
$$\psi(x,0) = \psi(x,b) = 0 \qquad \text{for all } x$$

and Equations 3 and 4 become

$$\psi_{n_x,n_y}(x,y) = A_x A_y \sin\frac{n_x\pi x}{a}\sin\frac{n_y\pi y}{b} \qquad \begin{array}{l} n_x = 1,\ 2,\ \dots \\ n_y = 1,\ 2,\ \dots \end{array}$$

and

$$\int_0^a dx \int_0^b dy\,\psi^*(x,y)\psi(x,y) = 1$$

The normalization constant is $A_x A_y = (4/ab)^{1/2}$, and so

$$\psi_{n_x,n_y}(x,y) = \left(\frac{4}{ab}\right)^{1/2}\sin\frac{n_x\pi x}{a}\sin\frac{n_y\pi y}{b}$$

Equation 6 becomes

$$E_{n_x n_y} = \frac{\hbar^2}{8m}\left(\frac{n_x^2}{a^2}+\frac{n_y^2}{b^2}\right)^{1/2}$$

6.

$$\int_{-1}^1 P_0^0(x)P_1^0(x)\,dx = \int_{-1}^1 x\,dx = 0$$
$$\int_{-1}^1 P_0^0(x)P_2^0(x)\,dx = \frac{1}{2}\int_{-1}^1 (3x^2-1)\,dx = \frac{1}{2}\left[x^3-1\right]_{-1}^1 = 0$$
$$\int_{-1}^1 P_1^1(x)P_2^1(x)\,dx = \int_{-1}^1 (1-x^2)^{1/2}3x(1-x^2)^{1/2}\,dx = 0 \quad \text{(odd function)}$$

and so on.

9. Equation 10 tells us that

$$\sin\theta\frac{d}{d\theta}\left(\sin\theta\frac{d\Theta_{lm}}{d\theta}\right) + [\,l(l+1)\sin^2\theta - m^2\,]\Theta_{lm}(\theta) = 0$$

Now substitute $Y_l^m(\theta,\phi) = \Theta_{lm}(\theta)\Phi_m(\phi)$ into the partial differential equation given in the problem and divide by $\Theta_{lm}(\theta)\Phi_m(\phi)$ to get

$$\frac{\sin\theta}{\Theta_{lm}(\theta)}\frac{d}{d\theta}\left(\sin\theta\frac{d\Theta_{lm}}{d\theta}\right) + l(l+1)\sin^2\theta = -\frac{1}{\Phi_m(\phi)}\frac{d^2\Phi_m(\phi)}{d\phi^2}$$

Set both sides equal to m^2 to obtain

$$\frac{d^2\Phi_m(\phi)}{d\phi^2} + m^2\Phi_m(\phi) = 0$$

12. $|Y_1^1(\theta, \phi)|^2 + |Y_1^0(\theta, \phi)|^2 + |Y_1^{-1}(\theta, \phi)|^2 = \dfrac{3}{8\pi} \sin^2\theta + \dfrac{3}{4\pi}\cos^2\theta + \dfrac{3}{8\pi}\sin^2\theta = \dfrac{3}{4\pi}$

15.

$$
\begin{aligned}
L_0(x) &= e^x(e^{-x}x^0) = 1 \\
L_1(x) &= e^x \frac{d}{dx}(xe^{-x}) = 1 - x \\
L_2(x) &= e^x \frac{d^2}{dx^2}(x^2 e^{-x}) = e^x \frac{d}{dx}(2xe^{-x} - x^2 e^{-x}) \\
&= e^x(2e^{-x} - 4xe^{-x} + x^2 e^{-x}) = 2 - 4x + x^2
\end{aligned}
$$

18. Starting with Equation 24, we have

$$
-\frac{\hbar^2}{2mr^2}[\,r^2 R''(r) + 2rR'(r)\,] + \left[\frac{\hbar^2 l(l+1)}{2mr^2}\right] R(r) - ER(r) = 0
$$

or

$$
r^2 R''(r) + 2rR'(r) + \left[\frac{2mEr^2}{\hbar^2} - l(l+1)\right] R(r) = 0
$$

Compare this equation to Equation 12.5.42 to get

$$
1 - 2\alpha = 2 \qquad \gamma = 1 \qquad \beta^2 \gamma^2 = \frac{2mE}{\hbar^2} \qquad \alpha^2 - \nu^2\gamma^2 = -l(l+1)
$$

or

$$
\alpha = -\frac{1}{2} \qquad \gamma = 1 \qquad \beta = \left(\frac{2mE}{\hbar^2}\right)^{1/2} \qquad \nu^2 = \frac{1}{4} + l(l+1) = \left(l + \frac{1}{2}\right)^2
$$

The general solution is (Equation 12.5.43)

$$
R_{l\beta}(r) = c_1 r^{-1/2} J_{l+\frac{1}{2}}(\beta r) + c_2 r^{-1/2} Y_{l+\frac{1}{2}}(\beta r)
$$

If $R_{l\beta}(r)$ is to be finite at $r = 0$, we must require that $c_2 = 0$. Now, the boundary condition $R(a) = 0$ tells us that $J_{l+\frac{1}{2}}(\beta a) = 0$. If β_{ln} is the nth zero of $J_{l+\frac{1}{2}}(\beta r)$, then the above equation $\beta = (2mE/\hbar^2)^{1/2}$ gives

$$
E_{ln} = \frac{\beta_{ln}\hbar^2}{2ma^2}
$$

Thus, the allowed energy levels of a particle in a spherical box are given in terms of the zeros of spherical Bessel functions.

16.7   The Classification of Partial Differential Equations

1. Start with $au_{xx} + bu_{xy} + cu_{yy} = 0$ and $\xi = x + vy$ and $\eta = x - vy$. Then

$$
\frac{\partial u}{\partial x} = \frac{\partial u}{\partial \xi}\frac{\partial \xi}{\partial x} + \frac{\partial u}{\partial \eta}\frac{\partial \eta}{\partial x} = \frac{\partial u}{\partial \xi} + \frac{\partial u}{\partial \eta}
$$

$$u_{xx} = \frac{\partial u}{\partial x}\left(\frac{\partial u}{\partial \xi} + \frac{\partial u}{\partial \eta}\right) = \frac{\partial^2 u}{\partial \xi^2} + 2\frac{\partial^2 u}{\partial \xi \partial \eta} + \frac{\partial^2 u}{\partial \eta^2}$$

$$\frac{\partial u}{\partial y} = \frac{\partial u}{\partial \xi}\frac{\partial \xi}{\partial y} + \frac{\partial u}{\partial \eta}\frac{\partial \eta}{\partial y} = \frac{\partial u}{\partial \xi}\alpha + \beta\frac{\partial u}{\partial \eta}$$

$$u_{yy} = \frac{\partial}{\partial y}\left(\alpha\frac{\partial u}{\partial \xi} + \beta\frac{\partial u}{\partial \eta}\right) = \alpha^2\frac{\partial^2 u}{\partial \xi^2} + 2\alpha\beta\frac{\partial^2 u}{\partial \xi \partial \eta} + \beta^2\frac{\partial^2 u}{\partial \eta^2}$$

and

$$u_{xy} = \frac{\partial}{\partial x}\left(\alpha\frac{\partial u}{\partial \xi} + \beta\frac{\partial u}{\partial \eta}\right) = \alpha\frac{\partial^2 u}{\partial \xi^2} + (\alpha + \beta)\frac{\partial^2 u}{\partial \xi \partial \eta} + \beta\frac{\partial^2 u}{\partial \eta^2}$$

Substituting these results into $au_{xx} + bu_{xy} + cu_{yy} = 0$ gives

$$
\begin{aligned}
au_{xx} + bu_{xy} + cu_{yy} &= au_{\xi\xi} + 2au_{\xi\eta} + au_{\eta\eta} + \alpha bu_{\xi\xi} + (\alpha + \beta)bu_{\xi\eta} + \beta bu_{\eta\eta} \\
&\quad + c\alpha^2 u_{\xi\xi} + 2\alpha\beta cu_{\xi\eta} + \beta^2 cu_{\eta\eta} = 0
\end{aligned}
$$

$$(a + \alpha b + \alpha^2 c)u_{\xi\xi} + [2a + (\alpha + \beta)b + 2\alpha\beta c]u_{\xi\eta} + (a + \beta b + \beta^2 c)u_{\eta\eta} = 0$$

3. We use the expressions for a', b', and c' in Problem 2 and the identities $\cos^2\theta - \sin^2\theta = \cos 2\theta$ and $2\sin\theta\cos\theta = \sin 2\theta$ in what follows:

$$
\begin{aligned}
(1)\ a' - c' &= a\cos^2\theta + b\cos\theta\sin\theta + c\sin^2\theta - a\sin^2\theta + b\cos\theta\sin\theta - c\cos^2\theta \\
&= a(\cos^2\theta - \sin^2\theta) + b(2\cos\theta\sin\theta) + c(\sin^2\theta - \cos^2\theta) \\
&= a\cos 2\theta + b\sin 2\theta - c\cos 2\theta \\
&= (a - c)\cos 2\theta + b\sin 2\theta
\end{aligned}
$$

(2) $b' = (c - a)\sin 2\theta + b\cos 2\theta$

(3) $a' + c' = a(\cos^2\theta + \sin^2\theta) + c(\cos^2\theta + \sin^2\theta) = a + c$

6. First recall that the equations of an ellipse and a hyperbola are

$$\frac{x^2}{\alpha^2} + \frac{y^2}{\beta^2} = 1 \qquad\qquad \frac{x^2}{\alpha^2} - \frac{y^2}{\beta^2} = 1$$
$$\text{(ellipse)} \qquad\qquad\qquad \text{(hyperbola)}$$

Thus, for an ellipse, the signs of the x^2 and y^2 terms are the same, and those of a hyperbola differ.

We need to use the result of Problem 4. (Problems 2 through 6 are sequential.) Using the result of Problem 3, it's easy to show that

$$
\begin{aligned}
(a' - c')^2 + b'^2 &= (a - c)^2\cos^2 2\theta + 2b(a - c)\cos 2\theta\sin 2\theta + b^2\sin^2 2\theta \\
&\quad + (c - a)^2\sin^2 2\theta + 2b(c - a)\cos 2\theta\sin 2\theta + b^2\cos^2 2\theta \\
&= (a - c)^2 + b^2
\end{aligned}
$$

Now subtract $(a' + c')^2$ from one side and $(a + c)^2$ from the other to get

$$(a' - c')^2 + b'^2 - (a' + c')^2 = b'^2 - 4a'c' = (a - c)^2 + b^2 - (a + c)^2 = b^2 - 4ac$$

The relation

$$b'^2 - 4a'c' = b^2 - 4ac$$

is key.

Now rotate axes so that $b' = 0$ in the equation $a'x'^2 + b'x'y' + c'y'^2 = 0$. If a' and c' have the same sign, then $a'c' > 0$ and we have an ellipse. If a' and c' have opposite signs, then $a'c' < 0$ and we have a hyperbola. If either a' or c' equals zero, then we have a parabola.

Finally, if $b' = 0$, then

$$-4a'c' = b^2 - 4ac$$

and the result given in the statement of the problem follows.

9. See the solution to Problem 16.3.21.

12. Use the identity $\cosh(x \pm iy) = \cos x \cosh y \mp i \sin x \sinh y$ to write

$$\cosh(x + iy) - \cosh(x - iy) = -2i \sin x \sinh y$$

and

$$\sin \frac{n\pi x}{a} \sinh \frac{n\pi y}{a} = \frac{i}{2} \left[\cosh(x + iy) \frac{n\pi}{a} - \cosh(x - iy) \frac{n\pi}{a} \right]$$

CHAPTER **17**

Integral Transforms

17.1 The Laplace Transform

1. Integrate by parts twice.

$$
\int_0^\infty e^{-st}\sin\omega t\,dt = \left[-e^{-st}\frac{\cos\omega t}{\omega}\right]_0^\infty - \frac{s}{\omega}\int_0^\infty e^{-st}\cos\omega t\,dt
$$

$$
= \frac{1}{\omega} - \frac{s}{\omega}\left[e^{-st}\frac{\sin\omega t}{\omega}\right]_0^\infty - \frac{s^2}{\omega^2}\int_0^\infty e^{-st}\sin\omega t\,dt
$$

Solve for $\int_0^\infty e^{-st}\sin\omega t\,dt$ to get

$$
\int_0^\infty e^{-st}\sin\omega t\,dt = \frac{\omega}{s^2+\omega^2}
$$

$\mathcal{L}\{\cos\omega t\}$ is evaluated the same way.

3. Use $\sinh at = (e^{at}-e^{-at})/2$ to write

$$
\int_0^\infty e^{-st}\sinh at\,dt = \frac{1}{2}\int_0^\infty e^{-(s-a)t}dt - \frac{1}{2}\int_0^\infty e^{-(s+a)t}dt
$$

$$
= \frac{1/2}{s-a} - \frac{1/2}{s+a} = \frac{a}{s^2-a^2}
$$

$\mathcal{L}\{\cosh at\}$ is evaluated in the same way.

6. If $f(t)$ is sectionally continuous, then $|f(t)| < M_1$, for $0 \le t \le t_0$. And if $f(t)$ is of exponential order, then $|f(t)| < M_2 e^{ct}$ for $t \ge t_0$. Then $|f(t)| < Me^{ct}$ for $t \ge 0$, where $M = \max(M_1, M_2)$. Then

$$
\left|\int_0^\infty e^{-st}f(t)\,dt\right| < M\int_0^\infty e^{-st}e^{ct}\,dt = \frac{M}{s-c} \qquad s > c
$$

which goes to zero as $s \to \infty$.

9.

$$\frac{(1-e^{-x})^2}{1-e^{-2x}} = \frac{[e^{-x/2}(e^{x/2}-e^{-x/2})]^2}{e^{-x}(e^x-e^{-x})} = \frac{(e^{x/2}-e^{-x/2})^2}{e^x-e^{-x}}$$

$$= \frac{(e^{x/2}-e^{-x/2})^2}{(e^{x/2}+e^{-x/2})(e^{x/2}-e^{-x/2})} = \frac{e^{x/2}-e^{-x/2}}{e^{x/2}+e^{-x/2}} = \tanh\frac{x}{2}$$

12. Start with $f(\alpha) = \displaystyle\int_0^\infty e^{-\alpha t}\sin t\, dt = \frac{1}{\alpha^2+1}$. Then

$$\frac{d^2 f}{d\alpha^2} = I(\alpha) = \frac{6\alpha^2-2}{(\alpha^2+1)^3}$$

15. (a) $\mathcal{L}\{e^{at}f(t)\} = \displaystyle\int_0^\infty e^{-(s-a)t}f(t)\,dt = \hat{F}(s-a)$

(b) Using $\mathcal{L}\{\cos\omega t\} = \dfrac{s}{s^2+\omega^2} = \hat{F}(s)$, we have

$$\mathcal{L}\{e^{2t}\cos\omega t\} = \hat{F}(s-2) = \frac{s-2}{s^2-4s+4+\omega^2}$$

18. $\hat{F}(s) = \mathcal{L}\{t^{-1/2}e^{-\alpha^2/t}\} = \displaystyle\int_0^\infty t^{-1/2}e^{-st-\alpha^2/t}dt$. Let $t = x^2$, so that $dt = 2x\,dx = 2t^{1/2}\,dx$ to get

$$\hat{F}(s) = 2\int_0^\infty e^{-sx^2-\alpha^2/x^2}dx = \left(\frac{\pi}{s}\right)^{1/2}e^{-2(\alpha^2 s)^{1/2}}$$

21. Starting with $J_0(t) = 1 - \dfrac{t^2}{2^2} + \dfrac{t^4}{2^2\cdot 4^2} - \dfrac{t^6}{2^2\cdot 4^2\cdot 6^2} + \cdots$, we have

$$\begin{aligned}
\mathcal{L}\{J_0(t)\} &= \frac{1}{s} - \frac{2!}{2^2 s^3} + \frac{4!}{2^2\cdot 4^2 s^5} - \frac{6!}{2^2\cdot 4^2\cdot 6^2 s^7} + \cdots \\
&= \frac{1}{s}\left(1 - \frac{1}{2}\frac{1}{s^2} + \frac{1\cdot 3}{2\cdot 4}\frac{1}{s^4} - \frac{1\cdot 3\cdot 5}{2\cdot 4\cdot 6}\frac{1}{s^6} + \cdots\right) \\
&= \frac{1}{s}\left(1+\frac{1}{s^2}\right)^{-1/2} = \frac{1}{\sqrt{s^2+1}}
\end{aligned}$$

24. First note that $H(t) - H(t-l/2) = \begin{cases} 1 & 0\le t\le l/2 \\ 0 & \text{otherwise} \end{cases}$. So

$$\hat{F}(s) = \int_0^{l/2} e^{-st}\,dt = \frac{1-e^{-ls/2}}{s}$$

27. $\displaystyle\lim_{s\to\infty}\int_0^\infty f'(t)e^{-st}\,dt = \lim_{s\to\infty}[s\hat{F}(s)-f(0+)]$. But $\displaystyle\lim_{s\to\infty}\int_0^\infty f'(t)e^{-st}\,dt = 0$, so

$$\lim_{t\to 0+}f(t) = \lim_{s\to\infty}s\hat{F}(s)$$

17.2   The Inversion of Laplace Transforms

1. Use partial fractions to write

$$\frac{1}{s^2 + s} = \frac{1}{s(s+1)} = \frac{\alpha}{s} + \frac{\beta}{s+1} = \frac{(\alpha+\beta)s + \alpha}{s(s+1)}$$

which gives us $\alpha = 1$ and $\beta = -1$. Therefore,

$$\frac{1}{s^2 + s} = \frac{1}{s} - \frac{1}{s+1} \qquad \text{and} \qquad \mathcal{L}^{-1}\left\{\frac{1}{s^2 + s}\right\} = 1 - e^{-t}$$

3. Use partial fractions to write

$$\frac{1}{(s^2+1)(s^2+2)} = \frac{\alpha s + \beta}{s^2 + 1} + \frac{\gamma s + \delta}{s^2 + 2} = \frac{(\alpha+\gamma)s^3 + (\beta+\delta)s^2 + (2\alpha+\gamma)s + 2\beta+\delta}{(s^2+1)(s^2+2)}$$

which gives us $\alpha = 0$, $\beta = 1$, $\gamma = 0$, and $\delta = -1$. Therefore,

$$\frac{1}{(s^2+1)(s^2+2)} = \frac{1}{s^2+1} - \frac{1}{s^2+2} \qquad \text{and} \qquad \mathcal{L}^{-1}\left\{\frac{1}{(s^2+1)(s^2+2)}\right\} = \sin t - \frac{\sin(\sqrt{2}\,t)}{\sqrt{2}}$$

6. Use partial fractions to write

$$\begin{aligned}
\frac{s}{(s+1)^4} &= \frac{\alpha}{s+1} + \frac{\beta}{(s+1)^2} + \frac{\gamma}{(s+1)^3} + \frac{\delta}{(s+1)^4} \\
&= \frac{\alpha s^3 + (3\alpha+\beta)s^2 + (3\alpha+2\beta+\gamma)s + (\alpha+\beta+\gamma+\delta)}{(s+1)^4}
\end{aligned}$$

which gives $\alpha = 0$, $\beta = 0$, $\gamma = 1$, and $\delta = -1$. Therefore,

$$\frac{s}{(s+1)^4} = \frac{1}{(s+1)^3} - \frac{1}{(s+1)^4} \qquad \text{and} \qquad \mathcal{L}^{-1}\left\{\frac{s}{(s+1)^4}\right\} = \frac{1}{2}t^2 e^{-t} - \frac{1}{6}t^3 e^{-t}$$

9. Express $\hat{F}(s)$ as $\hat{F}(s-3)$:

$$\frac{s+2}{s^2 - 6s + 8} = \frac{s+2}{(s-3)^2 - 1} = \frac{s-3}{(s-3)^2 - 1} + \frac{5}{(s-3)^2 - 1}$$

Therefore,

$$\begin{aligned}
\mathcal{L}^{-1}\left\{\frac{s+2}{s^2 - 6s + 8}\right\} &= e^{3t}(\cosh t + 5\sinh t) \\
&= \frac{e^{4t}}{2} + \frac{e^{2t}}{2} + \frac{5e^{4t}}{2} - \frac{5e^{2t}}{2} = 3e^{4t} - 2e^{2t}
\end{aligned}$$

12. Using Equations 3 and 4, we have $\mathcal{L}^{-1}\left\{\dfrac{e^{-5t}}{s^2}\right\} = (t-5)H(t-5)$.

15. First express $\hat{F}(s)$ as an expansion in powers of e^{-s}

$$\hat{F}(s) = \frac{1}{s}\sum_{n=0}^{\infty} e^{-2ns} - \frac{1}{s^2}\sum_{n=0}^{\infty} e^{-(2n+1)s} + \frac{1}{s^2}\sum_{n=0}^{\infty} e^{-2(n+1)s}$$

Using Equations 3 and 4, we have

$$f(t) = \sum_{n=0}^{\infty} H(t-2n) - \sum_{n=0}^{\infty}(t-2n-1)H(t-2n-1) + \sum_{n=0}^{\infty}(t-2n-2)H(t-2n-2)$$

The following figure shows $f(t)$ from $t = 0$ to $t = 10$.

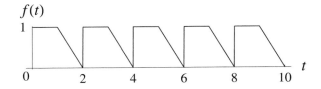

Notice that $f(t) = \begin{cases} 1 & 0 \le t \le 1 \\ 2-t & 1 \le t < 2 \end{cases}$ with $f(t) = f(t+2)$.

18. Use the convolution theorem and $\Gamma(x) = \displaystyle\int_0^{\infty} dz\, e^{-z} z^{x-1}$.

21. Using the fact that $\mathcal{L}\left\{\displaystyle\int_0^t f(u)\,du\right\} = \dfrac{\hat{F}(s)}{s}$, we see that

$$\mathcal{L}\{Si(t)\} = \mathcal{L}\left\{\int_0^t du\, \frac{\sin u}{u}\right\} = \frac{1}{s}\mathcal{L}\left\{\frac{\sin t}{t}\right\} = \frac{1}{s}\tan^{-1}\frac{1}{s}$$

24. We wish to show that

$$\frac{e^{-as^{1/2}}}{s^{1/2}} = \int_0^{\infty} e^{-st}\frac{e^{-a^2/4t}}{(\pi t)^{1/2}}\,dt$$

Assuming that we may differentiate both sides with respect to a, we have

$$e^{-as^{1/2}} = \frac{a}{(4\pi)^{1/2}}\int_0^{\infty} e^{-st}\frac{e^{-a^2/4t}}{t^{3/2}}\,dt$$

But Problem 22 shows that

$$e^{-as^{1/2}} = \frac{a}{(4\pi)^{1/2}}\int_0^{\infty} e^{-st}t^{-3/2}e^{-a^2/4t}\,dt$$

and so the above equality is valid.

17.3 Laplace Transforms and Ordinary Differential Equations

1. The substitution $y(t) = e^{\alpha t}$ gives the auxiliary equation $\alpha^2 + 3\alpha + 2 = 0$, or $\alpha = -1$ and -2. Therefore, the general solution is $y(t) = c_1 e^{-t} + c_2 e^{-2t}$. Now, $y(0) = y_0$ and $y'(0) = y_1$ give $y_0 = c_1 + c_2$ and $y_1 = -c_1 - 2c_2$, or $c_1 = 2y_0 + y_1$ and $c_2 = -(y_0 + y_1)$. Thus, we have

$$y(t) = (2y_0 + y_1)e^{-t} - (y_0 + y_1)e^{-2t}$$

3. The Laplace transform of $y''(t) - 4y(t) = 5\sin 2t$ with the initial conditions, $y(0) = 0$ and $y'(0) = 1$ is

$$s^2 \hat{Y}(s) - 1 - 4\hat{Y}(s) = \frac{10}{s^2 + 4}$$

Solving for $\hat{Y}(s)$:

$$\hat{Y}(s) = \frac{10}{(s^2 - 4)(s^2 + 4)} + \frac{1}{s^2 - 4}$$

Applying the method of partial fractions to the first term gives

$$\hat{Y}(s) = \frac{5/4}{s^2 - 4} + \frac{-5/4}{s^2 + 4} + \frac{1}{s^2 - 4} = \frac{9/4}{s^2 - 4} - \frac{5/4}{s^2 + 4}$$

or

$$y(t) = \frac{9}{8}\sinh 2t - \frac{5}{8}\sin 2t$$

6. The Laplace transform of $y'''(t) - 6y''(t) + 11y'(t) - 6y(t) = 1$ with the initial conditions $y(0) = y'(0) = y''(0) = 0$ gives

$$s^3 \hat{Y}(s) - 6s^2 \hat{Y}(s) + 11s\hat{Y}(s) - 6\hat{Y}(s) = \frac{1}{s}$$

or

$$\hat{Y}(s) = \frac{1}{s(s^3 - 6s^2 + 11s - 6)}$$

$s = 1$ is a root of $s^3 - 6s^2 + 11s - 6$, so $s^3 - 6s^2 + 11s - 6 = (s-1)(s^2 - 5s + 6) = (s-1)(s-2)(s-3)$. Apply the method of partial fractions to write

$$\hat{Y}(s) = \frac{1}{s(s-1)(s-2)(s-3)} = -\frac{1}{6s} + \frac{1}{2(s-1)} - \frac{1}{2(s-2)} + \frac{1}{6(s-3)}$$

The inverse is

$$y(t) = -\frac{1}{6} + \frac{1}{2}e^t - \frac{1}{2}e^{2t} + \frac{1}{6}e^{3t} = \frac{1}{6}(e^t - 1)^3$$

9. Let $\mathcal{L}\{x(t)\} = \hat{X}(s)$ and $\mathcal{L}\{y(t)\} = \hat{Y}(s)$. Then, taking the Laplace transform of the two differential equations and the initial conditions gives

$$s\hat{X}(s) - 2 - \hat{X}(s) - 2\hat{Y}(s) = \frac{1}{s^2}$$

and

$$-2\hat{X}(s) + s\hat{Y}(s) - 4 - \hat{Y}(s) = \frac{1}{s^2}$$

Solving for $\hat{X}(s)$ and $\hat{Y}(s)$ gives

$$\hat{X}(s) = \frac{(s-1)(1+2s^2) + 2(1+4s^2)}{s^2(s+1)(s-3)}$$

$$\hat{Y}(s) = \frac{(4s^2+1)(s-1) + 2(2s^2+1)}{s^2(s-3)(s+1)}$$

Using partial fractions gives

$$\hat{X}(s) = -\frac{1}{9s} - \frac{1}{3s^2} + \frac{28}{9(s-3)} - \frac{1}{s+1}$$

$$\hat{Y}(s) = -\frac{1}{9s} - \frac{1}{3s^2} + \frac{28}{9(s-3)} + \frac{1}{s+1}$$

Therefore,

$$x(t) = -\frac{1}{9} - \frac{t}{3} - e^{-t} + \frac{28}{9}e^{3t}$$

$$y(t) = -\frac{1}{9} - \frac{t}{3} + e^{-t} + \frac{28}{9}e^{3t}$$

12. The Laplace transform of the differential equation and its initial conditions gives $Ls^2\hat{Q}(s) + \frac{1}{C}\hat{Q}(s) = \frac{E_0}{s}$. Solving for $\hat{Q}(s)$ gives

$$\hat{Q}(s) = \frac{E_0/L}{s(s^2 + 1/LC)} = \frac{E_0 C}{s} - \frac{E_0 C s}{s^2 + 1/LC}$$

and so

$$q(t) = E_0 C \left[1 - \cos\frac{t}{(LC)^{1/2}}\right]$$

$$i(t) = E_0 \left(\frac{C}{L}\right)^{1/2} \sin\frac{t}{(LC)^{1/2}}$$

15. The Laplace transform of the differential equation $R\dot{q} + \frac{1}{C}q = E_0\sin\omega t$ with the initial conditions $q(0) = i(0) = 0$ is

$$Rs\hat{Q}(s) + \frac{1}{C}\hat{Q}(s) = \frac{E_0\omega}{s^2 + \omega^2}$$

Solving for $\hat{Q}(s)$ gives

$$\hat{Q}(s) = \frac{1}{Rs + 1/C} \cdot \frac{E_0\omega}{s^2 + \omega^2} = \frac{E_0\omega/R}{(s + 1/RC)(s^2 + \omega^2)}$$

$$= \frac{E_0 R}{\omega Z^2}\frac{1}{s + 1/RC} - \frac{E_0 R}{\omega Z^2}\frac{s}{s^2 + \omega^2} + \frac{E_0}{\omega C Z^2}\frac{1}{s^2 + \omega^2}$$

The inverse is

$$q(t) = \frac{E_0 R}{\omega Z^2} e^{-t/RC} - \frac{E_0 R}{\omega Z^2} \cos \omega t + \frac{E_0}{\omega^2 C Z^2} \sin \omega t$$

and so

$$i(t) = -\frac{E_0}{\omega C Z^2} e^{-t/RC} + \frac{E_0 R}{Z^2} \sin \omega t + \frac{E_0}{\omega C Z^2} \cos \omega t$$

18. The differential equation is $\dfrac{d^4 y}{dx^4} = -\dfrac{w_0 \delta(x - a)}{\gamma}$ and the boundary conditions in this case are $y(0) = y'(0) = 0$ and $y(2a) = y'(2a) = 0$. The Laplace transform gives

$$s^4 \hat{Y}(s) - s y''(0) - y'''(0) = -\frac{w_0 e^{-as}}{\gamma}$$

where $y''(0)$ and $y'''(0)$ are as yet unknowns. Now let $y''(0) = \alpha$ and $y'''(0) = \beta$. Then $\hat{Y}(s)$ becomes

$$\hat{Y}(s) = \frac{\alpha}{s^3} + \frac{\beta}{s^4} - \frac{w_0 e^{-as}}{s^4 \gamma}$$

and $y(x)$ is given by

$$y(x) = \frac{\alpha x^2}{2} + \frac{\beta x^3}{6} - \frac{w_0}{\gamma} \frac{(x - a)^3}{6} H(x - a)$$

For $a < x \le 2a$,

$$y(x) = \frac{\alpha x^2}{2} + \frac{\beta x^3}{6} - \frac{w_0}{\gamma} \frac{(x - a)^3}{6}$$

The boundary condition $y(2a) = 0$ gives $2\alpha a^2 + \dfrac{4\beta a^3}{3} - \dfrac{w_0}{\gamma} \dfrac{a^3}{6} = 0$ and $y'(2a) = 0$ gives $2\alpha a + 2\beta a^2 - \dfrac{w_0}{\gamma} \dfrac{a^2}{2} = 0$. Therefore, $\beta = \dfrac{w_0}{2\gamma}$ and $\alpha = -\dfrac{w_0 a}{4\gamma}$, and the solution is

$$y(x) = -\frac{w_0}{2\gamma} \left[\frac{a x^2}{4} - \frac{x^3}{6} + \frac{(x - a)^3}{3} H(x - a) \right]$$

21. The differential equation is $\gamma \dfrac{d^4 y}{dx^4} = -w_0$ and the boundary conditions in this case are $y(0) = y(2a) = y''(0) = y''(2a) = 0$. Taking the Laplace transform of the differential equations with its initial conditions gives

$$s^4 \hat{Y}(s) - s^2 y'(0) - y'''(0) = -\frac{w_0}{\gamma s}$$

Let $y'(0) = \alpha$ and $y'''(0) = \beta$ to write

$$\hat{Y}(s) = \frac{\alpha}{s^2} + \frac{\beta}{s^4} - \frac{w_0}{\gamma s^5}$$

The inverse of $\hat{Y}(s)$ is

$$y(x) = \alpha x + \frac{\beta x^3}{6} - \frac{\omega_0}{\gamma}\frac{x^4}{24}$$

The boundary conditions at $x = 2a$ gives

$$y(2a) = 0 = 2a\alpha + \frac{4}{3}\beta a^3 - \frac{2}{3}\frac{\omega_0}{\gamma}a^4 = 0$$

$$y''(2a) = 0 = 2a\beta - 2a^2\frac{\omega_0}{\gamma} = 0$$

and so $\beta = \dfrac{a\omega_0}{\gamma}$ and $\alpha = -\dfrac{a^3\omega_0}{\gamma}$. Therefore,

$$y(x) = -\frac{\omega_0}{24\gamma a}(x^4 - 4ax^3 + 8a^3 x)$$

We can also solve this problem by simply integrating the differential equation $\gamma\dfrac{d^4y}{dx^4} = -\omega_0$ four times. The general solution is

$$\gamma y(x) = -\frac{w_0 x^4}{24} + \frac{c_1 x^3}{6} + \frac{c_2 x^2}{2} + c_3 x + c_4$$

There are four boundary conditions to determine the four constants of integration, and it turns out that

$$c_4 = 0, \qquad c_2 = 0, \qquad -\frac{w_0(2a)^4}{24} + \frac{c_1(2a)^3}{6} + c_3(2a) = 0, \qquad -\frac{w_0(2a)^2}{2} + c_1(2a) = 0$$

from which we find that $c_1 = w_0 a$ and $c_3 = -\frac{1}{3}w_0 a^3$. Therefore, we obtain

$$\gamma y(x) = -\frac{\omega_0}{24}(x^4 - 4ax^3 + 8a^3 x)$$

in agreement with using the Laplace transform.

17.4   Laplace Transforms and Partial Differential Equations

1. $\dfrac{\partial}{\partial x}(te^{-xt}) = -t^2 e^{-xt}$ and $\mathcal{L}\{-t^2 e^{-xt}\} = -\dfrac{2}{(s+x)^3}$. Now

$$\mathcal{L}\{te^{-xt}\} = \frac{1}{(s+x)^2} \qquad \text{and} \qquad \frac{\partial}{\partial x}\mathcal{L}\{te^{-xt}\} = -\frac{2}{(s+x)^3}$$

3.

$$\frac{\partial u}{\partial x} = \frac{\beta_0}{\kappa}\left\{-\frac{2x}{(4\pi\alpha^2 t)^{1/2}}e^{-x^2/4\alpha^2 t} - \text{erfc}\left[\frac{x}{(4\alpha^2 t)^{1/2}}\right] + \frac{2x}{(4\pi\alpha^2 t)^{1/2}}e^{-x^2/4\alpha^2 t}\right\}$$

$$= -\frac{\beta_0}{\kappa}\text{erfc}\left[\frac{x}{(4\alpha^2 t)^{1/2}}\right]$$

$$\frac{\partial^2 u}{\partial x^2} = \frac{\beta_0}{\kappa}\frac{1}{(\pi\alpha^2 t)^{1/2}}e^{-x^2/4\alpha^2 t}$$

$$\frac{\partial u}{\partial t} = \frac{\beta_0}{\kappa}\left[\left(\frac{\alpha^2}{\pi t}\right)^{1/2}e^{-x^2/4\alpha^2 t} + \left(\frac{4\alpha^2 t}{\pi}\right)^{1/2}\frac{x^2}{4\alpha^2 t^2}e^{-x^2/4\alpha^2 t} - \frac{x^2}{(4\pi\alpha^2 t^3)^{1/2}}e^{-x^2/4\alpha^2 t}\right]$$

$$= \frac{\beta_0}{\kappa}\frac{\alpha^2}{(\pi\alpha^2 t)^{1/2}}e^{-x^2/4\alpha^2 t} = \alpha^2\frac{\partial^2 u}{\partial x^2}$$

6. The heat equation is $\dfrac{\partial^2 u}{\partial x^2} = \dfrac{1}{\alpha^2}\dfrac{\partial u}{\partial t}$ $\begin{array}{c} 0 \le x < \infty \\ 0 \le t \end{array}$ with the conditions $u(x,0) = 0$, $u(0,t) = f(t)$, and $\lim\limits_{x \to \infty} u(x,t) = 0$. Taking the Laplace transform with respect to t yields

$$\frac{d^2 \hat{U}}{ds^2} - \frac{s}{\alpha^2}\hat{U} = 0$$

whose general solution is

$$\hat{U}(x,s) = c_1(s)e^{x(s/\alpha^2)^{1/2}} + c_2(s)e^{-x(s/\alpha^2)^{1/2}}$$

The condition as $x \to \infty$ requires that $c_1(s) = 0$. The condition at $x = 0$ requires that $\hat{U}(0,s) = c_2(s) = \mathcal{L}\{f(t)\} = \hat{F}(s)$, so that

$$\hat{U}(x,s) = \hat{F}(s)e^{-x(s/\alpha^2)^{1/2}}$$

The inverse Laplace transform of $e^{-x(s/\alpha^2)^{1/2}}$ is (See Problem 17.2.22.)

$$\mathcal{L}^{-1}\{e^{-x(s/\alpha^2)^{1/2}}\} = \frac{1}{(4\pi\alpha^2 t^3)^{1/2}}e^{-x^2/4\alpha^2 t}$$

We now use the convolution theorem to write the inverse of $\hat{U}(x,s)$ as

$$u(x,t) = \int_0^t f(t-z)\frac{x}{(4\pi\alpha^2 z^3)^{1/2}}e^{-x^2/4\alpha^2 z}\,dz$$

9. The heat equation is $\dfrac{\partial^2 u}{\partial x^2} = \dfrac{1}{\alpha^2}\dfrac{\partial u}{\partial t}$ $\begin{array}{c} 0 \le x < \infty \\ 0 \le t \end{array}$ with the conditions $u(x,0) = 0$, $u_x(0,t) = -f(t)$, and $u(x,t) \to 0$ as $x \to \infty$. The Laplace transform with respect to t gives

$$\frac{d^2 \hat{U}}{ds^2} - \frac{s}{\alpha^2}\hat{U} = 0$$

whose general solution is

$$\hat{U}(x,s) = c_1(s)e^{x(s/\alpha^2)^{1/2}} + c_2(s)e^{-x(s/\alpha^2)^{1/2}}$$

The condition as $x \to \infty$ requires that $c_1(s) = 0$ and the boundary condition at $x = 0$ requires that

$$\left(\frac{\partial \hat{U}}{\partial x}\right)_{x=0} = \mathcal{L}\{u_x(0,t)\} = \mathcal{L}\{-f(t)\} = -\hat{F}(s) = -\left(\frac{s}{\alpha^2}\right)^{1/2}c_2(s)$$

and so we find that $c_2(s) = \left(\dfrac{\alpha^2}{s}\right)^{1/2}\hat{F}(s)$. Therefore,

$$\hat{U}(x,s) = \left(\frac{\alpha^2}{s}\right)^{1/2}\hat{F}(s)e^{-x(s/\alpha^2)^{1/2}}$$

The inverse Laplace transform of $e^{-x(s/\alpha^2)^{1/2}}/s^{1/2}$ is (Problem 17.2.24)

$$\mathcal{L}^{-1}\left\{\frac{e^{-x(s/\alpha^2)^{1/2}}}{s^{1/2}}\right\} = \frac{e^{-x^2/4\alpha^2 t}}{(\pi t)^{1/2}}$$

We now use the convolution theorem to write the inverse of $\hat{U}(x,t)$ as

$$u(x,t) = \left(\frac{\alpha^2}{\pi}\right)^{1/2}\int_0^t f(t-z)\frac{e^{-x^2/4\alpha^2 z}}{z^{1/2}}\,dz$$

12. The heat equation is $\dfrac{\partial^2 u}{\partial x^2} = \dfrac{1}{\alpha^2}\dfrac{\partial u}{\partial t}$ $\begin{array}{c}0 \le x < \infty \\ 0 \le t\end{array}$ with the conditions $u(x,0) = 0$, $u_x(0,t) = -\beta$, and $u(x,t) \to 0$ as $x \to \infty$. The Laplace transform with respect to t gives

$$\frac{d^2\hat{U}}{ds^2} - \frac{s}{\alpha^2}\hat{U} = 0$$

whose general solution is

$$\hat{U}(x,s) = c_1(s)e^{x(s/\alpha^2)^{1/2}} + c_2(s)e^{-x(s/\alpha^2)^{1/2}}$$

The condition as $x \to \infty$ requires that $c_1(s) = 0$ and the boundary condition at $x = 0$ requires that

$$\left(\frac{\partial\hat{U}}{\partial x}\right)_{x=0} = \mathcal{L}\{u_x(0,t)\} = -\frac{\beta}{s} = -c_2(s)\left(\frac{s}{\alpha^2}\right)^{1/2}$$

Therefore, $c_2(s) = \alpha\beta/s^{3/2}$, and

$$\hat{U}(x,s) = \frac{\alpha\beta}{s^{3/2}}e^{-x(s/\alpha^2)^{1/2}}$$

The inverse is

$$u(x,t) = \beta\left\{2\left(\frac{\alpha^2 t}{\pi}\right)^{1/2}e^{-x^2/4\alpha^2 t} - x\,\mathrm{erfc}\left[\frac{x}{(4\alpha^2 t)^{1/2}}\right]\right\}$$

We can check this answer by substituting $f = \beta$ into the result of Problem 9. This gives

$$u(x,t) = \beta\left(\frac{\alpha^2}{\pi}\right)^{1/2}\int_0^t\frac{e^{-x^2/4\alpha^2 z}}{z^{1/2}}\,dz$$

Let $z = 1/y^2$ to get

$$u(x,t) = \beta\left(\frac{4\alpha^2}{\pi}\right)^{1/2}\int_{1/t^{1/2}}^\infty\frac{e^{-(x^2/4\alpha^2)y^2}}{y^2}\,dy$$

Now integrate by parts, letting "u" $= e^{-(x^2/4\alpha^2)y^2}$ and "dv" $= dy/y^2$ to obtain

$$\begin{aligned}
u(x,t) &= \beta\left(\frac{4\alpha^2}{\pi}\right)^{1/2}\left\{\left[-\frac{e^{-(x^2/4\alpha^2)y^2}}{y}\right]_{1/t^{1/2}}^\infty - \frac{x^2}{2\alpha^2}\int_{1/t^{1/2}}^\infty e^{-(x^2/4\alpha^2)y^2}\,dy\right\} \\
&= \frac{2\alpha\beta}{\pi^{1/2}}\left[t^{1/2}e^{-x^2/4\alpha^2 t} - \frac{x\pi^{1/2}}{2\alpha}\,\mathrm{erfc}\left\{\frac{x}{(4\alpha^2 t)^{1/2}}\right\}\right]
\end{aligned}$$

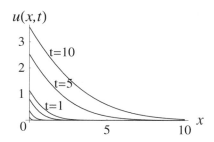

This is the same result as we obtained above. This solution is plotted against x for various values of t in the figure above.

15. The Laplace transform with respect to t of the wave equation with the initial conditions $u(x,0) = \sin n\pi x/l$ and $u_t(x,0) = 0$ gives

$$\frac{d^2\hat{U}}{dx^2} = \frac{s^2}{v^2}\hat{U} - \frac{s}{v^2}\sin\frac{n\pi x}{l}$$

The solution to the homogeneous equation is

$$\hat{U}_h(x,s) = c_1(s)e^{xs/v} + c_2(s)e^{-xs/v}$$

Notice by inspection that the particular solution is of the form $A\sin n\pi x/l$. Substituting this into the above nonhomogeneous equation for $\hat{U}$ gives

$$-\frac{n^2\pi^2 A}{l^2} - \frac{s^2}{v^2}A = -\frac{s}{v^2}$$

or $A = \dfrac{s}{s^2 + n^2\pi^2 v^2/l^2}$. Therefore, the particular solution is

$$\hat{U}_p(x,s) = \frac{s}{s^2 + n^2\pi^2 v^2/l^2}\sin\frac{n\pi x}{l}$$

and the complete solution is

$$\hat{U}(x,s) = c_1(s)e^{xs/v} + c_2(s)e^{-xs/v} + \frac{s\sin n\pi x/l}{s^2 + n^2\pi^2 v^2/l^2}$$

The two boundary conditions give $c_1(s) = c_2(s) = 0$, so

$$\hat{U}(x,s) = \frac{s\sin n\pi x/l}{s^2 + n^2\pi^2 v^2/l^2}$$

The inverse of $\hat{U}(x,s)$ is

$$u(x,t) = \sin\frac{n\pi x}{l}\cos\frac{n\pi vt}{l}$$

18. The Laplace transform with respect to t of the equation $u_{xx} = v^2 u_{tt}$ and the given initial conditions yields

$$\frac{d^2\hat{U}}{dx^2} - \frac{s^2}{v^2}\hat{U} = 0$$

The general solution to this equation is

$$\hat{U}(x,s) = c_1(s)e^{xs/v} + c_2(s)e^{-xs/v}$$

The condition $\hat{U}(0,s) = 0$ gives $c_1(s) + c_2(s) = 0$ and the condition $u_x(l,t) = \gamma\delta(t)$ gives

$$\hat{U}_x(x,s) = \mathcal{L}\{\gamma\delta(t)\} = \gamma = \frac{s}{v}[c_1(s)e^{ls/v} - c_2(s)e^{-ls/v}] = \frac{2sc_1(s)}{v}\cosh\frac{ls}{v}$$

Therefore,

$$\hat{U}(x,s) = \frac{\gamma v}{s}\frac{\sinh(xs/v)}{\cosh(ls/v)}$$

You can invert $\hat{U}(x,s)$ by expanding in powers of $e^{-xs/v}$ and $e^{-ls/v}$ to get

$$u(x,t) = \gamma\left\{\sum_{n=0}^{\infty}(-1)^n H[vt - (2n+1)l + x] - \sum_{n=0}^{\infty}(-1)^n H[vt - (2n+1)l - x]\right\}$$

This solution for the motion at the end of the bar is plotted below for $l = 2$.

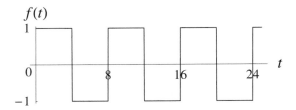

17.5   Fourier Transform

1.

$$\mathcal{F}\left\{\frac{1}{t^2 + a^2}\right\} = \hat{F}(\omega) = \frac{1}{(2\pi)^{1/2}}\int_{-\infty}^{\infty}\frac{e^{-i\omega t}}{t^2 + a^2}\,dt = \left(\frac{2}{\pi}\right)^{1/2}\int_0^{\infty}\frac{\cos\omega t}{t^2 + a^2}\,dt$$

$$= \left(\frac{\pi}{2a^2}\right)^{1/2}e^{-|a\omega|}$$

3.

$$\mathcal{F}\{te^{-a|t|}\} = \hat{F}(\omega) = \frac{1}{(2\pi)^{1/2}}\int_{-\infty}^{\infty}te^{-a|t|}e^{-i\omega t}\,dt = -\frac{2i}{(2\pi)^{1/2}}\int_{)}^{\infty}te^{-at}\sin\omega t\,dt$$

$$= -\left(\frac{2}{\pi}\right)^{1/2}\frac{2ia\omega}{(\omega^2 + a^2)^2}$$

6. We'll do Problems 4 and 5 first. For Problem 4,

$$\mathcal{F}\left\{\frac{1}{x}\right\} = \frac{1}{(2\pi)^{1/2}}\int_{-\infty}^{\infty}\frac{e^{-ikx}}{x}\,dx = -i\left(\frac{2}{\pi}\right)^{1/2}\int_0^{\infty}\frac{\sin kx}{x}\,dx$$

$$= -i\left(\frac{2}{\pi}\right)^{1/2}\begin{cases}\dfrac{\pi}{2} & k > 0 \\ 0 & k = 0 \\ -\dfrac{\pi}{2} & k < 0\end{cases} = -i\left(\frac{2}{\pi}\right)^{1/2}\operatorname{sgn}k$$

For Problem 5, we invert the result of Problem 4 to write

$$\mathcal{F}^{-1}\{\operatorname{sgn} k\} = \frac{1}{(2\pi)^{1/2}} \int_{-\infty}^{\infty} e^{ikx} \operatorname{sgn} k \, dk = \left(\frac{2}{\pi}\right)^{1/2} \frac{i}{x}$$

Replace x by $-x$ to write

$$\frac{1}{(2\pi)^{1/2}} \int_{-\infty}^{\infty} e^{-ikx} \operatorname{sgn} k \, dk = -\left(\frac{2}{\pi}\right)^{1/2} \frac{i}{x}$$

and then interchange x and k to obtain

$$\mathcal{F}\{\operatorname{sgn} x\} = -i \left(\frac{2}{\pi}\right)^{1/2} \frac{1}{k}$$

Finally, to solve Problem 6, we use the relation $H(x) = \frac{1}{2}[\,1 + \operatorname{sgn}(x)\,]$ to write

$$\mathcal{F}\{H(x)\} = \mathcal{F}\left\{\frac{1}{2}\right\} + \frac{1}{2}\mathcal{F}\{\operatorname{sgn}(x)\} = \left(\frac{\pi}{2}\right)^{1/2} \delta(k) - \frac{i}{(2\pi)^{1/2}k}$$

9.

$$
\begin{aligned}
\mathcal{F}\{e^{iax} f(x)\} &= \frac{1}{(2\pi)^{1/2}} \int_{-\infty}^{\infty} f(x) e^{iax} e^{-ikx} dx \\
&= \frac{1}{(2\pi)^{1/2}} \int_{-\infty}^{\infty} f(x) e^{-i(k-a)x} dx = \hat{F}(k-a)
\end{aligned}
$$

12.

$$
\begin{aligned}
\mathcal{F}\{e^{2it} e^{-|t|}\} &= \hat{F}(\omega) = \frac{1}{(2\pi)^{1/2}} \int_{-\infty}^{\infty} e^{2it} e^{-|t|} e^{-i\omega t} \, dt \\
&= \frac{1}{(2\pi)^{1/2}} \int_{-\infty}^{\infty} (\cos\omega t - i\sin\omega t)(\cos 2t + i\sin 2t) e^{-|t|} \, dt \\
&= \frac{2}{(2\pi)^{1/2}} \int_{0}^{\infty} (\cos\omega t \cos 2t + \sin\omega t \sin 2t) e^{-t} \, dt \\
&= \left(\frac{2}{\pi}\right)^{1/2} \int_{0}^{\infty} e^{-t} \cos\left[(\omega - 2)t\right] dt = \left(\frac{2}{\pi}\right)^{1/2} \frac{1}{(\omega - 2)^2 + 1} \\
&= \left(\frac{2}{\pi}\right)^{1/2} \frac{1}{\omega^2 - 4\omega + 5}
\end{aligned}
$$

15. Start with Equation 21,

$$
\begin{aligned}
\hat{F}(k) &= \frac{2\pi}{(2\pi)^{3/2}} \int_{0}^{\infty} dr \, r^2 f(r) \int_{0}^{\infty} d\theta \, \sin\theta e^{-ikr\cos\theta} \\
&= \frac{1}{(2\pi)^{1/2}} \int_{0}^{\infty} dr \, r^2 f(r) \left[\frac{e^{-ikr\cos\theta}}{ikr}\right]_{0}^{\pi} = \left(\frac{2}{\pi}\right)^{1/2} \frac{1}{k} \int_{0}^{\infty} dr \, r f(r) \sin kr
\end{aligned}
$$

18. We'll do the previous problem first. Using Equation 22, we have

$$\mathcal{F}\left\{\frac{e^{-\alpha r}}{r}\right\} = \hat{F}(k) = \left(\frac{2}{\pi}\right)^{1/2} \frac{1}{k} \int_0^\infty e^{-\alpha r} \sin kr\, dr = \left(\frac{2}{\pi}\right)^{1/2} \frac{1}{k^2 + \alpha^2}$$

Now let $\alpha \to 0$ to get

$$\mathcal{F}\left\{\frac{1}{r}\right\} = \left(\frac{2}{\pi}\right)^{1/2} \frac{1}{k^2}$$

21. The integral here is a three-dimensional convolution integral. Taking Fourier transforms gives

$$\hat{W}(k) = \left(\frac{2}{\pi}\right)^{1/2} \frac{1}{k^2} - \frac{\kappa^2}{4\pi}(2\pi)^{3/2} \hat{W}(k) \left(\frac{2}{\pi}\right)^{1/2} \frac{1}{k^2}$$

where we have used the result of Problem 18. Solving for $\hat{W}(k)$ gives

$$\hat{W}(k) = \left(\frac{2}{\pi}\right)^{1/2} \frac{1}{k^2 + \kappa^2}$$

Inverting $\hat{W}(k)$ yields $w(r) = e^{-\kappa r}/r$, which is called a screened Coulombic potential.

24. If $f(x)$ is an even function of t, then

$$\hat{F}_c(\omega) = \left(\frac{2}{\pi}\right)^{1/2} \int_0^\infty f(t) \cos \omega t\, dt = \frac{1}{(2\pi)^{1/2}} \int_{-\infty}^\infty f(t) e^{-i\omega t}\, dt$$

Now use Equation 9 to write

$$f(t) = \frac{1}{(2\pi)^{1/2}} \int_{-\infty}^\infty d\omega\, \hat{F}_c(\omega) e^{i\omega t} = \left(\frac{2}{\pi}\right)^{1/2} \int_0^\infty d\omega\, \hat{F}_c(\omega) \cos \omega t$$

where we have recognized that $\hat{F}_c(\omega)$ is an even function of ω.

27. The result of Example 6 and Equation 23 says that

$$\begin{aligned}
f(r) &= \frac{z^3}{\pi} e^{-2zr} = \left(\frac{2}{\pi}\right)^{1/2} \int_0^\infty \hat{F}(k) \frac{k \sin kr}{r}\, dk \\
&= \frac{2}{\pi} \frac{1}{r} \int_0^\infty \frac{4z^4 k \sin kr}{\pi(k^2 + 4z^2)^2}\, dk
\end{aligned}$$

or that

$$e^{-2zr} = \frac{8z}{\pi r} \int_0^\infty \frac{k \sin kr}{(k^2 + 4z^2)^2}\, dk$$

Now let $2z = \beta$, $k = x$, and $r = a$ to write

$$\int_0^\infty \frac{x \sin ax\, dx}{(x^2 + \beta^2)^2} = \frac{a\pi}{4\beta} e^{-\beta a}$$

17.6   Fourier Transforms and Partial Differential Equations

1. We use Equation 7 with $f(u) = c_0 H(u)$, or

$$
\begin{aligned}
c(x,t) &= \frac{c_0}{(4\pi Dt)^{1/2}} \int_0^\infty e^{-(x-u)^2/4Dt} du = \frac{c_0}{(4\pi Dt)^{1/2}} \int_{-x}^\infty e^{-z^2/4Dt} dz \\
&= \frac{c_0}{\pi^{1/2}} \int_{-x/(4Dt)^{1/2}}^\infty e^{-y^2} dy = \frac{c_0}{\pi^{1/2}} \left[\int_{-x/(4Dt)^{1/2}}^0 e^{-u^2} du + \int_0^\infty e^{-u^2} du \right] \\
&= \frac{c_0}{2} \left\{ 1 + \text{erf} \left[\frac{x}{(4Dt)^{1/2}} \right] \right\}
\end{aligned}
$$

This result is plotted in the figure below for several values of Dt.

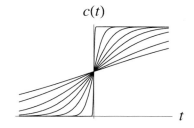

$c(t)$

t

3. Use Equation 7

$$
c(x,t) = \frac{c_0}{(4\pi Dt)^{1/2}} \int_{-\infty}^\infty e^{-(u-x)^2/4Dt} du = c_0
$$

This solution is what you would expect physically and is valid. Yet, $c(x,t)$ does not satisfy the condition $c(x,t) \to 0$ as $x \to \pm\infty$, which is that $c(x,t)$ have a Fourier transform in the usual sense.

6.

$$
\begin{aligned}
\mathcal{F}\{\nabla^2 c(x,y,z)\} &= \frac{1}{(2\pi)^{3/2}} \iiint_{-\infty}^\infty dxdydz\, e^{-ik_x x} e^{-ik_y y} e^{-ik_z z} \nabla^2 c(x,y,z) \\
&= \frac{1}{(2\pi)^{3/2}} \iiint_{-\infty}^\infty dxdydz\, e^{-ik_x x} e^{-ik_y y} e^{-ik_z z} \frac{\partial^2 c}{\partial x^2} + \text{similar terms in } y \text{ and } z
\end{aligned}
$$

Now use Equation 1 to write

$$
\begin{aligned}
\mathcal{F}\{\nabla^2 c(x,y,z)\} &= -k_x^2 \hat{C}(k_x, k_y, k_z) - k_y^2 \hat{C}(k_x, k_y, k_z) - k_z^2 \hat{C}(k_x, k_y, k_z) \\
&= -k^2 \hat{C}(k_x, k_y, k_z)
\end{aligned}
$$

9. The solution for $u(x,t)$ in Problem 8 is

$$
u(x,t) = \frac{c_0}{(4\pi Dt)^{1/2}} e^{-(x-x_0)^2/4Dt}
$$

and so

$$
\begin{aligned}
c(x,t) &= \frac{c_0}{(4\pi Dt)^{1/2}} e^{-(x-x_0)^2/4Dt} e^{(\mu x/2D - \mu^2 t)/4D} \\
&= \frac{c_0}{(4\pi Dt)^{1/2}} e^{-[(x-x_0)^2 - 2\mu x + \mu^2 t^2]/4Dt}
\end{aligned}
$$

This result differs only by a constant factor, $e^{\mu x_0/2D}$, from

$$
c(x,t) = \frac{c_0}{(4\pi Dt)^{1/2}} e^{-(x-x_0-\mu t)^2/4Dt}
$$

12.

$$
\begin{aligned}
\mathcal{F}_2\{e^{-(x^2+y^2)/4at}\} &= \frac{1}{2\pi} \int_{-\infty}^{\infty} dx \int_{-\infty}^{\infty} dy\, e^{i(kx+ly)} e^{-(x^2+y^2)/4at} \\
&= \left[\frac{1}{(2\pi)^{1/2}} \int_{-\infty}^{\infty} dx\, e^{-ikx} e^{-x^2/4at} \right] \left[\frac{1}{(2\pi)^{1/2}} \int_{-\infty}^{\infty} dy\, e^{-ly} e^{-y^2/4at} \right] \\
&= [(2at)^{1/2} e^{-ak^2 t}][(2at)^{1/2} e^{-al^2 t}] = 2at e^{-a(k^2+l^2)t}
\end{aligned}
$$

15. Start with

$$
c(r,t) = \frac{1}{2\pi} \iint_{-\infty}^{\infty} \hat{C}(k,t) e^{i\mathbf{k}\cdot\mathbf{r}}\, dk_x dk_y
$$

Now convert to plane polar coordinates to write

$$
\begin{aligned}
c(r,t) &= \frac{1}{2\pi} \int_0^{\infty} dk\, k\hat{C}(k,t) \int_0^{2\pi} d\theta\, e^{ikr\cos\theta} \\
&= \int_0^{\infty} dk\, k\hat{C}(k,t) J_0(kr)
\end{aligned}
$$

where we used the definition $\displaystyle\int_0^{2\pi} d\theta\, e^{ikr\cos\theta} = 2\pi J_0(kr)$.

18. In Problem 17, there is only one absorbing barrier, and so when we place an image sink at $x = 2a$ in order to make $c(a,t) = 0$, no other boundary conditions are violated. However, if there is another absorbing barrier at $x = -a$, then $c(-a,t)$ is given by

$$
c(-a,t) = \frac{1}{(4\pi Dt)^{1/2}} \left(e^{-a^2/4Dt} - e^{-9a^2/4Dt} \right)
$$

instead of zero. To make $c(-a,t) = 0$, we must add an image sink at $x = -2a$ and an image source at $x = -4a$ so that $c(x,t)$ now becomes

$$
c(x,t) = \frac{1}{(4\pi Dt)^{1/2}} \left[e^{-x^2/4Dt} - e^{-(x-2a)^2/4Dt} - e^{-(x+2a)^2/4Dt} + e^{-(x+4a)^2/4Dt} \right]
$$

The addition of these terms fixes up the boundary requirement at $x = -a$, but now $c(a,t) \neq 0$. We can make $c(a,t) = 0$ by adding an image source at $x = 4a$ and an image sink at $x = 6a$. By continuing in this manner, you generate the solution given in the statement of the problem.

21. The results differ by the sign of k.

CHAPTER 18

Functions of a Complex Variable: Theory

18.1   Functions, Limits, and Continuity

1. If $\text{Im } z > 0$, then z is in the upper half plane, Thus, the distance from z to $-i$, $|z - i|$, must be greater than the distance from z to i, $|z + i|$. Analytically, the distance from z to i is $(x^2 + (y - 1)^2)^{1/2}$ and the distance from z to $-i$ is $(x^2 + (y + 1)^2)^{1/2}$. Squaring both sides gives

$$x^2 + (y - 1)^2 < x^2 + (y + 1)^2$$

or $-2y < 2y$, which is so if $y > 0$.

3. The condition $|z - z_0| \le R$ specifies the disk; the condition $\text{Im } (z - z_0) > 0$ specifies its upper half; and $z_0 = x_0 + iy_0$ is the center of the disk.

6. $\sin iz = \dfrac{e^{i(iz)} - e^{-i(iz)}}{2i} = \dfrac{e^{-z} - e^{z}}{2i} = i\left(\dfrac{e^{z} - e^{-z}}{2}\right) = i \sinh z$

9. Use $1 + i = \sqrt{2}\, e^{i\pi/4} = \sqrt{2}\, e^{i\pi/4 + i2\pi n}$ for $n = 0, \pm 1, \pm 2, \ldots$ Therefore,

$$\ln(1 + i) = \frac{1}{2}\ln 2 + \frac{i\pi}{4} + i2\pi n \qquad \text{where} \quad n = 0, \pm 1, \pm 2, \ldots$$

12. Using the fact that $x \to 1$ and $y \to 2$ as $z \to 1 + 2i$, we have

$$\lim_{\substack{x \to 1 \\ y \to 2}} (x^2 + y^3) = 9 \qquad \lim_{\substack{x \to 1 \\ y \to 2}} \sin \frac{\pi y}{2} = 0 \qquad \text{and} \qquad \lim_{z \to 1 + 2i} \left(x^2 + y^3 + i \sin \frac{\pi y}{2}\right) = 9$$

15. $w(x, y) = z^2 + (z^*)^2 = (x + iy)^2 + (x^2 - iy)^2 = 2x^2 - 2y^2$

18. Everywhere except at $z = 0$.

18.2   Differentiation: The Cauchy-Riemann Equations

1. (a)

$$
\begin{aligned}
f'(z) &= \lim_{\Delta z \to 0} \frac{f(z + \Delta z) - f(z)}{\Delta z} = \lim_{\Delta z \to 0} \frac{(z + \Delta z)^n - z^n}{\Delta z} \\
&= \lim_{\Delta z \to 0} \frac{nz^{n-1}\Delta z + O(\Delta z)^2}{\Delta z} = nz^{n-1} \qquad \text{for all } z
\end{aligned}
$$

(b) $f'(z) = \lim\limits_{\Delta z \to 0} \dfrac{\dfrac{1}{z + \Delta z} - \dfrac{1}{z}}{\Delta z} = \lim\limits_{\Delta z \to 0}\left\{-\dfrac{1}{z(z + \Delta z)}\right\} = -\dfrac{1}{z^2} \quad (z \neq 0)$

(c) $f'(z) = \lim\limits_{\Delta z \to 0} \dfrac{\dfrac{1}{1 + z + \Delta z} - \dfrac{1}{1 + z}}{\Delta z} = \lim\limits_{\Delta z \to 0}\left\{-\dfrac{1}{(1 + z)(1 + z + \Delta z)}\right\} = -\dfrac{1}{(1 + z)^2} \quad (z \neq -1)$

3. First write $\sin z = \sin(x + iy) = \sin x \cosh y + i \cos x \sinh y$ and then use Equation 5:

$$
f'(z) = u_x + iv_x = \cos x \cosh y - i \sin x \sinh y = \cos z
$$

6. There are singularities at $z = \pm 2i$. Write $f(z)$ as $f(z) = \dfrac{z}{(z + 2i)^2(z - 2i)^2}$. Each singularity is a pole of order 2.

9. We need to show that $u(x, y)$ satisfies Laplace's equation.

$$
\frac{\partial u}{\partial x} = 3x^2y - y^3 \qquad\qquad \frac{\partial^2 u}{\partial x^2} = 6xy
$$

$$
\frac{\partial u}{\partial y} = x^3 - 3xy^2 \qquad\qquad \frac{\partial^2 u}{\partial y^2} = -6xy
$$

and so $\dfrac{\partial^2 u}{\partial x^2} + \dfrac{\partial^2 u}{\partial y^2} = 0$.

12. The second of Equations 4 gives us

$$
\frac{\partial v}{\partial x} = -\frac{\partial u}{\partial y} = -\frac{\partial}{\partial y}(x^3y - xy^3) = -x^3 + 3xy^2
$$

Now integrate partially with respect to x to obtain

$$
v = -\frac{x^4}{4} + \frac{3x^2y^2}{2} + f(y)
$$

where $f(y)$ is to be determined. Now, the first of Equations 4, $\dfrac{\partial v}{\partial y} = \dfrac{\partial u}{\partial x}$, gives us

$$
3x^2y + \frac{df}{dy} = 3x^2y - y^3
$$

or $f(y) = -\dfrac{y^4}{4} + c$. Therefore,

$$v(x,y) = -\frac{x^4}{4} - \frac{y^4}{4} + \frac{3x^2y^2}{2} + c$$

is a harmonic conjugate to $u(x,y) = x^3y - xy^3$.

15. First note that $f(z) = r^2 e^{i\theta} = r^2\cos\theta + ir\sin\theta = u(r,\theta) + iv(r,\theta)$. The first of Equations 14 gives

$$u_r(r,\theta) = \frac{\partial}{\partial r}(r^2\cos\theta) = 2r\cos\theta \neq \frac{1}{r}v_\theta(r,\theta) = r\cos\theta$$

and so $f(z) = r^2 e^{i\theta}$ is not differentiable. Note, however, that $f(z) = r^2 e^{2i\theta}$ is differentiable because

$$u_r(r,\theta) = \frac{\partial}{\partial r}(r^2\cos 2\theta) = 2r\cos 2\theta = \frac{1}{r}\frac{\partial v}{\partial \theta} = 2r\cos 2\theta$$

18. We'll use l'Hôpital's rule:

$$\lim_{z\to n\pi}\frac{z - n\pi}{\sin(z - n\pi)} = \lim_{z\to n\pi}\frac{1}{\cos(z - n\pi)} = 1$$

21. Starting with Equation 1 with $z = re^{i\theta_0}$, we have

$$f'(z_0) = \lim_{z\to z_0}\frac{f(z) - f(z_0)}{z - z_0} = \lim_{r\to r_0}\left\{e^{-i\theta_0}\left[\frac{u(r,\theta_0) - u(r_0,\theta_0) + i[v(r,\theta_0) - v(r_0,\theta_0)]}{r - r_0}\right]\right\}$$

$$= \left(\frac{\partial u}{\partial r} + i\frac{\partial v}{\partial r}\right)e^{-i\theta_0}$$

Now, letting $z = r_0 e^{i\theta}$, we have

$$f'(z_0) = \lim_{\theta\to\theta_0}\left\{\frac{u(r_0,\theta) - u(r_0,\theta_0) + i[v(r_0,\theta) - v(r_0,\theta_0)]}{r(e^{i\theta} - e^{i\theta_0})}\right\}$$

$$= \frac{1}{r}\lim_{\theta\to\theta_0}\left\{\frac{u(r_0,\theta) - u(r_0,\theta_0) + i[v(r_0,\theta) - v(r_0,\theta_0)]}{e^{i\theta_0}i(\theta - \theta_0) + O[(\theta - \theta_0)^2]}\right\}$$

$$= -\frac{i}{r}(u_\theta + iv_\theta)e^{-i\theta_0} = \left(\frac{v_\theta}{r} - \frac{iu_\theta}{r}\right)e^{-i\theta_0}$$

These two results for $f'(z_0)$ must be equal, so $u_r = \dfrac{v_\theta}{r}$ and $v_r = -\dfrac{u_\theta}{r}$.

18.3 Complex Integration: Cauchy's Theorem

1. Given that $\mathbf{v}(t) = (3x^2 + y)\mathbf{i} - 6xy\,\mathbf{j}$ and $x = t$ and $y = t^2$ along the path of integration, we have

$$\int_C \mathbf{v}\cdot d\mathbf{r} = \int_C [(3x^2 + y)\,dx - 6xy\,dy] = \int_0^2 (4t^2\,dt - 12t^4\,dt)$$

$$= \left[\frac{4t^3}{3} - \frac{12t^5}{5}\right]_0^2 = \frac{32}{3} - \frac{384}{5} = -\frac{992}{15}$$

3. Given that $\mathbf{v} = xy\,\mathbf{i} - x\,\mathbf{j}$, we have

$$
\begin{aligned}
\int_C \mathbf{v} \cdot d\mathbf{r} &= \int_C (xy\,dx - x\,dy) \\
&= \int_{-a}^{a} -ax\,dx + \int_{-a}^{a} -a\,dy + \int_{a}^{-a} ax\,dx + \int_{a}^{-a} a\,dy = 0 - 2a^2 + 0 - 2a^2 = -4a^2
\end{aligned}
$$

It's a nice exercise to show that Stokes's theorem in the form of Equation 7.5.1 gives the same result.

6.

$$
\begin{aligned}
\int_C f(z)\,dz &= \int_0^1 e^{(1-i)t}(1+i)\,dt = (1+i)\left[\frac{e^{(1-i)t}}{1-i}\right]_0^1 \\
&= \frac{(1+i)^2}{2}[e^{1-i} - 1] = i(e^{1-i} - 1)
\end{aligned}
$$

9.

$$
\begin{aligned}
\int_C z^*\,dz &= \int_C (x - iy)(dx + i\,dy) = \int_C [(x\,dx + y\,dy) + i(x\,dy - y\,dx)] \\
&= \int_0^1 (x\,dx + 2x^3\,dx) + i\int_0^1 (2x^2\,dx - x^2\,dx) = 1 + \frac{i}{3}
\end{aligned}
$$

12.

$$
\begin{aligned}
\left|\int_C \frac{dz}{1+z^2}\right| \le \left|\frac{1}{1+z^2}\right| |2\pi R| &= \frac{2\pi R}{|1 + R^2\cos 2\theta + iR^2\sin 2\theta|} \\
&= \frac{2\pi R}{[(1 + R^2\cos 2\theta)^2 + R^4\sin^2 2\theta]^{1/2}} \\
&= \frac{2\pi R}{[1 + R^4 + 2R^2\cos 2\theta]^{1/2}} \longrightarrow 0
\end{aligned}
$$

as $R \to \infty$.

15. According to the Cauchy-Goursat theorem, we can choose any convenient contour that encircles the point $z = a$. We'll choose a unit circle centered at $z = a$, so that $z - a = e^{i\theta}$, and

$$
\oint_C \frac{dz}{(z-a)^n} = i\int_0^{2\pi} e^{i(1-n)\theta}\,d\theta = \begin{cases} 0 & n \ne 1 \\ 2\pi i & n = 1 \end{cases}
$$

18. Use partial fractions to write

$$
I = \oint_C \frac{2z - 1}{z(z-1)}\,dz = \oint_C \frac{dz}{z} + \oint_C \frac{dz}{z-1}
$$

Now that the integrand has been written as two separate integrands, we evaluate each one separately, getting $2\pi i$ for each for a total of $4\pi i$.

21. (a) The integrand has a simple pole at $z = -2$. The path does not enclose the pole, so $I = 0$.

(b) The path encloses the pole, so let's deform the path such that $z + 2 = \rho e^{i\theta}$. Then

$$I = \int_0^{2\pi} \frac{(\rho e^{i\theta} - 2)\rho i e^{i\theta}}{\rho e^{i\theta}} \, d\theta = -4\pi i \qquad \text{as} \qquad \rho \to 0$$

(c) The path does not enclose the pole, so $I = 0$.

(d) The path encloses the pole, so we deform the path and get the same result as in (b), $-4\pi i$.

24. A parametric representation for z is given by $z = a \cos t + i b \sin t$. Now,

$$
\begin{aligned}
\frac{dz}{z} &= \frac{-a \sin t \, dt + i b \cos t \, dt}{a \cos t + i b \sin t} \\[2mm]
&= \frac{(-a \sin t + i b \cos t)(a \cos t - i b \sin t) \, dt}{a^2 \cos^2 t + b^2 \sin^2 t} \\[2mm]
&= \frac{\left[-a^2 \sin t \cos t + b^2 \sin t \cos t + i a b (\cos^2 t + \sin^2 t) \right] dt}{a^2 \cos^2 t + b^2 \sin^2 t} \\[2mm]
&= \frac{(b^2 - a^2) \sin t \cos t \, dt + i a b \, dt}{a^2 \cos^2 t + b^2 \sin^2 t}
\end{aligned}
$$

and

$$\int_0^{2\pi} \frac{(b^2 - a^2) \sin t \cos t \, dt + i a b \, dt}{a^2 \cos^2 t + b^2 \sin^2 t} = 0 + i a b \cdot \frac{2\pi}{ab} = 2\pi i$$

18.4 Cauchy's Integral Formula

1. $f(x) = x^3 \sin(1/x)$. $f'(x) = 3x^2 \sin(1/x) - x \cos(1/x) = 0$ when $x = 0$.

$f''(x) = 6x \sin(1/x) - 3 \cos x(1/x) - \cos(1/x) - \dfrac{1}{x} \sin(1/x)$ does not exist at $x = 0$.

3. The singularity (a simple pole) is at $z = \pi/4$.

(a) $I = 0$ because the singular point lies outside the circle $|2z| = 1$.

(b) Write

$$I = \oint \frac{\tan z}{4z - \pi} \, dz = \frac{1}{4} \oint \frac{\tan z}{z - \pi/4} \, dz$$

and use Equation 1 with $a = \pi/4$ and $f(z) = \tan z$ to get $I = \dfrac{2\pi i}{4} \tan \dfrac{\pi}{4} = \dfrac{2\pi i}{4} = \dfrac{\pi i}{2}$.

6. First write I as $\displaystyle \oint_C \frac{\sin z}{(z - \pi/2)(z + \pi/2)} \, dz$

(a) Only the singular point at $z = \pi/2$ is enclosed by $|z - 1| = 2$. Using Equation 1 with $a = \pi/2$ and $f(z) = \sin z/(z + \pi/2)$, we get $I = 2i \sin \pi/2 = 2i$.

(b) Both singular points are enclosed by the contour $|z - 1| = 3$ and so

$$I = 2\pi i \left[\frac{\sin \pi/2}{\pi} - \frac{\sin(-\pi/2)}{\pi} \right] = 4i$$

9. There is a pole of order $n + 1$ $(n \geq 0)$ at $z = 0$, and the contour encloses the pole. If $n = 0$, there is no pole $(\sin z/z \to 1$ as $z \to 0)$ and the value of the integrand is zero. If $n \neq 0$, Equation 7 gives

$$I = \oint_C \frac{\sin z}{z^{n+1}} \, dz = \frac{2\pi i}{n!} \left[\frac{d^n \sin z}{dz^n} \right]_{z=0} = \begin{cases} 0 & n \text{ even} \\ \dfrac{2\pi i}{n!} (-1)^{(n-1)/2} & n \text{ odd} \end{cases}$$

12. The zeros of $\cos z$, $\pm\pi/2$, $\pm 3\pi/2$, $\ldots$, all lie outside C and there is a simple pole at $z = 0$. From Equation 1, $f(0) = \lim\limits_{z \to 0} \dfrac{(z+1)^2}{\cos z} = 1$ and so $I = 2\pi i$.

15. Subtract Equation 9 from Equation 1 to obtain

$$\frac{1}{2\pi i} \left[\oint_C \frac{f(z)\,dz}{z - \zeta_1} - \oint_C \frac{f(z)\,dz}{z - \zeta_2} \right] = \frac{1}{2\pi i} \int_{-R}^{R} \frac{(\zeta_1 - \zeta_2)f(x,0)\,dx}{(x - \zeta_1)(x - \zeta_2)} + \frac{1}{2\pi i} \int_{C_1} \frac{(\zeta_1 - \zeta_2)f(z)\,dz}{(z - \zeta_1)(z - \zeta_2)}$$
$$= I_1 + I_2$$

where C_1 is shown in Figure 18.26. Now let's look at the second integral as $R \to \infty$. Let $z = Re^{i\theta}$ to write

$$I_2 = \frac{\zeta_1 - \zeta_2}{2\pi i} \int_0^\pi \frac{f(z)Rie^{i\theta}\,d\theta}{(Re^{i\theta} - \zeta_1)(Re^{i\theta} - \zeta_2)} \leq \frac{|\zeta_1 - \zeta_2|}{2\pi} \left| \frac{f(z)Rie^{i\theta}}{(Re^{i\theta} - \zeta_1)(Re^{i\theta} - \zeta_2)} \right|$$

Now let M be the maximum value of $f(z)$ on C_1 to write

$$I_2 \leq \frac{|\zeta_1 - \zeta_2|}{2\pi} M \frac{R}{(R + \cdots)(R + \cdots)} \longrightarrow \quad \text{as} \quad R \to \infty$$

Let's prove now that the quantity $(\zeta_1 - \zeta_2)/(x - \zeta_1)(x - \zeta_2)$ is purely imaginary if $\zeta_1 = \zeta_2^*$ (remember that z lies along the x axis in I_1 above.) If $\zeta_1 = a + ib$, then $\zeta_2 = \zeta_1^* = a - ib$ and

$$\frac{\zeta_1 - \zeta_2}{(z - \zeta_1)(z - \zeta_2)} = \frac{2ib}{(x - a)^2 + b^2} = \text{pure imaginary}$$

Now change the notation by letting $\zeta_1 = x + iy$, $\zeta_2 = x - iy$, and $x = \xi$ to write

$$f(x,y) = \frac{y}{\pi} \int_{-\infty}^{\infty} \frac{f(\xi, 0)\,d\xi}{(\xi - x)^2 + y^2}$$

Now let $f(z) = f(x,y) = u(x,y) + iv(x,y)$ and equate real parts to obtain

$$u(x,y) = \frac{y}{\pi} \int_{-\infty}^{\infty} \frac{u(\xi, 0)\,d\xi}{(\xi - x)^2 + y^2}$$

18.

$$u(x, y) = \frac{y}{\pi} \int_{-\infty}^{-1} \frac{u_0 \, d\xi}{(x - \xi)^2 + y^2} + \frac{y}{\pi} \int_{-1}^{1} \frac{u_1 \, d\xi}{(x - \xi)^2 + y^2} + \frac{y}{\pi} \int_{1}^{\infty} \frac{u_2 \, d\xi}{(x - \xi)^2 + y^2}$$

$$= \frac{u_0}{\pi} \left[\tan^{-1} \frac{\xi - x}{y} \right]_{-\infty}^{-1} + \frac{u_1}{\pi} \left[\tan^{-1} \frac{\xi - x}{y} \right]_{-1}^{1} + \frac{u_2}{\pi} \left[\tan^{-1} \frac{\xi - x}{y} \right]_{1}^{\infty}$$

$$= \frac{u_0}{\pi} \left[\frac{\pi}{2} - \tan^{-1} \frac{x + 1}{y} \right] + \frac{u_1}{\pi} \left[\tan^{-1} \frac{1 - x}{y} + \tan^{-1} \frac{1 + x}{y} \right] + \frac{u_2}{\pi} \left[\frac{\pi}{2} - \tan^{-1} \frac{1 - x}{y} \right]$$

$$= \frac{u_0}{\pi} \left[\tan^{-1} \frac{y}{x + 1} \right] + \frac{u_1}{\pi} \left[\frac{\pi}{2} - \tan^{-1} \frac{y}{x + 1} - \frac{\pi}{2} + \tan^{-1} \frac{y}{x - 1} \right]$$

$$\quad + \frac{u_2}{\pi} \left[\frac{\pi}{2} + \frac{\pi}{2} - \tan^{-1} \frac{y}{x - 1} \right]$$

$$= \frac{u_0 - u_1}{\pi} \tan^{-1} \frac{y}{x + 1} + \frac{u_1 - u_2}{\pi} \tan^{-1} \frac{y}{x - 1} + u_2$$

Problem 19 asks you to plot this result. The following is a list of a few *Mathematica* commands to give the plot for $u_0 = 1$, $u_1 = 10$, and $u_2 = 5$ for various values of y.

$$u(x, y) = -\frac{9}{\pi} \tan^{-1} \frac{y}{x + 1} + \frac{5}{\pi} \tan^{-1} \frac{y}{x - 1} + 5$$

```
Clear [y]
y = plot1 = plot[5 + (5/Pi)*ArcTan[y/(x-1)]-(9/Pi)*ArcTan[y/(x+1)],{x,1,3}]
plot2 = plot[5+(5/Pi)*(Pi-ArcTan[y/Abs[(x-1)]])- (9/Pi)*ArcTan[y/(x+1)],{x,-1,1}]

plot3 = plot[5+(5/Pi)*(Pi-ArcTan[y/Abs[(x-1)]])-
    (9/Pi)*(Pi-ArcTan[y/Abs[(x+1)]]),{x,-3,-1}]
Show[plot1,plot2,plot3]
```

21. Start with Equation 7

$$f^{(n)}(a) = \frac{n!}{2\pi i} \oint_C \frac{f(z)}{(z - a)^{n+1}} \, dz \qquad n = 0, \, 1, \, 2, \dots$$

and use the *ML* inequality to write

$$|f^{(n)}(a)| = \frac{n!}{2\pi} \left| \oint_C \frac{f(z)}{(z - a)^{n+1}} \, dz \right| \leq \frac{n!}{2\pi} \frac{M}{R^{n+1}} \cdot 2\pi R = \frac{n! M}{R^n}$$

24. (a) Using $z = e^{i\theta}$ on $|z| = 1$, we see that $z^* = e^{-i\theta} = 1/z$, so that

$$I_1 = \int_{|z|=1} \frac{dz}{2 + z^*} = \int_{|z|=1} \frac{z\,dz}{2z + 1} = \frac{1}{2} \int_{|z|=1} \frac{z\,dz}{z + \frac{1}{2}}$$

There is a simple pole at $z = -1/2$, which is enclosed by $|z| = 1$. $f(-1/2) = -1/4$, and so $I_1 = -i\pi/2$.

(b) In this case, $z = 3e^{i\theta}$, and $z^* = 3e^{-i\theta} = 9/z$ on $|z| = 3$. Therefore, $I_3 = \int_{|z|=3} \frac{z\,dz}{2z + 9}$.

There is a simple pole at $z = -9/2$, which lies outside $|z| = 3$. Therefore, $I_3 = 0$.

18.5   Taylor Series and Laurent Series

1. Let $z = x + iy$ to write $\sum_{n=0}^{\infty} e^{nx}e^{iny}$. This series converges only if $x < 0$, or $\mathrm{Re}\,z < 0$.

3. The series is a geometric series with $r = (1 - i)/3$, and so

$$S = \frac{1}{1 - \dfrac{1 - i}{3}} - 1 = \frac{3}{2 + i} - 1 = \frac{1 - i}{2 + i} = \frac{1 - 3i}{5}$$

6. We'll use the ratio test in each case.

(a) $\lim_{n\to\infty} \left| \dfrac{z^{2n+2}}{a^{n+1}} \cdot \dfrac{a^n}{z^{2n}} \right| = \left| \dfrac{z^2}{a} \right|; \quad |z| = \sqrt{a}$

(b) $\lim_{n\to\infty} \left| \dfrac{(z - 1)^{n+1}}{(n + 1)^2} \cdot \dfrac{n^2}{(z - 1)^n} \right| = |z - 1|; \quad |z - 1| = 1$

(c) $\lim_{n\to\infty} \left| \dfrac{(n + 1 + 2^{n+1})z^{n+1}}{(n + 2^n)z^n} \right| = |2z| = 1, \text{ or } |z| = 1/2$

(d) $\lim_{n\to\infty} \left| \dfrac{(n + 1)!z^{n+1}}{(n + 1)^{n+1}} \cdot \dfrac{n^n}{n!z^n} \right| = \lim_{n\to\infty} \left| \dfrac{(n + 1)z^n}{(n + 1)\left(1 + \dfrac{1}{n}\right)^n} \right| = \left| \dfrac{z}{e} \right|, \text{ or } |z| = e$

9. Use $\cosh z = \frac{1}{2}(e^z + e^{-z})$ to write

$$\cosh z = \frac{1}{2} \sum_{n=0}^{\infty} \left(\frac{z^n}{n!} + \frac{(-1)^n z^n}{n!} \right) = \sum_{n=0}^{\infty} \frac{z^{2n}}{(2n)!}$$

12. We'll first do Problem 11 here. The various derivatives of $\mathrm{Ln}\,(1 + z)$ are

$$f(z) = \mathrm{Ln}\,(1 + z) \qquad f'(x) = \frac{1}{1 + z} \qquad f''(z) = -\frac{1}{(1 + z)^2}$$

$$f^{(3)}(z) = \frac{2}{(1 + z)^3} \qquad f^{(4)}(z) = -\frac{2 \cdot 3}{(1 + z)^4} \qquad \text{and so on}$$

Notice that $f^{(n)}(0) = (-1)^{n+1}(n+1)!$ for $n \geq 1$ and so $f^{(n)}(0)/n! = (-1)^{n+1}/n$ for $n \geq 1$. Thus Equation 10 says that

$$\mathrm{Ln}\,(1+z) = z - \frac{z^2}{2} + \frac{z^3}{3} - \frac{z^4}{4} + \cdots$$

Now write $\mathrm{Ln}\,z = \mathrm{Ln}\,(1 + z - 1)$ and use the above result to obtain

$$\mathrm{Ln}\,z = (z-1) - \frac{1}{2}(z-1)^2 + \frac{1}{3}(z-1)^3 + \cdots$$

15. Refer to Figure 18.27. Because ζ lies in the region between $|z| = R$ and $|z| = \rho$, we have that $\dfrac{|z-a|}{|\zeta - a|} = \alpha < 1$. Let M be the maximum value of $f(\zeta)$ on the contour $|\zeta - a| = R$ and m be the minimum value of $|\zeta - z|$. Then

$$|R_n(z)| \leq \frac{M}{2\pi}\frac{\alpha^n}{m} = \frac{M}{2\pi m}\frac{\alpha^n}{R} \longrightarrow 0 \qquad \text{as} \quad n \longrightarrow \infty$$

18.

$$\begin{aligned}
f(z) &= \frac{1}{1-z}\cdot\frac{1}{2-z}\\
&= -\frac{1}{z}\left(1+\frac{1}{z}+\frac{1}{z^2}+\frac{1}{z^3}+\cdots\right)\left(-\frac{1}{z}\right)\left(1+\frac{2}{z}+\frac{4}{z^2}+\frac{8}{z^3}+\cdots\right)\\
&= \frac{1}{z^2}\left[1+\frac{2}{z}+\frac{4}{z^2}+\frac{8}{z^3}+O(z^{-4})\right.\\
&\quad +\frac{1}{z}+\frac{2}{z^2}+\frac{4}{z^3}+O(z^{-4})\\
&\quad +\frac{1}{z^2}+\frac{2}{z^3}+O(z^{-4})\\
&\quad \left.+\frac{1}{z^3}+O(z^{-4})\right]\\
&= \frac{1}{z^2}+\frac{3}{z^3}+\frac{7}{z^4}+\frac{15}{z^5}+O(z^{-6})
\end{aligned}$$

21.

$$\begin{aligned}
f(z) &= \frac{1}{1+z^2} = \frac{1}{z^2}\left(\frac{1}{1+1/z^2}\right) = \frac{1}{z^2}\sum_{n=0}^{\infty}(-1)^n\left(\frac{1}{z^2}\right)^n = \sum_{n=0}^{\infty}(-1)^n\frac{1}{z^{2n+2}}\\
&= \frac{1}{z^2}-\frac{1}{z^4}+\frac{1}{z^6}+\cdots \qquad |z|>1
\end{aligned}$$

24.

$$\begin{aligned}
f(z) &= z^3\sum_{n=0}^{\infty}(-1)^n\frac{(1/z)^{2n+1}}{(2n+1)!} = \sum_{n=0}^{\infty}(-1)^n\frac{z^{2-2n}}{(2n+1)!}\\
&= z^2 - \frac{1}{6} + \frac{1}{120z^2} - \frac{1}{5040z^4} + \cdots \qquad 0<|z|
\end{aligned}$$

27. We shall do Problem 26 first. We need to show that $\sum\limits_{n=0}^{\infty}\left(\dfrac{z-a}{\zeta-a}\right)$ is uniformly convergent. Referring to Problem 15, use $\dfrac{|z-a|}{|\zeta-a|}=\alpha<1$, or $M_n=\alpha^n$, because z lies in the region between C_1 and C_2 and ζ lies on C_1 in Figure 18.29. According to the Weierstrass M-test, the series converges uniformly in ζ because $\sum\limits_{n=0}^{\infty}M_n$ converges.

For Problem 27, we simply take $\left|\dfrac{\zeta-a}{z-a}\right|=\beta<1$ because ζ lies on C_2 in Figure 18.29.

18.6   Residues and the Residue Theorem

1. The singular points occur at $z=2\pi in$ for $n=0,\ \pm1,\ \pm2,\ \ldots$ because $e^{2\pi in}=1$ for $n=0,\pm1,\ldots$. You can use l'Hôpital's rule to show that they are all simple poles;

$$\lim_{z\to 2\pi in}\frac{z-2\pi in}{e^z-1}=\frac{1}{e^z}=1.$$

3. The point $z=1$ is a pole of order 2; $\lim\limits_{z\to 1}(z-1)^2\dfrac{\sin z}{(z-1)^2}=\sin 1$.

6. The point $z=0$ is a pole of order 2:

$$\frac{1-\cos z}{z^4}=\frac{-z^2/2+O(z^4)}{z^4}=-\frac{1}{2z^2}+O(1)$$

9. By writing $f(z)$ as $\dfrac{z^3}{(z+2)^3(z-2)}$, we see that there is a pole of order 3 at $z=-2$ and a simple pole at $z=2$.

12. There is an essential singularity at $z=0$. Expand $f(z)$ to obtain

$$f(z)=1-\frac{1}{z}+\frac{1}{2!}\frac{1}{z^2}-\frac{1}{3!}\frac{1}{z^3}+\cdots$$

The residue at $z=0$ is -1.

15. There is a pole of order 3 at $z=0$ and a simple pole at $z=i$. The residue at $z=0$ is given by

$$\text{Res}\,(z{=}0)=\frac{1}{2}\lim_{z\to 0}\left[\frac{d^2}{dz^2}\frac{e^z}{z-i}\right]=1-\frac{i}{2}$$

The residue at $z=i$ is given by

$$\text{Res}\,(z{=}i)=\lim_{z\to i}\frac{e^z}{z^3}=ie^i$$

18. There are simple poles at $z=n\pi i$, $n=0,\ \pm1,\ \pm2,\ldots$

$$\text{Res}\,(z{=}n\pi i)=\lim_{z\to n\pi i}\frac{z-n\pi i}{\sinh z}=\lim_{z\to n\pi i}\frac{1}{\cosh n\pi i}=\begin{cases}1 & \text{when } n \text{ is even}\\ -1 & \text{when } n \text{ is odd}\end{cases}$$

21. The integrand has simple poles at $z = 2n\pi i$, $n = 0, \pm 1, \ldots$. There are three poles ($n = 0, \pm 1$) within C and the residue equals unity at each one. Therefore, $\displaystyle\oint_C \frac{dz}{e^z - 1} = 6\pi i$.

24. There are poles at $z = n\pi$, when $n = 0, \pm 1, \pm 2, \ldots$. The poles at $z = 0$ and $z = \pm \pi$ lie within C, and the residues at those points are $1/2$, $-1/2$, and $-1/2$. Therefore,

$$\oint_C \frac{1 - \cos z}{\sin^3 z}\, dz = \left(\frac{1}{2} - \frac{1}{2} - \frac{1}{2}\right) 2\pi i = -\pi i$$

27. The point $z = 0$ is an essential singularity. The residue of the integrand at $z = 0$ is 0. Therefore, the integral is equal to zero.

CHAPTER 19

Functions of a Complex Variable: Applications

19.1 The Inversion Formula for Laplace Transforms

1.

$$\mathcal{L}\left\{\frac{\sin 2t}{4} + \frac{t\cos 2t}{2}\right\} = \frac{1}{4}\int_0^\infty e^{-st}\sin 2t\,dt + \frac{1}{2}\int_0^\infty te^{-st}\cos 2t\,dt$$

$$= \frac{1}{4}\cdot\frac{2}{s^2+4} + \frac{1}{2}\cdot\frac{s^2-4}{(s^2+4)^2} = \frac{s^2}{(s^2+4)^2}$$

3. The zeros of $\cosh z$ are at $z = in\pi/2$, where $n = \pm1,\ \pm3,\dots$. Therefore, the zeros of $\cosh\sqrt{s}$ are at $\sqrt{s} = in\pi/2$, $n = \pm1,\ \pm3,\dots$ or at $s = -n^2\pi^2/4$, where $n = \pm1,\ \pm3,\dots$.

6. There is a pole of order 3 at $s = -2$. The residue is

$$\text{Res}\,(s=-2) = \frac{1}{2!}\lim_{s\to-2}\left[\frac{d^2}{ds^2}s^2e^{st}\right] = \frac{1}{2}(2e^{-2t} - 8te^{-2t} + 4t^2e^{-2t})$$

$$\mathcal{L}^{-1}\left\{\frac{s^2}{(s+2)^3}\right\} = e^{-2t}(1 - 4t + 2t^2)$$

9. The function $f(s) = s/(s^4+1)$ has simple poles at $s = e^{i\pi/4}$, $e^{i3\pi/4}$, $e^{i5\pi/4}$, and $e^{i7\pi/4}$, or at $\dfrac{1+i}{\sqrt{2}}, \dfrac{-1+i}{\sqrt{2}}, \dfrac{-1-i}{\sqrt{2}}$, and $\dfrac{1-i}{\sqrt{2}}$. The residues are

$$\text{Res}\left(s=\frac{1+i}{\sqrt{2}}\right) = \lim_{s\to(1+i)/\sqrt{2}}\frac{se^{st}}{\left(s+\dfrac{1-i}{\sqrt{2}}\right)\left(s+\dfrac{1+i}{\sqrt{2}}\right)\left(s-\dfrac{1-i}{\sqrt{2}}\right)} = -\frac{i}{4}e^{(1+i)t/\sqrt{2}}$$

$$\text{Res}\left(s=\frac{-1+i}{\sqrt{2}}\right) = \lim_{s\to(-1+i)/\sqrt{2}}\frac{se^{st}}{\left(s-\dfrac{1+i}{\sqrt{2}}\right)\left(s+\dfrac{1+i}{\sqrt{2}}\right)\left(s-\dfrac{1-i}{\sqrt{2}}\right)} = \frac{i}{4}e^{(-1+i)t/\sqrt{2}}$$

$$\text{Res}\left(s=\frac{-1-i}{\sqrt{2}}\right) = \lim_{s\to(-1-i)/\sqrt{2}} \frac{se^{st}}{\left(s-\dfrac{1+i}{\sqrt{2}}\right)\left(s+\dfrac{1-i}{\sqrt{2}}\right)\left(s-\dfrac{1-i}{\sqrt{2}}\right)} = -\frac{i}{4}e^{(-1-i)t/\sqrt{2}}$$

$$\text{Res}\left(s=\frac{1-i}{\sqrt{2}}\right) = \lim_{s\to(1-i)/\sqrt{2}} \frac{se^{st}}{\left(s-\dfrac{1+i}{\sqrt{2}}\right)\left(s+\dfrac{1-i}{\sqrt{2}}\right)\left(s+\dfrac{1+i}{\sqrt{2}}\right)} = \frac{i}{4}e^{(1-i)t/\sqrt{2}}$$

$$\mathcal{L}^{-1}\left\{\frac{s}{s^4+1}\right\} = \frac{e^{it/\sqrt{2}}i}{4}(-e^{t/\sqrt{2}}+e^{-t/\sqrt{2}}) + \frac{e^{-it/\sqrt{2}}i}{4}(e^{t/\sqrt{2}}-e^{-t/\sqrt{2}})$$

$$= \sinh\frac{t}{\sqrt{2}}\left(\frac{e^{it/\sqrt{2}}-e^{-it/\sqrt{2}}}{2i}\right) = \sinh\frac{t}{\sqrt{2}}\sin\frac{t}{\sqrt{2}}$$

12. There is no branch point. There are simple poles at $s=0$ and at $as^{1/2}=n\pi i$ for $n=\pm1,\ \pm2,\dots$ or at $s=-n^2\pi^2/a^2$ for $n=0,\ 1,\ 2,\dots$. The residues are

$$\text{Res}\,(s=0) = \lim_{s\to0}\left[\frac{se^{st}\cosh xs^{1/2}}{s^{1/2}\sinh as^{1/2}}\right] = \frac{1}{a}$$

$$\text{Res}\left(s=-\frac{n^2\pi^2}{a^2}\right) = \lim_{s\to-n^2\pi^2/a^2}\left[\frac{s+n^2\pi^2/a^2}{s^{1/2}\sinh as^{1/2}}\right]e^{-n^2\pi^2t/a^2}\cosh\frac{n\pi xi}{a}$$

$$= \left(\frac{2}{a\cosh n\pi i}\right)e^{-n^2\pi^2t/a^2}\cos\frac{n\pi x}{a}$$

$$= \frac{(-1)^n 2}{a}e^{-n^2\pi^2t/a^2}\cos\frac{n\pi x}{a} \qquad n=1,\ 2,\dots$$

$$\mathcal{L}^{-1}\left\{\frac{\cosh xs^{1/2}}{s^{1/2}\sinh as^{1/2}}\right\} = \frac{1}{a}+\frac{2}{a}\sum_{n=1}^{\infty}(-1)^n e^{-n^2\pi^2t/a^2}\cos\frac{n\pi x}{a}$$

15. There is a branch point at $s=0$, so we use the contour shown in Figure 19.2. We let $s=xe^{i\pi}$ along C_2 and $s=xe^{-i\pi}$ along C_3, and write

$$\int_{c-i\infty}^{c+i\infty}\frac{e^{-as^{1/2}}e^{st}}{s^{1/2}}\,ds = -\int_{C_2}\frac{e^{st}e^{-as^{1/2}}}{s^{1/2}}\,ds - \int_{C_3}\frac{e^{st}e^{-as^{1/2}}}{s^{1/2}}\,ds$$

$$= -\int_{\infty}^{0}\frac{e^{-xt}e^{-aix^{1/2}}}{ix^{1/2}}(-dx) - \int_{0}^{\infty}\frac{e^{-xt}e^{aix^{1/2}}}{-ix^{1/2}}(-dx)$$

$$= 2i\int_{0}^{\infty}\frac{e^{-xt}\cos ax^{1/2}}{x^{1/2}}\,dx$$

Let $u=x^{1/2}$. Then

$$\mathcal{L}^{-1}\left\{\frac{e^{-as^{1/2}}}{s^{1/2}}\right\} = \frac{2i}{2\pi i}\left[2\int_{0}^{\infty}e^{-u^2t}\cos au\,du\right] = \frac{2}{\pi}\cdot\left(\frac{\pi}{4t}\right)^{1/2}e^{-a^2/4t} = \frac{1}{(\pi t)^{1/2}}e^{-a^2/4t}$$

18. The negative real axis is a branch cut for $1/(s^{1/2} + a)$. Therefore, referring to Figure 19.2, let $s = xe^{i\pi}$ along C_2 and $s = xe^{-i\pi}$ along C_3, and write

$$\int_{c-i\infty}^{c+i\infty} \frac{e^{st}}{s^{1/2} + a} \, ds = -\int_{C_2} \frac{e^{st}}{s^{1/2} + a} \, ds - \int_{C_2} \frac{e^{st}}{s^{1/2} + a} \, ds$$

$$= -\int_{\infty}^{0} \frac{e^{-xt}\,(-dx)}{ix^{1/2} + a} - \int_{0}^{\infty} \frac{e^{-xt}\,(-dx)}{-ix^{1/2} + a}$$

$$= \int_{0}^{\infty} e^{-xt} \left[\frac{1}{a - ix^{1/2}} - \frac{1}{a + ix^{1/2}} \right] dx$$

$$= 2i \int_{0}^{\infty} e^{-xt} \frac{x^{1/2}}{a^2 + x} \, dx$$

$$= 2i \int_{0}^{\infty} \frac{e^{-xt}}{x^{1/2}} \, dx - 2ia^2 \int_{0}^{\infty} e^{-xt} \frac{dx}{x^{1/2}(a^2 + x)}$$

$$= 2i \frac{\Gamma(1/2)}{t^{1/2}} - 2ia^2 \cdot \frac{\pi}{a} e^{a^2 t} \text{erfc}\,(at^{1/2})$$

Divide by $2\pi i$ to get the desired result.

21. Simply let $s - a = p$ in the inversion formula.

19.2   Evaluation of Real, Definite Integrals

1. Let $z = e^{i\theta}$, or $\cos\theta = \dfrac{z^2 + 1}{2z}$ and $d\theta = \dfrac{dz}{iz}$, to write the integral as $I = \displaystyle\oint \frac{2\,dz}{i(z^2 + 6z + 1)}$.
There are simple poles at $z = -3 \pm 2\sqrt{2}$, only one of which $(-3 + 2\sqrt{2})$ lies within $|z| = 1$.
Therefore,

$$I = \int_0^{2\pi} \frac{d\theta}{3 + \cos\theta} = \oint \frac{2\,dz}{i(z^2 + 6z + 1)} = \frac{2}{i} 2\pi i \lim_{z \to -3 + 2\sqrt{2}} \frac{1}{z + 3 + 2\sqrt{2}} = \frac{\pi}{\sqrt{2}}$$

3. Let $z = e^{i\theta}$, or $\sin\theta = \dfrac{z^2 - 1}{2iz}$ and $d\theta = \dfrac{dz}{iz}$, to write the integral as

$$I = \oint \frac{2\,dz}{bz^2 + 2aiz - b} = \frac{2}{b} \oint \frac{dz}{z^2 + \frac{2a}{b}iz - 1}$$

The integrand has simple poles at $(-a \pm (a^2 - b^2)^{1/2})i/b$. Only $(-a + (a^2 - b^2)^{1/2})i/b$ lies within $|z| = 1$ if $a > b$. Therefore,

$$I = \int_0^{2\pi} \frac{d\theta}{a + b\sin\theta} = \frac{2}{b} \oint \frac{dz}{z^2 + \frac{2a}{b}iz - 1}$$

$$= \left(\frac{4\pi i}{b}\right) \lim_{z \to (-a + (a^2 - b^2)^{1/2})i/b} \frac{1}{z + (a + (a^2 - b^2)^{1/2})i/b} = \frac{2\pi}{(a^2 - b^2)^{1/2}}$$

6. Simply let $\theta = 2\pi - \phi$.

9. Use the contour in Figure 19.5. The singular points of the integrand are poles of order 2 at $z = \pm i$. Only the pole at $z = i$ lies within the contour. The residue at $z = i$ is

$$\text{Res}\,(z{=}i) = \lim_{z \to i} \frac{d}{dz} \frac{z^2}{(z+i)^2} = -\frac{i}{4}$$

Therefore,

$$\int_{-\infty}^{\infty} \frac{x^2}{(1+x^2)^2}\, dx = 2\pi i \left(-\frac{i}{4}\right) = \frac{\pi}{2}$$

12. If the degree of the denominator is at least two greater than the degree in the numerator, then the argument on page 931 shows that the asymptotic form of the integrand will be $\dfrac{1}{R^\alpha}$ with $\alpha \geq 1$, and so the integral along the arc will be zero.

15. Start with $\displaystyle\int_C \frac{e^{i\pi z}}{z^2 - 2z + 2}\, dz$ (we will subsequently take the real and imaginary parts), where C is the contour in Figure 19.5. The singular points of the integrand are at $z = 1 \pm i$, and only $z = 1 + i$ lies within the contour. The residue at this point is

$$\text{Res}\,(z{=}1+i) = \lim_{z \to 1+i} \frac{e^{i\pi z}}{z - 1 + i} = -\frac{e^{-\pi}}{2i}$$

Therefore,

$$\int_{-\infty}^{\infty} \frac{e^{i\pi x}}{x^2 - 2x + 2} = 2\pi i \left(-\frac{e^{-\pi}}{2i}\right) = -\pi e^{-\pi}$$

and we see then that

$$\int_{-\infty}^{\infty} \frac{\cos \pi x\, dx}{x^2 - 2x + 2} = -\pi e^{-\pi} \qquad \text{and} \qquad \int_{-\infty}^{\infty} \frac{\sin \pi x\, dx}{x^2 - 2x + 2} = 0$$

18. Start with $I = \displaystyle\oint_C \frac{e^{iz}}{z}\, dz$ (we will subsequently use the imaginary part.) There are no singularities within the contour shown in Figure 19.8, and so

$$
\begin{aligned}
I &= 0 = \lim_{\epsilon \to 0} \int_{-\infty}^{-\epsilon} \frac{e^{ix}}{x}\, dx + \lim_{\epsilon \to 0} \oint_{C_\epsilon} \frac{e^{iz}}{z}\, dz + \lim_{\epsilon \to 0} \int_{\epsilon}^{\infty} \frac{e^{ix}}{x}\, dx \\
&= \int_{-\infty}^{\infty} \frac{e^{ix}}{x}\, dx + \lim_{\epsilon \to 0} \int_{\pi}^{0} e^{i\epsilon e^{i\theta}}\, i\, d\theta
\end{aligned}
$$

or

$$\int_{-\infty}^{\infty} \frac{e^{ix}}{x}\, dx - i\pi = 0$$

The imaginary part gives $\displaystyle\int_{-\infty}^{\infty} \frac{\sin x}{x}\, dx = \pi$.

21. First note that

$$\int_0^\infty \frac{dx}{(1+x^2)^N} = \frac{1}{2}\int_{-\infty}^\infty \frac{dx}{(1+x^2)^N} = \frac{1}{2}\int_{-\infty}^\infty \frac{dx}{(x+i)^N(x-i)^N}$$

If we use the contour in Figure 19.5, only the pole at $x = i$ lies within the contour. The residue at $z = i$ is given by Equation 18.6.10. The $(N-1)$th derivative of $1/(x+i)^N$ is $\dfrac{(-1)^{N-1}(2N-2)!}{(N-1)!(2i)^{2N-1}}$. Substitute this result into Equation 18.6.10 to find that

$$b_1 = \frac{(-1)^{n-1}(2n-2)!}{(n-1)!(n-1)!(2i)^{2n-1}(2n-1)} = -\frac{i\Gamma(2n)}{\Gamma^2(n)2^{2n-1}(2n-1)}$$

Now use Equation 3.1.8 for $\Gamma(2n)$:

$$b_1 = -\frac{i\,\Gamma(n+1/2)}{\sqrt{\pi}\,\Gamma(n)(2n-1)}$$

Therefore,

$$\int_0^\infty \frac{dx}{(1+x^2)^n} = \frac{1}{2}(2\pi i)\left[-\frac{i(n-1/2)\Gamma(n-1/2)}{\sqrt{\pi}\Gamma(n)(2n-1)}\right]$$
$$= \frac{\sqrt{\pi}}{2}\frac{\Gamma(n-1/2)}{\Gamma(n)}$$

24. The integral, $I = \displaystyle\int_0^\infty \frac{x^4\,dx}{1+x^2}$, with $-1 < a < 1$, has a branch cut so we'll use the contour in Figure 19.7. The integral along Γ vanishes as $R \to \infty$ because $a < 1$ and the integral along C_ϵ vanishes as $\epsilon \to 0$ because $a > -1$. Use $z = x$ along the upper part of the branch cut and $z = xe^{2\pi i}$ along the lower part. The integrand has simple poles at $z = \pm i$, with residues

$$\text{Res}\,(z=i) = \lim_{z\to i}\frac{z^a}{z+i} = \frac{i^a}{2i} = \frac{e^{i\pi a/2}}{2i}$$
$$\text{Res}\,(z=-i) = \lim_{z\to -i}\frac{z^a}{z-i} = \frac{e^{3i\pi a/2}}{-2i}$$

The sum of the residues is $-e^{\pi a i}\sin(\pi a/2)$.

So far, then, we have

$$\int_0^\infty \frac{x^a\,dx}{1+x^2} + \int_\infty^0 \frac{e^{2\pi i a}x^a\,dx}{1+x^2} = 2\pi i\left[-e^{\pi a i}\sin(\pi a/2)\right]$$

and so

$$\int_0^\infty \frac{x^a\,dx}{1+x^2} = -2\pi i\frac{e^{\pi a i}\sin(\pi a/2)}{1-e^{2\pi i a}}$$
$$= \pi\frac{\sin(\pi a/2)}{\sin(\pi a)} = \frac{\pi}{2\cos(\pi a/2)}$$

27. Use the method outlined in the statement of the problem. This procedure leads to

$$\int_0^\infty \frac{x\,dx}{(1+x^2)^2} = -\text{sum of the residues of } \frac{z\,\text{Ln}\,z}{(1+z^2)^2}$$

There are poles of order 2 at $z = \pm i$.

$$
\begin{aligned}
\text{Res}\,(z=i) &= \lim_{z\to i} \frac{d}{dz} \frac{z\,\text{Ln}\,z}{(z+i)^2} = \lim_{z\to i}\left[\frac{1+\text{Ln}\,z}{(z+i)^2} - \frac{2z\,\text{Ln}\,z}{(z+i)^3}\right] \\
&= \frac{1+i\pi/2}{-4} - \frac{2i(i\pi/2)}{8i^3} = -\frac{1}{4} \\
\text{Res}\,(z=-i) &= \lim_{z\to -i}\left[\frac{1+\text{Ln}\,z}{(z-i)^2} - \frac{2z\,\text{Ln}\,z}{(z-i)^3}\right] \\
&= \frac{1+3i\pi/2}{-4} - \frac{2i(3i\pi/2)}{8i^3} = -\frac{1}{4}
\end{aligned}
$$

Therefore,

$$\int_0^\infty \frac{x\,dx}{(1+x^2)^2} = -\left(-\frac{1}{4}-\frac{1}{4}\right) = \frac{1}{2}$$

30. We'll use the method that is outlined in Problem 27. We'll consider $\oint \dfrac{z\,\text{Ln}\,z}{1+z^3}\,dz$ on the contour in Figure 19.7. There are three simple poles, at $z = -1$, $e^{i\pi/3} = \frac{1}{2}(1+\sqrt{3}\,i)$, and $e^{5i\pi/3} = \frac{1}{2}(1-\sqrt{3}\,i)$. This leads to

$$\int_\infty^0 \frac{x\,dx}{1+x^3} = \text{sum of the residues of } \frac{z\,\text{Ln}\,z}{1+z^3}$$

$$
= \frac{-\text{Ln}\,(-1)}{\left(-\frac{3}{2}-\frac{\sqrt{3}}{2}i\right)\left(-\frac{3}{2}+\frac{\sqrt{3}}{2}i\right)} + \frac{\frac{1}{2}(1+\sqrt{3}\,i)\,\text{Ln}\,e^{i\pi/3}}{\left(\frac{3}{2}+\frac{\sqrt{3}}{2}i\right)(\sqrt{3}\,i)} + \frac{\frac{1}{2}(1-\sqrt{3}\,i)\,\text{Ln}\,e^{5i\pi/3}}{\left(\frac{3}{2}-\frac{\sqrt{3}}{2}i\right)(-\sqrt{3}\,i)}
$$

$$
= -\frac{i\pi}{3} + \frac{\pi}{3\sqrt{3}}\frac{1+\sqrt{3}\,i}{(3+\sqrt{3}\,i)} - \frac{5\pi}{3\sqrt{3}}\frac{1-\sqrt{3}\,i}{(3-\sqrt{3}\,i)}
$$

$$
= -\frac{i\pi}{3} + \frac{\pi}{3\sqrt{3}}(-2+\sqrt{3}\,i) = -\frac{2\pi}{3\sqrt{3}}
$$

Therefore,

$$\int_0^\infty \frac{x\,dx}{1+x^3} = \frac{2\pi}{3\sqrt{3}}$$

19.3   Summation of Series

1. First expand $\sin \pi z$ and $\cos \pi z$ about the point $z = n$:

$$
\begin{aligned}
\sin \pi z &= \sin n\pi + \left[\frac{d\sin \pi z}{dz}\right]_{z=n}(z-n) + \frac{1}{2}\left[\frac{d^2 \sin \pi z}{dz^2}\right]_{z=n}(z-n)^2 + \cdots \\
&= (-1)^n \pi (z-n) - \frac{(-1)^n \pi^3}{6}(z-n)^3 + \cdots \\
\cos \pi z &= (-1)^n - \frac{(-1)^n \pi^2}{2}(z-n)^2 + \cdots
\end{aligned}
$$

The ratio of these two series gives

$$\cot \pi z = \frac{(-1)^n - (-1)^n \pi^2 (z-n)^2/2 + \cdots}{(-1)^n \pi (z-n) + O((z-n)^3)} = \frac{1 + O((z-n)^2)}{\pi(z-n)[1 + O((z-n)^2)]}$$

$$= \frac{1}{\pi(z-n)} + O(z-n)$$

3. The residue of $\pi \cot \pi z/(z+a)^2$ at $z = -a$, a pole of order 2, is

$$\mathrm{Res} \left\{ \frac{\pi \cot \pi z}{(z+a)^2} \text{ at } z = -a \right\} = \pi \lim_{z \to -a} \frac{d}{dz} \cot \pi z$$

$$= -\pi^2 \csc^2 a = -\frac{\pi^2}{\sin^2 \pi a}$$

and so $\displaystyle\sum_{n=-\infty}^{\infty} \frac{1}{(n+a)^2} = \frac{\pi^2}{\sin^2 \pi a}$.

6. We first write $f(z) = 1/(z^2 + a^2)^2$ as $f(z) = \dfrac{1}{(z+ia)^2(z-ia)^2}$. Thus, we have poles of order 2 at $\pm ia$. The residues of $\pi \cot \pi z/(z^2 + a^2)^2$ at these poles are given by

$$\mathrm{Res} \left\{ \frac{\pi \cot \pi z}{(z^2+a^2)^2} \text{ at } z = ia \right\} = \pi \lim_{z \to ia} \frac{d}{dz} \frac{\cot \pi z}{(z+ia)^2}$$

$$= +\frac{\pi^2 \mathrm{cosec}^2 \pi ia}{4a^2} - i\frac{\pi \cot \pi ia}{4a^3} = -\frac{\pi^2 \mathrm{cosech}^2 \pi a}{4a^2} - \frac{\pi \coth \pi a}{4a^3}$$

$$\mathrm{Res} \left\{ \frac{\pi \cot \pi z}{(z^2+a^2)^2} \text{ at } z = -ia \right\} = \frac{\pi^2 \mathrm{cosec}^2 \pi ia}{4a^2} - \frac{i\pi \cot \pi ia}{4a^3} = -\frac{\pi^2 \mathrm{cosech}^2 \pi a}{4a^2} - \frac{\pi \coth \pi a}{4a^3}$$

Therefore,

$$\sum_{n=-\infty}^{\infty} \frac{1}{(n^2+a^2)^2} = \frac{\pi^2 \mathrm{cosech}^2 \pi a}{2a^2} + \frac{\pi \coth \pi a}{2a^3}$$

9. Following Example 2,

$$\sum_{n=1}^{\infty} \frac{1}{n^6} = -\frac{1}{2} \mathrm{Res} \left\{ \frac{\pi \cot \pi z}{z^6} \text{ at } z = 0 \right\} = \frac{\pi^6}{945}$$

12. First write $\mathrm{cosec}\,\pi z = \dfrac{1}{\sin \pi z}$ and then expand $\sin \pi z$ about $z = n$ to obtain

$$\mathrm{cosec}\,\pi z = \frac{1}{(\pi \cos \pi n)(z-n) + O[(z-n)^3]} = \frac{(-1)^n}{\pi(z-n)} + O[(z-n)]$$

Therefore, $\displaystyle\lim_{z \to n}(z-n)\pi f(z)\mathrm{cosec}\,\pi z = (-1)^n f(n)$.

15. First write $\mathrm{cosec}\,x$ as

$$\mathrm{cosec}\,x = \frac{1}{\sin x} = \frac{2i}{e^{ix} - e^{-ix}} = \frac{2ie^{ix}}{e^{2ix} - 1}$$

Now use the Maclaurin series for e^{ix} and the equation given in the statement of the problem to write ($B_0 = 1$, $B - 1 = -1/2$, $B_2 = 1/6$, $B_3 = 0$, $B_4 = -1/30$)

$$\operatorname{cosec} x = 2i \left[1 + ix - \frac{x^2}{2} - \frac{ix^3}{6} + \frac{x^4}{24} + O(x^5) \right] \cdot \frac{1}{2ix} \left[1 - ix - \frac{x^2}{3} - \frac{x^4}{45} + O(x^5) \right]$$

$$= \frac{1}{x} \left[1 - ix - \frac{x^2}{3} - \frac{x^4}{45} + O(x^5) + ix + x^2 - \frac{ix^3}{3} + O(x^5) \right.$$

$$\left. - \frac{x^2}{2} + \frac{ix^3}{2} + \frac{x^4}{6} + O(x^5) - \frac{ix^3}{6} - \frac{x^4}{6} + O(x^5) + \frac{x^4}{24} + O(x^5) \right]$$

$$= \frac{1}{x} + \frac{x}{6} + \frac{7x^3}{360} + O(x^5)$$

18. Let $a = 1/2$ in $\displaystyle\sum_{n=-\infty}^{\infty} \frac{1}{(n+a)^2} = \frac{\pi^2}{\sin^2 \pi a}$ to obtain $\displaystyle\sum_{n=-\infty}^{\infty} \frac{1}{(n+1/2)^2} = \pi^2$. Now,

$$\sum_{n=-\infty}^{\infty} \frac{1}{(n+1/2)^2} = 2 \sum_{n=0}^{\infty} \frac{1}{\left(\dfrac{2n+1}{2}\right)^2} = 8 \sum_{n=0}^{\infty} \frac{1}{(2n+1)^2} = \pi^2$$

or $\displaystyle\sum_{n=0}^{\infty} \frac{1}{(2n+1)^2} = \frac{\pi^2}{8}$.

19.4   Location of Zeros

1. First note that

$$\frac{b}{s^2 - 2as + a^2 + b^2} = \frac{b}{(s-a)^2 + b^2}$$

The inverse Laplace transform of this expression is (Table 17.1 and Equation 17.1.4) $f(t) = e^{at} \sin bt$. This result diverges with increasing time if $a > 0$ and decays to zero if $a < 0$.

3. $f(z) = z^2/(2z^2 + 1)$ has a zero of multiplicity two ($z = 0$) and two simple poles ($z = \pm i/\sqrt{2}$). Now

$$\frac{f'(z)}{f(z)} = \frac{2}{z} - \frac{4z}{2z^2 + 1}$$

and

$$\frac{1}{2\pi i} \oint_{|z|=1} \left(\frac{2}{z} - \frac{4z}{2z^2 + 1} \right) dz = 2 \left(\text{Residue of } \frac{1}{z} \text{ at } z = 0 \right)$$

$$- 2 \left(\text{sum of the Residues of } \frac{z}{z^2 + 1/2} \text{ at } z = \frac{\pm i}{\sqrt{2}} \right)$$

$$= 2 - 2 \left(\frac{i/\sqrt{2}}{2i/\sqrt{2}} + \frac{i/\sqrt{2}}{2i/\sqrt{2}} \right) = 2 - 2 \left(\frac{1}{2} + \frac{1}{2} \right) = 0$$

6. Start with

$$
\begin{aligned}
w &= z^4 + z + 1 = (x+iy)^4 + (x+iy) + 1 \\
&= x^4 - 6x^2y^2 + y^4 + x + 1 + 4i(x^3y - xy^3) + iy \\
&= u(x,y) + iv(x,y)
\end{aligned}
$$

Let $x = a\cos\theta$ and $y = a\sin\theta$ and plot $v(\theta)$ against $u(\theta)$. The following figures show that two roots lie within $|z| = 1$ and four roots lie within $|z| = 2$.

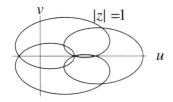   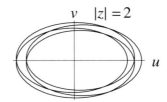

9. Start with

$$
\frac{1}{2\pi i} \oint_C \frac{f'(z) + g'(z)}{f(z) + g(z)} = \text{number of zeros of } \{f(z) + g(z)\} \text{ within } C
$$

Now,

$$
\frac{f'(z) + g'(z)}{f(z) + g(z)} = \frac{f'(z) + g'(z)}{f(z)\left[1 + \dfrac{g(z)}{f(z)}\right]}
$$

The method of partial fractions gives

$$
\frac{1}{f(z)\left[1 + \dfrac{g(z)}{f(z)}\right]} = \frac{\alpha}{f(z)} + \frac{\beta}{1 + \dfrac{g(z)}{f(z)}}
$$

and

$$
\alpha + \alpha\frac{g(z)}{f(z)} + \beta f(z) = f'(z) + g'(z)
$$

Let $\alpha = f'(z)$ and $\beta = \dfrac{g'(z)}{f(z)} - \dfrac{f'(z)g(z)}{f^2(z)} = \left(\dfrac{g(z)}{f(z)}\right)'$. Therefore,

$$
\frac{f'(z) + g'(z)}{f(z) + g(z)} = \frac{f'(z)}{f(z)} + \frac{(g(z)/f(z))'}{1 + g(z)/f(z)} = \frac{f'(z)}{f(z)} + \frac{\left(1 + \dfrac{g(z)}{f(z)}\right)'}{1 + \dfrac{g(z)}{f(z)}}
$$

Using Equation 4, we have

$$
\begin{pmatrix} \text{number of zeros of} \\ f(z) + g(z) \text{ within } C \end{pmatrix} = \begin{pmatrix} \text{number of zeros of} \\ f(z) \text{ within } C \end{pmatrix} + \frac{1}{2\pi i} \oint_C \frac{\left(1 + \dfrac{g(z)}{f(z)}\right)'}{1 + \dfrac{g(z)}{f(z)}}\, dz
$$

Because $|f(z)| > |g(z)|$ on C,

$$\left|\left(1 + \frac{g(z)}{f(z)}\right) - 1\right| = \frac{|g(z)|}{|f(z)|} < 1$$

Thus, $1 + \dfrac{g(z)}{f(z)}$ lies *inside* the circle of unit radius centered at $z = 1$, but this circle can never enclose the origin.

12. Take $f(z) = -4z^3$, so that $|f(z)| = 4$ on $|z| = 1$. Similarly, $|g(z)| = |z^7 + z - 1| \le |z^7| + |z| + 1 = 3$ on $|z| = 1$. Thus, $|f(z)| > |g(z)|$ on $|z| = 1$. Because $f(z)$ has three roots within $|z| = 1$, $f(z) + g(z) = z^7 - 4z^3 + z - 1 = 0$ has three roots within $|z| = 1$.

15. Take $f(z) = az^n$ and $g(z) = -e^z$. Then $|f(z)| = a$ and $|g(z)| = e^x$, where $-1 \le x \le 1$ on $|z| = 1$, and

$$\frac{|g(z)|}{|f(z)|} = \frac{e^x}{a} < 1 \quad \text{if} \quad a > e$$

But $f(z) = az^n$ has n zeros, so $e^z - az^n$ has n zeros. In Example 3, $n = 2$, and, indeed, there are two roots of $e^z = 3z^2$ that lie within the unit circle.

18. Let $s = i\omega$ and write

$$\begin{aligned} Q(\omega) &= i\omega^5 + 3\omega^4 - 5i\omega^3 - 5\omega^2 + 3i\omega + 1 \\ &= 3\omega^4 - 5\omega^2 + 1 + i(\omega^5 - 5\omega^3 + 3\omega) \\ &= u(\omega) + iv(\omega) \end{aligned}$$

So

$$u(\omega) = 3\omega^4 - 5\omega^2 + 1 \quad \text{and} \quad v(\omega) = \omega(\omega^4 - 5\omega^2 + 3)$$

We see that $v = 0$ at $\omega = 0$, $\pm 0.83\ldots$, and $\pm 2.07\ldots$, where $u = 1$, -1.02, and 35.

   The figure below shows the image curve in the ω plane. As $\omega \to \pm\infty$,
$\dfrac{v}{u} = \dfrac{\omega^5 - 5\omega^3 + 3\omega}{3\omega^4 - 5\omega^2 + 1} \longrightarrow \dfrac{\omega}{3} \longrightarrow \pm\infty$. This says that the curve in the figure is vertical at its two extremes ($\omega = \pm\infty$). Therefore, Arg ω varies from $5\pi/2$ to $-5\pi/2$, for a change of -5π. This added to the 5π from the semicircular arc gives a total of 0. Thus, there are no zeros of $Q(s) = s^5 + 3s^4 + 5s^3 + 5s^2 + 3s + 1$ lying in the right half plane.

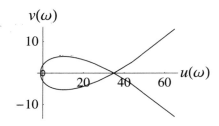

21. Recall that the principal branch of the arctangent is such that

$$-\frac{\pi}{2} \le \arctan \frac{v(\omega)}{u(\omega)} < \frac{\pi}{2}$$

For $\dfrac{v(\omega)}{u(\omega)} = \dfrac{10\omega - \omega^3}{6(1 - \omega^2)}$, this occurs for $-1 \le \omega \le 1$, and so we use $\arctan \dfrac{10\omega - \omega^3}{6(1 - \omega^2)}$

for $-1 \le \omega \le 1$. For $\omega \ge 1$, we use $\pi + \arctan \dfrac{10\omega - \omega^3}{6(1 - \omega^2)}$, and for $\omega \le -1$, we use

$-\pi + \arctan \dfrac{10\omega - \omega^3}{6(1 - \omega^2)}$. The combination of the three plots is shown below.

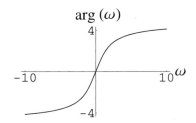

19.5   Conformal Mapping

1. Start with the Cauchy-Riemann equations

$$\frac{\partial u}{\partial x} = \frac{\partial v}{\partial y} \quad \text{and} \quad \frac{\partial u}{\partial y} = -\frac{\partial v}{\partial x}$$

Now

$$\frac{\partial^2 u}{\partial x^2} = \frac{\partial^2 v}{\partial x \partial y} \quad \text{and} \quad \frac{\partial^2 u}{\partial y^2} = -\frac{\partial^2 v}{\partial x \partial y}$$

and so $\dfrac{\partial^2 u}{\partial x^2} + \dfrac{\partial^2 v}{\partial y^2} = 0$.

3. Simply scale all distances by a factor of b to obtain $|w - bz_0| \le ba$.

6. Consider the case $c \ne 0$. The first step is a scaling-multiplication by c: $w_1 = cz$. The next step is a translation by d: $w_2 = w_1 + d = cz + d$. The next step is an inversion: $w_3 = \dfrac{1}{w_2} = \dfrac{1}{cz + d}$, and the next step is a scaling-multiplication by $(bc - ad)/c$:

$w_4 = \left(\dfrac{bc - ad}{c}\right) w_3 = \dfrac{bc - ad}{c(cz + d)}$. The final step is a translation by a/c:

$$w = w_4 + \frac{a}{c} = \frac{bc - ad + acz + ad}{c(cz + d)} = \frac{az + b}{cz + d}$$

If $c = 0$, then we have only two steps:

$$w_1 = \frac{a}{d}z \quad \text{and} \quad w_2 = w_1 + \frac{b}{d} = \frac{az + b}{d}$$

9. Start with

$$\begin{aligned} w &= \cosh \frac{\pi z}{a} = \cosh \frac{\pi x}{a} \cos \frac{\pi y}{a} + i \sinh \frac{\pi x}{a} \sin \frac{\pi y}{a} \\ &= u(x, y) + iv(x, y) \end{aligned}$$

The positive real axis in the z-plane ($y = 0$) maps into ($u = \cosh \pi x/a$, $v = 0$) the interval $(1, \infty)$ of the u axis. The positive y axis ($x = 0$) maps into ($u = \cos \pi y/a$, $v = 0$) the interval $(-1, 1)$ of the u axis. The line at $y = a$ ($x \geq 0$) in the z-plane maps into ($u = -\cosh \pi x/a$, $v = 0$) the interval $(-\infty, -1)$ of the u axis. Any point within the shaded region in the z-plane maps into the upper half plane in the w-plane.

12.

$$w = \frac{1}{z} = \frac{1}{x + iy} = \frac{x}{x^2 + y^2} - \frac{iy}{x^2 + y^2}$$
$$= u(x, y) + iv(x, y)$$

The y axis in the z-plane ($x = 0$) maps into ($u = 0$, $v = -1/y$) the v axis in the w-plane. Note that the point 0 in the z-plane maps into $\pm\infty$ in the w-plane; the point B ($x = 0$, $y = 0$) maps into B' ($u = 0$, $v = -1$), and so on.

The equation of the circle in the z-plane, $\left| z - \frac{1}{2} \right| = \frac{1}{2}$, gives $\left(x - \frac{1}{2} \right)^2 + y^2 = \frac{1}{4}$, or $x^2 + y^2 = x$. Substitute this result into $u(x, y) = x/(x^2 + y^2)$ and $v(x, y) = -y/(x^2 + y^2)$ to get $u = 1$ and $v = -y/x$. Therefore, the circle in the z-plane maps into a vertical line at $u = 1$ in the w-plane. Note that the point E ($x = 1$, $y = 0$) maps onto the point E' ($u = 1$, $v = 0$), the point F ($x = 1/2$, $y = 1/2$) maps onto F' ($u = 1$, $v = -1$), and so on. Points outside the circle in the z-plane map onto points between the vertical lines $u = 0$ and $u = 1$ in the w-plane.

15. $w = \left(\frac{1+z}{1-z} \right)^2 = \left(\frac{1+x+iy}{1-x-iy} \right)^2 = \left[\frac{(1-x^2) - y^2 + 2iy}{(1-x)^2 + y^2} \right]^2$ so that

$$u(x, y) = \frac{(1-x)^2 - 2y^2(1-x) + y^4 - 4y^2}{((1-x)^2 + y^2)^2} \quad \text{and} \quad v(x, y) = \frac{4y\left[(1-x)^2 - y^2 \right]}{((1-x)^2 + y^2)^2}$$

Let's look at how the x axis ($y = 0$) maps into the w-plane. In this case, $u = ((1+x)/(1-x))^2$ and $v = 0$. The point O ($x = 0$, $y = 0$) maps onto O'; the point A ($x = 1$, $y = 0$) maps onto A' ($u = -\infty$, $v = 0$) and C ($x = -1$, $y = 0$) maps onto C' ($u = 0$, $v = 0$). The point B ($x = 0$, $y = 1$) maps onto B' ($-1, 0$).

18. Because $f'(z) = 0$ at $z = 0$.

21. If the point (x_0, y_0) is on C_z in the z-plane, then the point $x_0(u_0, v_0), y_0(u_0, v_0)$ is on C_w in the w-plane, and so $\Phi(u, v) = g\{x(u, v), y(u.v)\}$ satisfies the Dirichlet condition in C_w.

19.6 Conformal Mapping and Boundary Value Problems

1. The solution depends upon only v, and so

$$\Phi = \alpha v + \beta = \frac{\Phi_1 - \Phi_0}{a} v + \Phi_0$$

3. For Entry 5 of Table 19.1, $w = u + iv = \text{Ln} \, z = \ln(x^2 + y^2)^{1/2} + i \tan^{-1} \frac{y}{x}$. The x axis ($y = 0$) transforms to $u = \ln x$. Because $0 \leq x < \infty$, we see that $-\infty < u < \infty$. So the origin in the z-plane (A) transforms to the point $(-\infty, 0)$ (A') in the w-plane; the point $(x = 1, y = 0)$ transforms to the point $(0,0)$. The y axis ($x = 0$) transforms to $\ln y + i\pi/2$, and so we see that the points C and D are mapped onto the points C' and D'.

6. Poisson's integral formula for a disk (Problem 16.2.13) in this case is

$$
\begin{aligned}
\phi(r, \theta) &= \frac{V_0}{2\pi} \int_0^\pi \frac{(1 - r^2)\, d\phi}{1 + r^2 - 2r\cos(\theta - \phi)} + \frac{V_1}{\pi} \int_\pi^{2\pi} \frac{(1 - r^2)\, d\phi}{1 + r^2 - 2r\cos(\theta - \phi)} \\
&= \frac{V_0(1 - r^2)}{2\pi} \int_0^\pi \frac{d\phi}{1 + r^2 - 2r\cos\theta\cos\phi - 2r\sin\theta\sin\phi} \\
&\quad + \frac{V_1(1 - r^2)}{2\pi} \int_\pi^{2\pi} \frac{d\phi}{1 + r^2 - 2r\cos\theta\cos\phi - 2r\sin\theta\sin\phi} \\
&= \frac{V_0}{\pi} \left[\tan^{-1} \frac{-2r\sin\theta + (1 + r^2 - 2r\cos\theta)\tan(\phi/2)}{1 - r^2} \right]_0^\pi \\
&\quad + \frac{V_1}{\pi} \left[\tan^{-1} \frac{-2r\sin\theta + (1 + r^2 - 2r\cos\theta)\tan(\phi/2)}{1 - r^2} \right]_\pi^{2\pi} \\
&= \frac{V_0}{\pi} \left(\frac{\pi}{2} - \tan^{-1} \frac{-2r\sin\theta}{1 - r^2} \right) + \frac{V_1}{\pi} \left(\tan^{-1} \frac{-2r\sin\theta}{1 - r^2} - \frac{\pi}{2} \right) \\
&= \frac{V_0}{\pi} \left(\frac{\pi}{2} + \frac{\pi}{2} - \tan^{-1} \frac{1 - r^2}{2r\sin\theta} \right) + \frac{V_1}{\pi} \left(\tan^{-1} \frac{1 - r^2}{2r\sin\theta} \right) \\
&= V_0 + \frac{V_1 - V_0}{\pi} \tan^{-1} \frac{1 - r^2}{2r\sin\theta} \\
&= V_0 + \frac{V_1 - V_0}{\pi} \tan^{-1} \frac{1 - x^2 - y^2}{2y} \qquad x^2 + y^2 < 1
\end{aligned}
$$

9. Use Entry 14 of Table 19.1 with $a = 1$, $b = \infty$, $\theta_1 = 0$, and $\theta_2 = \pi/2$. The corresponding region in the w-plane is shown in the accompanying figure.

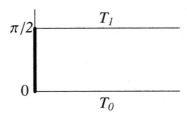

The solution in the w-plane is $\Phi(u, v) = T_0 + \dfrac{2}{\pi}(T_1 - T_0)v$. The transformation

$$
w = u + iv = \ln z = \ln(x^2 + y^2)^{1/2} + i\tan^{-1}\frac{y}{x}
$$

gives

$$
T(x, y) = T_0 + \frac{2}{\pi}(T_1 - T_0)\tan^{-1}\frac{y}{x} \qquad x^2 + y^2 \geq 1
$$

The following figure shows a contour plot of $\Phi(x, y)$.

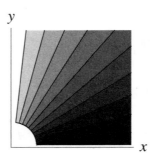

12. The mapping is

$$w = \sin z = \sin x \cosh y + i \cos x \sinh y = u(x,y) + iv(x,y)$$

The vertical line $x = \pi/2$ is mapped onto $u = \cosh y$, $v = 0$, or onto the u axis for $u \geq 1$. The vertical line $x = -\pi/2$ is mapped onto the u axis for $u \leq -1$. The line segment $(-\pi/2 \leq x \leq \pi/2, y = 0)$ is mapped into $u = \sin x$, $v = 0$, or onto the u axis for $-1 \leq u \leq 1$. Therefore, the mapping maps the region shown in Figure 19.31 onto the upper half plane in the w-plane.

15. The potential is given by

$$\phi(x,y) = \frac{V_2 - V_1}{2 \ln 5} \ln \frac{(x-3)^2 + y^2}{(3x-1)^2 + 9y^2} + V_2$$

Let's consider the potential on the unit circle surrounding the origin in Figure 19.35, where $x^2 + y^2 = 1$. Now,

$$\ln \frac{(x-3)^2 + y^2}{(3x-1)^2 + 9y^2} = \frac{x^2 + y^2 - 6x + 9}{9x^2 + 9y^2 + 1 - 6x} = \frac{1 - 6x + 9}{9 + 1 - 6x} = 1$$

and so $\phi(x,y) = V_2$.

The equation of the other equipotential circle in Figure 19.35 is $\left(x - \dfrac{9}{2}\right)^2 + y^2 = \dfrac{25}{4}$ or $x^2 + y^2 = 9x - 14$. In this case,

$$\ln \frac{(x-3)^2 + y^2}{(3x-1)^2 + 9y^2} = \frac{9x - 14 - 6x + 9}{81x - 126 + 1 - 6x} = \frac{3x - 5}{75x - 125} = \frac{1}{25}$$

and so $\phi(x,y) = -(V_2 - V_1) + V_2 = V_1$.

18. We use Entry 12 in Table 19.1 with $x_2 = 0$ and $x_1 = 1/2$. According to Entry 12, $a = \dfrac{1 + \sqrt{3}/2}{1/2} = 2 + \sqrt{3}$ and $R = 2 + \sqrt{3}$. The solution in the annulus in the w-plane is (see Problem 2)

$$\Phi(u,v) = \frac{T_1 - T_2}{\ln R} \ln \rho + T_2 = \frac{T_1 - T_2}{\ln R} \ln |w| + T_2 = \frac{T_1 - T_2}{\ln R} \ln \left| \frac{z - a}{az - 1} \right| + T_2$$

and so

$$T(x,y) = \frac{T_1 - T_2}{2 \ln R} \ln \left[\frac{(x-a)^2 + y^2}{(ax-1)^2 + a^2 y^2} \right] + T_2$$

The logarithm term is

$$\ln\left[\frac{(x-a)^2+y^2}{(ax-1)^2+a^2y^2}\right] = \ln\left[\frac{x^2+y^2-2ax+a^2}{a^2(x^2+y^2)-2ax+1}\right]$$

The equation of the inner circle in Figure 19.43 is $\left(x-\frac{1}{4}\right)^2+y^2=\left(\frac{1}{4}\right)^2$, or $x^2+y^2=\frac{x}{2}$ and so

$$
\begin{aligned}
\ln\left[\frac{(x-a)^2+y^2}{(ax-1)^2+a^2y^2}\right] &= \frac{\dfrac{x}{2}-2ax+a^2}{\dfrac{a^2x}{2}-2ax+1} = \frac{x-4ax+2a^2}{a^2x-4ax+2} \\[2mm]
&= \frac{14+8\sqrt{3}-7x-4\sqrt{3}\,x}{2-x} = \frac{(7+4\sqrt{3})(2-x)}{2-x} \\[2mm]
&= 7+4\sqrt{3} = R^2
\end{aligned}
$$

Therefore

$$T(x,y) = \frac{T_1-T_2}{2\ln R}\ln R^2 + T_2 = T_1 \qquad \text{on the inner cylinder}$$

The equation of the outer circle in Figure 19.43 is $x^2+y^2=1$, and so

$$\ln\left[\frac{(x-a)^2+y}{(ax-1)^2+a^2y^2}\right] = \ln\left[\frac{1-2ax+a^2}{a^2-2ax+1}\right] = 0$$

Therefore,

$$T(x,y)) = T_2 \qquad \text{on the outer cylinder}$$

19.7   Conformal Mapping and Fluid Flow

1. Start with the continuity equation (Equation 7.1.10), $\dfrac{\partial\rho}{\partial t} = \boldsymbol{\nabla}\cdot(\rho\mathbf{v})$. If $\rho=$ constant, then div $\mathbf{v}=0$.

3. If $\Omega(z) = v_0 x\cos\alpha + v_0 y\sin\alpha + i(v_0 y\cos\alpha - v_0 x\sin\alpha)$, then

$$v_x = \frac{\partial\phi}{\partial x} = v_0\cos\alpha \qquad \text{and} \qquad v_y = \frac{\partial\phi}{\partial y} = v_0\sin\alpha$$

which represents uniform flow that makes an angle α with the x axis.

6. We'll do Problem 5 first. Start with $\Omega(z) = \Omega(x,y) = \phi(x,y) + i\,\psi(x,y)$. Then

$$\frac{d\Omega}{dz} = \frac{\partial\phi}{\partial x} + i\frac{\partial\psi}{\partial x} = \frac{\partial\phi}{\partial x} - i\frac{\partial\psi}{\partial y} = v_x - iv_y$$

For uniform flow, $\Omega(z) = v_0 e^{-i\alpha}z$ and $\dfrac{d\Omega}{dz} = v_0\cos\alpha - iv_0\sin\alpha$.

Using the result of Problem 5, $|v(z)| = \left|\dfrac{d\Omega}{dz}\right| = |\Omega'(z)|$.

9. The field depends only upon the radial coordinate r. Use Gauss's law to get $2\pi r E_r = \dfrac{\lambda}{\epsilon_0}$, or $\phi = \dfrac{\lambda}{2\pi\epsilon_0}\ln r + \text{constant}$.

12. We use the map $w = z^2$ (Entry 1 of Figure 19.1.) The solution in the w-plane is $\Phi(u, v) = v_0 w = v_0 z^2 = v_0(x^2 - y^2) + 2iv_0 xy$. The streamlines are given by $\psi(x, y) = 2v_0 xy = \text{constant}$. These are plotted in Figure 19.45a.

15. Because $f'(z) = 0$ at $z = 1$, the mapping is not conformal there.

CHAPTER 20

Calculus of Variations

20.1 The Euler Equation

1. Start with

$$
\begin{aligned}
\phi &= c_1 \int \frac{d\theta}{(a^2 \sin^4 \theta - c_1^2 \sin^2 \theta)^{1/2}} \\
&= \int \frac{\csc^2 \theta \, d\theta}{(a^2/c_1^2 - \csc^2 \theta)^{1/2}} = \int \frac{\csc^2 \theta \, d\theta}{\left(\dfrac{a^2}{c_1^2} - 1 - \cot^2 \theta \right)^{1/2}} \\
&= \frac{1}{[(a^2/c_1^2) - 1]^{1/2}} \int \frac{\csc^2 \theta \, d\theta}{\left[1 - \dfrac{\cot^2 \theta}{(a^2/c_1^2) - 1} \right]^{1/2}}
\end{aligned}
$$

Now let $u = \cot\theta/[(a^2/c_1^2) - 1]^{1/2}$ to write $du = -\dfrac{\csc^2 \theta \, d\theta}{[(a^2/c_1^2) - 1]^{1/2}}$. With this substitution, ϕ becomes

$$
\begin{aligned}
\phi &= -\int \frac{du}{(1 - u^2)^{1/2}} = -\sin^{-1} u + c_2 \\
&= -\sin^{-1} \frac{\cot \theta}{[(a^2/c_1^2) - 1]^{1/2}} + c_2
\end{aligned}
$$

3. Start with

$$
I = \int_a^b (py'^2 - qy^2) \, dx
$$

and use Equation 8 to write

$$
\frac{d}{dx}(2py') + 2qy = 0
$$

or $[p(x)y'(x)]' + q(x)y(x) = 0$. This is a form of a Sturm-Liouville equation.

6. Use Equation 8 to write

$$\frac{d}{dx}[2xy' - y] + y' - 1 = 0$$

or $2xy'' + 2y' - 1 = 0$.

9. Start with $I = \int (r'^2 + r^2 \sin^2 \alpha)^{1/2} d\phi$. Equation 10 (with $r \leftrightarrow y$, $\phi \leftrightarrow x$) gives $-r^2 \sin^2 \alpha = c(r^2 \sin^2 \alpha + r'^2)^{1/2}$, or

$$r^4 \sin^4 \alpha - cr^2 \sin^2 \alpha = c\left(\frac{dr}{d\phi}\right)^2$$

Let $z = r \sin \alpha$, so that we have

$$\phi = \frac{c}{\sin \alpha} \int \frac{dz}{z(z^2 - c)^{1/2}} = \frac{1}{\sin \alpha} \sec^{-1} \frac{z}{c} + c_1$$

or

$$r \sin \alpha = c \sec(\phi \sin \alpha + c_2)$$

12. The substitution $y = c_1 \cosh u$ and $dy = c_1 \sinh u \, du$ gives

$$
\begin{aligned}
x &= c_1 \int \frac{dy}{(y^2 - c_1^2)^{1/2}} = \int \frac{c_1 \sinh u \, du}{c_1 \sinh u} = \int du \\
&= c_1 u + c_2 = c_1 \cosh^{-1} \frac{y}{c_1} + c_2
\end{aligned}
$$

20.2   Two Laws of Physics in Variational Form

1. Start with the Maclaurin expansion of $\int_{t_1}^{t_2} L(Q_1, Q_2, \dot{Q}_1, \dot{Q}_2, t) \, dt$ about $\epsilon = 0$:

$$
\begin{aligned}
I(\epsilon) &= \int_{t_1}^{t_2} L(Q_1, Q_2, \dot{Q}_1, \dot{Q}_2, t) \, dt \\
&= \int_{t_1}^{t_2} L(q_1, q_2, \dot{q}_1, \dot{q}_2, t) \, dt + \epsilon \int_{t_1}^{t_2} \left[\frac{\partial L}{\partial Q_1} \frac{\partial Q_1}{\partial \epsilon} + \frac{\partial L}{\partial \dot{Q}_1} \frac{\partial \dot{Q}_1}{\partial \epsilon} \right]_{\epsilon=0} dt \\
&\quad + \epsilon \int_{t_1}^{t_2} \left[\frac{\partial L}{\partial Q_2} \frac{\partial Q_2}{\partial \epsilon} + \frac{\partial L}{\partial \dot{Q}_2} \frac{\partial \dot{Q}_2}{\partial \epsilon} \right]_{\epsilon=0} dt + O(\epsilon^2) \\
&= \int_{t_1}^{t_2} L(q_1, q_2, \dot{q}_1, \dot{q}_2, t) \, dt + \epsilon \int_{t_1}^{t_2} \left(\frac{\partial L}{\partial q_1} \eta_1 + \frac{\partial L}{\partial \dot{q}_1} \dot{\eta}_1 \right) dt \\
&\quad + \epsilon \int_{t_1}^{t_2} \left(\frac{\partial L}{\partial q_2} \eta_2 + \frac{\partial L}{\partial \dot{q}_2} \dot{\eta}_2 \right) dt + O(\epsilon^2)
\end{aligned}
$$

The condition $dI/d\epsilon = 0$ when $\epsilon = 0$ gives

$$\int_{t_1}^{t_2} \left(\frac{\partial L}{\partial q_1} \eta_1 + \frac{\partial L}{\partial \dot{q}_1} \dot{\eta}_1 \right) dt + \int_{t_1}^{t_2} \left(\frac{\partial L}{\partial q_2} \eta_2 + \frac{\partial L}{\partial \dot{q}_2} \dot{\eta}_2 \right) dt = 0$$

Now integrate by parts to obtain Equation 8.

3. Extremize $I = \int_{t_1}^{t_2} (\dot{x}^2 + \dot{y}^2 + \dot{z}^2)^{1/2} \, dt$ to obtain $\ddot{x} = \ddot{y} = \ddot{z} = 0$. Integration of these expressions with $x(t_1) = x_1$, $x(t_2) = x_2$, etc., gives

$$x = \left(\frac{x_2 - x_1}{t_2 - t_1} \right) t + \frac{x_1 t_2 - x_2 t_1}{t_2 - t_1} = \frac{(x_2 - x_1)t + x_1 t_2 - x_2 t_1}{t_2 - t_1}$$

with similar equations in y and z. These are the parametric equations of a straight line.

6. Let $x_1 = l \sin\theta$ and $y_1 = l \cos\theta$. The kinetic energy and potential energy are given by

$$K = \frac{m}{2}(\dot{x}^2 + \dot{y}^2) = \frac{ml^2}{2}(\cos^2\theta + \sin^2\theta)\dot{\theta}^2 = \frac{ml^2 \dot{\theta}^2}{2}$$

and

$$V = mg(l - y) = mgl(1 - \cos\theta)$$

The Lagrangian for this system is $L = \frac{ml^2}{2}\dot{\theta}^2 - mgl(1 - \cos\theta)$. The Euler-Lagrange equation, $\frac{d}{dt}\frac{\partial L}{\partial \dot{\theta}} - \frac{\partial L}{\partial \theta} = 0$, leads to

$$ml^2 \ddot{\theta} + mgl \sin\theta = 0 \qquad \text{or} \qquad \ddot{\theta} + \frac{g}{l}\sin\theta = 0$$

9. The kinetic and potential energies are given by

$$K = \frac{m}{2}\dot{x}^2 = \frac{m}{2}\dot{\xi}^2$$

and

$$V = \frac{k}{2}(x - l - a \sin\omega t)^2 = \frac{k}{2}(\xi - a \sin\omega t)^2$$

where $\xi = x - l$. The Lagrangian for this system is $L(\xi, \dot{\xi}, t) = \frac{m}{2}\dot{\xi}^2 - \frac{k}{2}(\xi - a \sin\omega t)^2$. The corresponding equation of motion is

$$m\ddot{\xi} + k(\xi - a \sin\omega t) = 0$$

$$m\ddot{\xi} + k\xi = ka \sin\omega t$$

12. Use Equation 12 with $n(y) = a/y$, which yields $y^2(1 + y'^2) = c^2$, or

$$x = \int \frac{y \, dy}{(c^2 - y^2)^{1/2}} = -(c^2 - y^2)^{1/2} + c_1$$

or $y^2 = c^2 - (x - c_1)^2$.

15. Let $\alpha = \theta$ and $\beta = z$ in cylindrical coordinates. In this case, $h_\alpha = a^2$ and $h_\beta = 1$, and we must minimize

$$I = \int_a^b \left[a^2 + \left(\frac{dz}{d\theta} \right)^2 \right]^{1/2} d\theta = \int_a^b (a^2 + z'^2)^{1/2} d\theta$$

The Euler equation in this case is $\dfrac{d}{d\theta} \left[\dfrac{z'}{(a^2 + z'^2)^{1/2}} \right] = 0$, which gives

$$z' = \frac{dz}{d\theta} = \text{constant} \qquad \text{or} \qquad z = c_1 \theta + c_2$$

which is the equation of a helix. (See Problem 20.1.8.)

20.3   Variational Problems with Constraints

1. Replace $y(x)$ by $y(x) + \epsilon \eta(x)$ in F and G in Equation 3, expand in a Taylor series about $\epsilon = 0$ to $O(\epsilon^2)$, and then set $dK/d\epsilon$ (in other words, the term linear in ϵ) equal to zero when $\epsilon = 0$. You also need to integrate by parts once.

3. Substitute $\rho g y - \lambda = c_1 \cosh z$, and $dy = \dfrac{c_1}{\rho g} \sinh z \, dz$ into Equation 7 to obtain

$$\begin{aligned} x &= \frac{c_1^2}{\rho g} \int \frac{\sinh z \, dz}{(c_1^2 \cosh^2 z - c_1^2)^{1/2}} = \frac{c_1}{\rho g} \int dz \\ &= \frac{c_1 z}{\rho g} + c_2 \end{aligned}$$

or $z = \dfrac{\rho g (x - c_2)}{c_1}$, or $\cosh^{-1} \dfrac{\rho g y - \lambda}{c_1} = \dfrac{\rho g (x - c_2)}{c_1}$, or finally

$$\rho g y - \lambda = c_1 \cosh \frac{\rho g (x - c_2)}{c_1}$$

6. According to Problem 5, extremizing the curve for a given area is equivalent to extremizing the area for a given curve, which gives the same result as that in Example 2.

9. In this case, $F(y, y') = p(x) y'^2(x) - q(x) y^2(x)$ and $G(y, y') = r(x) y^2(x)$. Equation 4 (with $\lambda = -\lambda$) gives

$$\frac{\partial (F - \lambda G)}{\partial y} = -2q(x)y(x) - 2\lambda r(x)y(x) = \frac{d}{dx} \frac{\partial (F - \lambda G)}{\partial y'} = \frac{d}{dx} [\, 2p(x)y'(x) \,]$$

or

$$\frac{d}{dx} \left(p(x) \frac{dy}{dx} \right) + [\, q(x) + \lambda r(x)y(x) \,] = 0$$

This is a Sturm-Liouville equation.

20.4   Variational Formulation of Eigenvalue Problems

1. Using Equation 4 directly gives

$$\lambda = \frac{\int_0^1 x(1-x)(2)\,dx}{\int_0^1 x^2(1-x)^2\,dx} = \frac{2B(2,2)}{B(3,3)} = \frac{2\Gamma(2)\Gamma(2)}{\Gamma(4)} \cdot \frac{\Gamma(6)}{\Gamma(3)\Gamma(3)} = 10$$

3. Substitute $\phi(x) = \sum_{n=1}^{\infty} a_n u_n(x)$ into $\lambda_\phi = \dfrac{\int \phi(x)\mathcal{L}\phi(x)\,dx}{\int r(x)\phi^2(x)\,dx}$ and use the orthonormality of the $u_n(x)$ with respect to $r(x)$ to obtain

$$\lambda_\phi = \frac{\int \left[\sum_{n=1}^{\infty} a_n u_n(x) \sum_{m=1}^{\infty} a_m \lambda_m u_m(x) \right] dx}{\int \left[\sum_{n=1}^{\infty} a_n u_n(x) \sum_{m=1}^{\infty} a_m u_m(x) \right] dx}$$

$$= \frac{\sum_{n=1}^{\infty}\sum_{m=1}^{\infty} \lambda_m a_n a_m \delta_{nm}}{\sum_{n=1}^{\infty}\sum_{m=1}^{\infty} a_n a_m \delta_{nm}} = \frac{\sum_{n=1}^{\infty} a_n^2 \lambda_n}{\sum_{n=1}^{\infty} a_n^2}$$

6. See the solution to 20.3.9.

9. Use $\mathcal{L} = -\dfrac{d^2}{dx^2}$ and $r(x) = 1/x$ in Equation 4 to write

$$\lambda_\phi = \frac{\int_0^1 2x(1-x)\,dx}{\int_0^1 x(1-x)^2\,dx} = \frac{2B(2,2)}{B(2,3)} = 4$$

12. Using $\mathcal{L} = -\dfrac{d^2}{dx^2}$ and $r(x) = 1$ in Equation 4 gives

$$L_{11} = 2\int_0^1 x(1-x)\,dx = 2B(2,2) = \frac{1}{3}$$

$$S_{11} = \int_0^1 x^2(1-x)^2\,dx = B(3,3) = \frac{1}{30}$$

$$L_{12} = L_{21} = 2\int_0^1 x^2(1-x)\,dx = 2B(3,2) = \frac{1}{6}$$

$$S_{12} = S_{21} = \int_0^1 x^3(1-x)^2\,dx = B(4,3) = \frac{1}{60}$$

$$L_{22} = 6\int_0^1 x^3(1-x)^2\,dx - 2\int_0^1 x^2(1-x)\,dx = 6B(4,2) - 2B(3,2) = \frac{2}{15}$$

$$S_{22} = \int_0^1 x^4(1-x)^2\,dx = B(5,3) = \frac{1}{105}$$

The corresponding secular determinantal equation is

$$
\begin{vmatrix} \dfrac{1}{3} - \dfrac{\lambda}{30} & \dfrac{1}{6} - \dfrac{\lambda}{60} \\[2mm] \dfrac{1}{6} - \dfrac{\lambda}{60} & \dfrac{2}{15} - \dfrac{\lambda}{105} \end{vmatrix} = \dfrac{1}{3} \cdot \dfrac{1}{3} \begin{vmatrix} 1 - \dfrac{\lambda}{10} & \dfrac{1}{2} - \dfrac{\lambda}{20} \\[2mm] \dfrac{1}{2} - \dfrac{\lambda}{20} & \dfrac{2}{5} - \dfrac{\lambda}{35} \end{vmatrix} = 0
$$

or

$$
\frac{\lambda^2}{2800} - \frac{13\lambda}{700} + \frac{3}{20} = 0
$$

or

$$
\lambda^2 - 52\lambda + 420 = 0
$$

or

$$
\lambda = 26 \pm \frac{1}{2}(32) = 10 \quad \text{and} \quad 42
$$

The exact value is $\pi^2 = 9.86960\ldots$. The solution to Problem 5 is $9.86975\ldots$.

15. Use

$$
\hat{H} = -\frac{h^2}{8\pi^2 \mu r^2} \frac{d}{dr}\left(r^2 \frac{d}{dr}\right) - \frac{e^2}{4\pi\epsilon_0 r}
$$

so that

$$
\hat{H}(e^{-ar}) = -\frac{h^2}{8\pi^2 \mu r^2}(-2ar + a^2 r^2)e^{-ar} - \frac{e^2 e^{-ar}}{4\pi\epsilon_0 r}
$$

Therefore,

$$
\begin{aligned}
4\pi \int dr\, r^2 e^{-ar} \hat{H}(e^{-ar}) &= 4\pi \int_0^\infty dr\, r^2 \left[-\frac{h^2}{8\pi^2 \mu r^2}(a^2 r^2 - 2ar)e^{-2ar} - \frac{e^2 e^{-2ar}}{4\pi\epsilon_0 r} \right] \\
&= -\frac{h^2 a^2}{2\pi\mu} \int_0^\infty dr\, r^2 e^{-2ar} + \frac{h^2 a}{\pi\mu} \int_0^\infty dr\, r e^{-2ar} - \frac{e^2}{\epsilon_0} \int_0^\infty dr\, r e^{-2ar} \\
&= -\frac{h^2 a^2}{2\pi\mu}\left(\frac{1}{4a^3}\right) + \frac{h^2 a}{\pi\mu}\left(\frac{1}{4a^2}\right) - \frac{e^2}{\epsilon_0}\left(\frac{1}{4a^2}\right)
\end{aligned}
$$

and $4\pi \displaystyle\int_0^\infty dr\, r^2 e^{-2ar} = \dfrac{\pi}{a^3}$. Then

$$
E(a) = -\frac{h^2 a^2}{8\pi^2 \mu} + \frac{h^2 a^2}{4\pi^2 \mu} - \frac{e^2 a}{4\pi\epsilon_0}
$$

and $\dfrac{dE}{da} = 0$ gives $a = \dfrac{\pi e^2 \mu}{\epsilon_0 h^2}$ and $E = -\dfrac{e^4 \mu}{8\epsilon_0^2 h^2} = -\dfrac{1}{2}\left(\dfrac{\mu e^4}{4\epsilon_0^2 h^2}\right)$, which is the exact ground-state electronic energy of a hydrogen atom.

The reason that we obtain the exact value is because e^{-ar} is the exact ground-state wave function.

18. Start with

$$E(\beta) = \frac{\int_{-\infty}^{\infty} \phi(x) \left[-\frac{\hbar^2}{2\mu} \frac{d^2}{dx^2} + \frac{k}{2} x^2 \right] \phi(x)\, dx}{\int_{-\infty}^{\infty} \phi^2(x)\, dx}$$

with $\phi(x) = 1/(1 + \beta x^2)$. The numerator is

$$
\begin{aligned}
N(\beta) &= -\frac{\hbar^2}{2\mu} \int_{-\infty}^{\infty} \frac{dx\, \beta(6\beta x^2 - 2)}{(\beta x^2 + 1)^4} + \frac{k}{2} \int_{-\infty}^{\infty} \frac{dx\, x^2}{(\beta x^2 + 1)^2} \\
&= -\frac{\hbar^2}{2\mu} \left(\frac{12\pi \beta^{1/2}}{32} - \frac{20\pi \beta^{1/2}}{32} \right) + \frac{k}{2} \frac{\pi}{2\beta^{3/2}}
\end{aligned}
$$

and the denominator is

$$D(\beta) = 2\beta^2 \int_{-\infty}^{\infty} \frac{dx}{(\beta x^2 + 1)^2} = \frac{\pi}{2\beta^{1/2}}$$

Therefore,

$$E(\beta) = \frac{N(\beta)}{D(\beta)} = \frac{\hbar^2 \beta}{4\mu} + \frac{k}{2\beta}$$

and $\dfrac{dE}{d\beta} = 0 = \dfrac{\hbar^2}{4\mu} - \dfrac{k}{2\beta^2}$ or $\beta = (2\mu k)^{1/2}/\hbar$. The ground-state energy is given by

$$E = \frac{\hbar}{4\mu}(2\mu k)^{1/2} + \frac{k}{2} \frac{\hbar}{(2\mu k)^{1/2}} = \frac{1}{\sqrt{2}} \hbar (k/\mu)^{1/2} = 0.707\, \hbar (k/\mu)^{1/2}$$

The exact ground-state energy is $\frac{1}{2}\hbar(k/\mu)^{1/2}$.

21. We need the result of Problem 5 first. We use Equation 21 to obtain

$$
\begin{aligned}
L_{11} &= 2B(2,2) = \frac{1}{3} \quad \text{(from Problem 1)} \\
S_{11} &= B(3,3) = \frac{1}{30} \quad \text{(from Problem 1)} \\
L_{12} &= L_{21} = 2B(3,3) = \frac{1}{15} \\
S_{12} &= S_{21} = B(4,4) = \frac{1}{140} \\
L_{22} &= -2B(3,3) + 12B(4,3) - 12B(5,3) = \frac{2}{105} \\
S_{22} &= B(5,5) = \frac{1}{630}
\end{aligned}
$$

Using these values, Equation 23 becomes

$$
\begin{vmatrix}
\dfrac{1}{3} - \dfrac{\lambda}{30} & \dfrac{1}{15} - \dfrac{\lambda}{140} \\[2mm]
\dfrac{1}{15} - \dfrac{\lambda}{140} & \dfrac{2}{105} - \dfrac{\lambda}{630}
\end{vmatrix} = 0
$$

The corresponding secular equation is

$$\lambda^2 - 112\lambda + 1008 = 0$$

which yields $\lambda = 9.86975$ and 102.13.

Now substitute $\lambda = 9.86975$ into the original algebraic equations (Equations 21 and 22) to find that

$$c_1\left(\frac{1}{3} - \frac{9.86975}{30}\right) + c_2\left(\frac{1}{15} - \frac{9.86975}{140}\right) = 0$$

or that $\dfrac{c_2}{c_1} = 1.13314$. So far, then, we have

$$\phi(x) = c_1\left[\,x(1-x) + 1.13314x^2(1-x)^2\right]$$

We now determine c_1 by requiring that $\phi(x)$ be normalized:

$$
\begin{aligned}
\int_0^1 \phi^2(x)\,dx &= 1 \\
&= c_1^2\left[\int_0^1 x^2(1-x)^2\,dx + 2(1.13314)\int_0^1 x^3(1-x)^3 + (1.13314)^2\int_0^1 x^4(1-x)^4\,dx\right.\\
&= c_1^2\left[\frac{1}{30} + 2(1.13314)\frac{1}{140} + (1.13314)^2\frac{1}{630}\right] \\
&= 0.0515592\,c_1^2
\end{aligned}
$$

or $c_1 = 4.404$.

20.5   Multidimensional Variational Problems

1. Expand

$$I = \iint_R F(U, U_x, U_y, x, y)\,dxdy$$

where $U(x,y) = u(x,y) + \epsilon\eta(x,y)$ in a Maclaurin series about ϵ to $O(\epsilon)$.

$$
\begin{aligned}
I(\epsilon) &= \iint_R F(u, u_x, u_y, x, y)\,dxdy + \epsilon\iint_R\left[\frac{\partial F}{\partial U}\frac{\partial U}{\partial\epsilon} + \frac{\partial F}{\partial U_x}\frac{\partial U_x}{\partial\epsilon} + \frac{\partial F}{\partial U_y}\frac{\partial U_y}{\partial\epsilon}\right]_{\epsilon=0}dxdy + O(\epsilon^2) \\
&= \iint_R F(u, u_x, u_y, x, y)\,dxdy + \epsilon\iint_R\left[\frac{\partial F}{\partial U}\eta + \frac{\partial F}{\partial U_x}\eta_x + \frac{\partial F}{\partial U_y}\eta_y\right]dxdy + O(\epsilon^2)
\end{aligned}
$$

Minimizing $I(\epsilon)$ with respect to ϵ gives Equation 3.

3. This problem is very similar to Example 1.

6. The first term represents potential energy due to tension, the second term represents kinetic energy, and the third term represents the potential energy due to the impressed load.

9. The trial function $\phi(r,\theta) = y(a-r) = r(a-r)\sin\theta$ is fairly simple and satisfies the boundary conditions. Substitute $\phi(r,\theta)$ into Equation 12:

$$
\begin{aligned}
\nabla^2\phi &= \frac{\partial^2\phi}{\partial r^2} + \frac{1}{r}\frac{\partial\phi}{\partial r} + \frac{1}{r^2}\frac{\partial^2\phi}{\partial\theta^2} \\
&= -2\sin\theta + \left(\frac{a}{r}-2\right)\sin\theta - \left(\frac{a}{r}-1\right)\sin\theta \\
&= -3\sin\theta
\end{aligned}
$$

Thus,

$$
\begin{aligned}
\int \phi\,\nabla^2\phi\,dv &= \int_0^a dr\, r \int_0^\pi d\theta\,\phi\,\nabla^2\phi = -3\int_0^\pi \sin^2\theta\,d\theta \int_0^a r^2(a-r)\,dr \\
&= -3\cdot\frac{\pi}{2}\cdot a^4 B(3,2) = -\frac{\pi a^4}{8}
\end{aligned}
$$

and

$$
\int \phi^2\,dv = \frac{\pi}{2}\cdot B(4,3) = \frac{\pi a^6}{120}
$$

Therefore, Equation 12 gives us $\dfrac{\omega^2}{v^2} = \dfrac{15}{a^2}$, or $\omega = \sqrt{15}\,v/a$.

CHAPTER 21

Probability Theory and Stochastic Processes

21.1 Discrete Random Variables

1. Each toss is an independent event, or the probability of 10 heads followed by heads is exactly the same as the probability of 10 tails followed by heads.

3. If A and B are independent events, then the probability that either occurs is given by

$$p(A + B) = p(A) + p(B) - p(A, B) = p(A) + p(B) - p(A)p(B) = 0.507$$

The probability that both occur is given by $p(A, B) = p(A)p(B) = 0.0630$. Finally, if A and B are mutually exclusive, then the probability that both occur is $p(A, B) = 0$ and the probability that either occurs is $p(A + B) = p(A) + p(B) = 0.570$.

6. The expectation operator is a linear operator (Equation 22.)

9.
$$\begin{aligned} E\left[(X - \mu_x + Y - \mu_y)^3\right] &= E\left[(X - \mu_x)^3\right] + 3E\left[(X - \mu_x)^2\right]E\left[Y - \mu_y\right] \\ &\quad + 3E\left[X - \mu_x\right]E\left[(Y - \mu_y)^2\right] + E\left[(Y - \mu_y)^3\right] \end{aligned}$$

But $E\left[X - \mu_x\right] = E\left[Y - \mu_y\right] = 0$, so $E\left[(X - \mu_x + Y - \mu_y)^3\right] = E\left[(X - \mu_x)^3\right] + E\left[(Y - \mu_y)^3\right]$.

12. The average value of n is given by

$$E\left[m\right] = \sum_{m=-n}^{n} \frac{1}{2^n} \frac{m\, n!}{\left(\dfrac{n+m}{2}\right)! \left(\dfrac{n-m}{2}\right)!} = \frac{n!}{2} \sum_{m=-n}^{n} \frac{m}{\left(\dfrac{n+m}{2}\right)! \left(\dfrac{n-m}{2}\right)!}$$

The figure on the following page shows $P(n, m)$ for $n = 10$ and 20. Notice from the figure or analytically that $P(n, m)$ is an even function of m, and so the summation over $mP(n, m)$ equals zero.

We can also show analytically that the sum is equal to zero. Start with

$$E\left[m\right] = \frac{n!}{2} \sum_{m=-n}^{n} \frac{m}{\left(\dfrac{n+m}{2}\right)! \left(\dfrac{n-m}{2}\right)!}, \text{ where } n \text{ and } m \text{ must be even. Therefore, let } n =$$

$2n'$ and $m = 2m'$, where n' and m' are positive and negative integers (including 0.) Now,

$$E\left[m\right] = \frac{(2n')!}{4^{n'}} \sum_{m'=-n'}^{+n'} \frac{2m'}{(n' + m')!(n' - m')!}$$

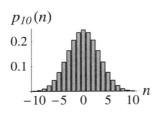

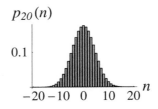

Let $n' + m' = l$, where $l = 0, 1, 2, \ldots, 2n'$ to write

$$
\begin{aligned}
E[m] &= \frac{2(2n')!}{4^{n'}} \sum_{l=0}^{2n'} \frac{l - n'}{l!(2n' - l)!} \\
&= \frac{2}{4^{n'}} \sum_{l=0}^{2n'} \frac{l(2n')!}{l!(2n' - l)!} - \frac{2n'}{4^{n'}} \sum_{l=0}^{2n'} \frac{(2n')!}{l!(2n' - l)!}
\end{aligned}
$$

To evaluate these two summations, we use $\displaystyle\sum_{k=0}^{N} \frac{N!x^k}{k!(N - k)!} = (1 + x)^N$, which, upon dif-

ferentiation, gives $\displaystyle\sum_{k=0}^{N} \frac{N!kx^{k-1}}{k!(N - k)!} = N(1 + x)^{N-1}$. Letting $x = 1$ gives the two required

summations above. Therefore,

$$
E[m] = \frac{2}{4^{n'}} \cdot 2n' 2^{2n'-1} - \frac{2n'}{4^{n'}} 2^{2n'} = 2n' - 2n' = 0
$$

15. (a) The probability that more than three customers will enter the store in any given minute is

$$
\sum_{n=4}^{\infty} \frac{a^n}{n!} e^{-a} = 1 - \sum_{n=0}^{3} \frac{a^n}{n!} e^{-a} = 1 - e^{-2}\left(1 + \frac{2}{1!} + \frac{2^2}{2!} + \frac{2^3}{3!}\right) = 0.1429
$$

(b) The probability that no customers enter is $p_0 = e^{-2} = 0.1352$

18.

$$
\begin{aligned}
p(D|+) &= \frac{(0.001)(0.98)}{(0.001)(0.98) + (0.999)(0.01)} = 0.0893 \\
p(D|-) &= \frac{(0.001)(0.02)}{(0.001)(0.02) + (0.999)(0.99)} = 1.97 \times 10^{-5} \\
p(\overline{D}|-) &= \frac{(0.999)(0.99)}{(0.999)(0.99) + (0.001)(0.02)} \approx 1.00 \\
p(\overline{D}|+) &= \frac{(0.999)(0.01)}{(0.999)(0.01) + (0.001)(0.98)} = 0.9107
\end{aligned}
$$

$p(+) = 0.01097$ and $p(-) = 0.98903$.

21.2   Continuous Random Variables

1. The expectation operator is linear (Equation 11), so $E[aX + bY] = aE[X] + bE[Y]$.

$$
\begin{aligned}
\text{Var}[X + Y] &= E[(X - \mu_x + Y - \mu_y)^2] \\
&= E[(X - \mu_x)^2] + 2E[(X - \mu_x)(Y - \mu_y)] + E[(Y - \mu_y)^2]
\end{aligned}
$$

If X and Y are independent random variables, then

$$
E[(X - \mu_x)(Y - \mu_y)] = E[(X - \mu_x)]E[Y - \mu_y)] = 0
$$

and so $\text{Var}[X + Y] = \text{Var}[X] + \text{Var}[Y]$.

3. We'll first prove a slightly more general version of the inequality in Problem 2. Start with

$$
\begin{aligned}
E[\{X - \mu_x + \lambda(Y - \mu_y)\}\{X - \mu_x + \lambda(Y - \mu_y)\}] &= E[(X - \mu_x)^2] + 2\lambda E[(X - \mu_x)(Y - \mu_y)] \\
&\quad + \lambda^2 E[(Y - \mu_y)^2] \geq 0
\end{aligned}
$$

Now let $\lambda = -E[(X - \mu_x)(Y - \mu_y)]/E[(Y - \mu_y)^2]$ to obtain

$$
E[(X - \mu_x)^2] - \frac{2E[(X - \mu_x)(Y - \mu_y)]^2}{E[(Y - \mu_y)^2]} + \frac{E[(X - \mu_x)(Y - \mu_y)]^2}{E[(Y - \mu_y)^2]} \geq 0
$$

or $E[(X - \mu_x)(Y - \mu_y)]^2 \leq E[(X - \mu_x)^2]E[(Y - \mu_y)^2]$.

We write this result as $\sigma_{xy}^2 \leq \sigma_x \sigma_y$, and so we see that $\rho_{xy}^2 = \dfrac{\sigma_{xy}^2}{\sigma_x \sigma_y} \leq 1$, or $-1 \leq \rho_{xy} \leq 1$.

6. Differentiate $\hat{P}(s) = \displaystyle\int_{-\infty}^{\infty} e^{isx} p(x)\, dx$ with respect to s n times and then let $s = 0$ to obtain

$$
\left[\frac{d^n \hat{P}(s)}{ds^n}\right]_{s=0} = \left[\int_{-\infty}^{\infty} (ix)^n e^{isx} p(x)\, dx\right]_{s=0} = i^n \int_{-\infty}^{\infty} x^n p(x)\, dx = i^n E[X^n]
$$

9. If S is the normalized modal matrix of A^{-1}, so that $\mathsf{S}^\mathsf{T} \mathsf{A}^{-1} \mathsf{S} = \mathsf{D}$ and $\mathbf{x}^\mathsf{T} \mathsf{A}^{-1} \mathbf{x} = \mathbf{x}'^\mathsf{T} \mathsf{D} \mathbf{x}'$, where $\mathbf{x} = \mathsf{S}\mathbf{x}'$, then $\mathbf{x}^\mathsf{T} = \mathbf{x}'^\mathsf{T} \mathsf{S}^\mathsf{T}$ and

$$
\mathbf{x}^\mathsf{T} \mathsf{A}^{-1} \mathbf{x} = \mathbf{x}'^\mathsf{T} \mathsf{S}^\mathsf{T} \mathsf{A}^{-1} \mathsf{S} \mathbf{x}' = \mathbf{x}'^\mathsf{T} \mathsf{D} \mathbf{x}'
$$

Now let $\mathbf{u} = \mathsf{S}^{-1} \mathbf{t} = \mathsf{S}^\mathsf{T} \mathbf{t}$ so that $\mathbf{t} = \mathsf{S}\mathbf{u}$ and $\mathbf{t}^\mathsf{T} \mathbf{x} = (\mathsf{S}\mathbf{u})^\mathsf{T} \mathbf{x} = \mathbf{u}^\mathsf{T} \mathsf{S}^\mathsf{T} \mathbf{x} = \mathbf{u}^\mathsf{T} \mathbf{x}'$. To verify Equation 1, note that $\mathbf{u}^\mathsf{T} \mathsf{D}^{-1} \mathbf{u} = \mathbf{t}^\mathsf{T} \mathsf{S} \mathsf{D}^{-1} \mathsf{S}^\mathsf{T} \mathbf{t}$.

To finish we need to prove that $\mathsf{S} \mathsf{D}^{-1} \mathsf{S}^\mathsf{T} = \mathsf{A}$. Start with $\mathsf{S}^\mathsf{T} \mathsf{A}^{-1} \mathsf{S} = \mathsf{S}^{-1} \mathsf{A}^{-1} \mathsf{S} = \mathsf{D}$, which gives $\mathsf{A}^{-1} = \mathsf{S} \mathsf{D} \mathsf{S}^{-1} = \mathsf{S} \mathsf{D} \mathsf{S}^\mathsf{T}$. Multiply from the left by A to get $\mathsf{I} = \mathsf{A} \mathsf{S} \mathsf{D} \mathsf{S}^\mathsf{T} = \mathsf{A} \mathsf{S} \mathsf{D} \mathsf{S}^{-1}$. Now multiply from the right by S to get $\mathsf{S} = \mathsf{A} \mathsf{S} \mathsf{D}$ and then from the right by D^{-1} to get $\mathsf{S} \mathsf{D}^{-1} = \mathsf{A} \mathsf{S}$ and finally from the right by $\mathsf{S}^{-1} = \mathsf{S}^\mathsf{T}$ to get $\mathsf{S} \mathsf{D}^{-1} \mathsf{S}^\mathsf{T} = \mathsf{A}$.

12. Start with

$$
\text{Prob}\{-u_{x0} \leq U_x \leq u_{x0}\} = \left(\frac{m}{2\pi k_\mathrm{B} T}\right)^{1/2} \int_{-u_{x0}}^{u_{x0}} e^{-mu_x^2/2k_\mathrm{B} T}\, du_x
$$

Let $w = (m/2k_{\mathrm{B}}T)^{1/2}u_x$ to write

$$
\begin{aligned}
\mathrm{Prob}\{-u_{x0} \le U_x \le u_{x0}\} &= \mathrm{Prob}\{-w_0 \le W \le w_0\} \\
&= \frac{1}{\pi}\int_{-w_0}^{w_0} e^{-w^2}\,dw = \frac{2}{\sqrt{\pi}}\int_0^{w_0} e^{-w^2}\,dw = \mathrm{erf}\,(w_0)
\end{aligned}
$$

Therefore,

$$
\mathrm{Prob}\{-(2k_{\mathrm{B}}T/m)^{1/2} \le U_x \le (2k_{\mathrm{B}}T/m)^{1/2}\} = \mathrm{erf}\,(1) = 0.84270
$$

15. We convert the cartesian coordinate expression $f(u_x)f(u_y)f(u_z)$ to a spherical coordinate expression by writing $u^2 = u_x^2 + u_y^2 + u_z^2$ and $du_x du_y du_z = 4\pi u^2 du$ to get

$$
F(u)\,du = 4\pi\left(\frac{m}{2\pi k_{\mathrm{B}}T}\right)^{3/2} u^2 e^{-mu^2/2k_{\mathrm{B}}T}\,du \qquad 0 \le u < \infty
$$

Using this result,

$$
\begin{aligned}
E[U] &= 4\pi\left(\frac{m}{2\pi k_{\mathrm{B}}T}\right)^{3/2}\int_0^\infty u^3 e^{-mu^2/2k_{\mathrm{B}}T}\,du = \left(\frac{8k_{\mathrm{B}}T}{\pi m}\right)^{1/2} \\
E[U^2] &= 4\pi\left(\frac{m}{2\pi k_{\mathrm{B}}T}\right)^{3/2}\int_0^\infty u^4 e^{-mu^2/2k_{\mathrm{B}}T}\,du = \frac{3k_{\mathrm{B}}T}{m} \\
E\left[\tfrac{1}{2}mU^2\right] &= \frac{m}{2}E[U^2] = \frac{3}{2}k_{\mathrm{B}}T
\end{aligned}
$$

18. Just let $c = \mu$ and $\epsilon = \sigma_x a$ in the result of the previous problem.

21.3   Characteristic Functions

1. The probability density function of X is $p(x)\,dx = \begin{cases} dx & 0 < x < 1 \\ 0 & \text{otherwise} \end{cases}$. Therefore, the probability density of Y is obtained by letting $x = (y - b)/a$ and $dx = dy/a$;

$$
q(y)\,dy = \frac{1}{a}p\left(\frac{y-b}{a}\right)dy = \begin{cases} \dfrac{dy}{a} & b < y < a+b \\ 0 & \text{otherwise} \end{cases}
$$

3. The probability density function of $Z = X + Y$ is given by the convolution integral

$$
r(z)\,dz = \frac{1}{2\pi}\int_{-\infty}^\infty e^{-x^2/2}e^{-(z-x)^2/2}dx = \frac{1}{2\pi}\int_{-\infty}^\infty e^{-(z^2-2xz+2x^2)/2}dx
$$

Let $z^2 - 2xz + 2x^2 = z^2 + (2x^2 - 2xz) = z^2 + \left(\sqrt{2}\,x - \dfrac{z}{\sqrt{2}}\right)^2 - \dfrac{z^2}{2} = \dfrac{z^2}{2} + \left(\sqrt{2}\,x + \dfrac{z}{\sqrt{2}}\right)^2$ to write

$$
r(z) = \frac{1}{2\pi}e^{-z^2/4}\int_{-\infty}^\infty e^{-(\sqrt{2}\,x - z/\sqrt{2})^2/2}\,dx
$$

Now let $\sqrt{2}\,x - \dfrac{z}{\sqrt{2}} = u$ ($\sqrt{2}\,dx = du$) to write

$$r(z) = \frac{1}{2^{3/2}\pi}e^{-z^2/4}\int_{-\infty}^{\infty} e^{-u^2/2}\,du = \frac{1}{(4\pi)^{1/2}}e^{-z^2/4}$$

6. Let $p(x) = f(x)/(2\pi)^{1/2}$ in Equation 1 to write

$$\phi(s) = \frac{1}{(2\pi)^{1/2}}\int_{-\infty}^{\infty} e^{isx} f(x)\,dx$$

The inverse transform is

$$f(x) = \frac{1}{(2\pi)^{1/2}}\int_{-\infty}^{\infty} e^{-isx}\phi(s)\,ds$$

which is Equation 4 if we replace $f(x)$ by $(2\pi)^{1/2}p(x)$.

9. Equation 11 is $\phi_X(s) = e^{-a}e^{ae^{is}}$ and so

$$\phi_X(0) = 1 \qquad \phi_X'(0) = ia \qquad \phi_X''(0) = i^2(a^2 + a) \qquad \phi_X'''(0) = i^3(a^3 + 3a^2 + a) \qquad \cdots$$

Verification:

$$E[X] = \sum_{x=o}^{\infty}\frac{xa^x}{x!}e^{-a} = e^{-a}a\sum_{x=o}^{\infty}\frac{xa^{x-1}}{x!} = e^{-a}a\frac{de^a}{da} = a$$

$$E[X^2] = e^{-a}\sum_{x=o}^{\infty}\frac{x^2 a^x}{x!} = e^{-a}a\sum_{x=o}^{\infty}\frac{[x(x-1)-x]a^x}{x!} = e^{-a}a^2\frac{d^2 e^a}{da^2} + e^{-a}a\frac{de^a}{da}$$

$$= a^2 + a$$

$$E[X^3] = e^{-a}\sum_{x=o}^{\infty}\frac{x^3 a^x}{x!} = e^{-a}a\sum_{.x=o}^{\infty}\frac{[x(x-1)(x-2)+3x^2-2x]a^x}{x!}$$

$$= a^3 + 3(a^2 + a) - 2a = a^3 + 3a^2 + a$$

12. We'll first do Problem 11. Starting with the binomial distribution, $p(m) = \dfrac{n!}{m!(n-m)!}$ $p^m q^{n-m}$ for $m = 0,\ 1,\ 2,\ \ldots, n$. We write the characteristic function

$$\phi_M(s) = E[e^{isM}] = \sum_{m=0}^{n}\frac{n!}{m!(n-m)!}(pe^{is})^m q^{n-m} = (pe^{is} + q)^n$$

with $p + q = 1$.

   Example 4 of the previous section shows that $E[M] = np$ and $E[M^2] = npq + (np)^2$. Using $\phi_M(s)$, we get

$$\phi_M(0) = 1 \qquad \phi_M'(s) = n(pe^{is} + q)^{n-1}ipe^{is} \qquad \text{or} \qquad \phi_M'(0) = inp$$

$$\phi_M''(s) = n(n-1)(pe^{is} + q)^{n-2}(ip)^2 e^{is} + n(pe^{is} + q)^{n-1}i^2 pe^{is}$$

or

$$\phi_M''(0) = [n(n-1)p^2 + np]i^2 = i^2[n^2p^2 + npq] = -n^2p^2 - npq$$

15. We'll do Problem 14 first. If X_j is uniformly distributed in the interval $(0, T)$, then

$$E[X_j] = \frac{1}{T}\int_0^T x\, dx = \frac{T}{2} \quad \text{and} \quad \text{Var}[X_j] = \frac{1}{T}\int_0^T \left(x - \frac{T}{2}\right)^2 dx = \frac{T^2}{12}$$

For $X = X_1 + X_2$, $E[X] = T$ and $\text{Var}[X] = T/6$, so the Gaussian approximation for p_X is

$$p_{G2}(x) \approx \frac{1}{T}\left(\frac{3}{\pi}\right)^{1/2} e^{-3(x-T)^2/T^2}$$

The exact density is given by

$$p_2(x) = \int_0^T p_1(u)p_1(x-u)\, du = \begin{cases} \dfrac{1}{T^2}\displaystyle\int_0^T du = \dfrac{x}{T^2} & 0 < x < T \\[2ex] \dfrac{1}{T^2}\displaystyle\int_{x-T}^T du = \dfrac{2T-x}{T^2} & T < x < 2T \\[2ex] 0 & \text{otherwise} \end{cases}$$

Using $E[X] = E[X_1 + X_2 + X_3] = 3T/2$ and $\text{Var}[X] = T^2/4$, the Gaussian approximation is

$$p_{G3}(x) \approx \frac{1}{T}\left(\frac{2}{\pi}\right)^{1/2} e^{-2(x-3T/2)^2/T^2}$$

The exact density function is given by the convolution integral $p_3(x) = \displaystyle\int_0^{2T} p_2(z)p_1(x-z)\, dz$ where

$$p_2(x) = \begin{cases} \dfrac{x}{T^2} & 0 < x < T \\[2ex] \dfrac{2T-x}{T^2} & T < x < 2T \\[2ex] 0 & \text{otherwise} \end{cases} \quad \text{and} \quad p_1(x) = \begin{cases} \dfrac{1}{T} & 0 < x < T \\[2ex] 0 & \text{otherwise} \end{cases}$$

For $0 < x < T$, we have

$$\frac{1}{T}\int_0^x p_2(z)\, dz = \frac{x^2}{2T^3}$$

For $T < x < 2T$, we have

$$\begin{aligned} \frac{1}{T}\int_{x-T}^x p_2(z)\, dz &= \frac{1}{T^3}\int_{x-T}^T z\, dz + \frac{1}{T^3}\int_T^x (2T - z)\, dz \\[2ex] &= \frac{1}{2T} - \frac{(x-T)^2}{2T^3} + \frac{2(x-T)}{T^2} - \frac{x^2}{2T^3} + \frac{1}{2T} = \frac{3x}{T^2} - \frac{x^2}{T^3} - \frac{3}{2T} \end{aligned}$$

For $2T < x < 3T$,

$$\frac{1}{T}\int_{x-T}^{2T} p_2(z)\,dz = \frac{1}{T^3}\int_{x-T}^{2T}(2T - z)\,dz$$

$$= \frac{1}{T^3}\left[2T(3T - x) - 2T^2 + \frac{(x - T)^2}{2}\right] = \frac{4}{T} - \frac{2x}{T^2} + \frac{(x - T)^2}{2T^3}$$

Thus,

$$p_3(x) = \begin{cases} \dfrac{x^2}{2T^3} & 0 < x < T \\[2mm] \dfrac{3x}{T^2} - \dfrac{x^2}{T^3} - \dfrac{3}{2T} & T < x < 2T \\[2mm] \dfrac{4}{T} - \dfrac{2x}{T^2} + \dfrac{(x - T)^2}{2T^3} & 2T < x < 3T \\[2mm] 0 & \text{otherwise} \end{cases}$$

The following figures show $p_2(x)$ and its Gaussian approximation, $p_{G2}(x)$ and $p_3(x)$ and its Gaussian approximation, $p_{G3}(x)$, for $T = 2$.

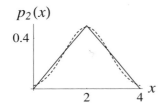

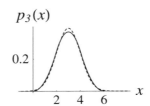

21.4   Stochastic Processes-General

1. Some simulations of a one-dimensional random walk are shown below.

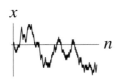

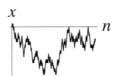

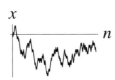

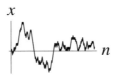

3. Some realizations of the stochastic process $a \cos \omega t$, where a is a uniformly distributed random variable:

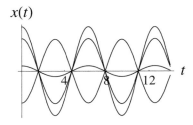

$x(t)$

6. Use the fact that $E[X(t)] = \lambda t$ for a Poisson process.

$$E[Y(t)] = \frac{1}{\epsilon} E[X(t+\epsilon)] - \frac{1}{\epsilon} E[X(t)] = \frac{\lambda(t+\epsilon) - \lambda t}{\epsilon} = \lambda$$

9. The ensemble average is

$$\langle X \rangle_{\text{ensemble}} = \frac{\langle a \rangle}{2\pi} \int_0^{2\pi} \cos(t+\phi)\, d\phi = 0$$

and the time average is

$$\langle X(t) \rangle_{\text{time}} = \lim_{T \to \infty} \frac{1}{2T} \int_{-T}^{T} a \cos(t+\phi)\, dt = 0$$

because the integral is finite.

12. $R(\tau) = \langle x(t)x(t+\tau) \rangle = \langle x(t-\tau)x(t) \rangle = \langle x(t)x(t-\tau) \rangle = R(-\tau)$

15. Take the Fourier transform of $m\ddot{y}(t) + \gamma \dot{y}(t) + y(t) = F(t)$ to get

$$\hat{Y}(\omega) = \frac{F(\omega)}{-\omega^2 m + i\gamma\omega + k} = \frac{F(\omega)/k}{1 - \left(\dfrac{\omega}{\omega_0}\right)^2 + i\xi\left(\dfrac{\omega}{\omega_0}\right)}$$

where $\omega_0 = (k/m)^{1/2}$ and $\xi = \gamma/(mk)^{1/2}$. Therefore, the transfer function,

$$\hat{H}(\omega) = \frac{1/k}{1 - \left(\dfrac{\omega}{\omega_0}\right)^2 + i\xi\left(\dfrac{\omega}{\omega_0}\right)}$$

and

$$|\hat{H}(\omega)| = \frac{1/k}{\left\{\left[1 - \left(\dfrac{\omega}{\omega_0}\right)^2\right]^2 + \xi^2\left(\dfrac{\omega}{\omega_0}\right)^2\right\}^{1/2}}$$

is plotted in the figure on the next page against ω/ω_0 for $k = 1$ and $\xi = 0.25$.

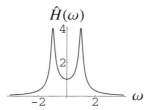

18 Start with $y(t) = \int_{-\infty}^{\infty} f(u)h(t-u)\,du$ or $Y(t) = \int_{-\infty}^{\infty} F(u)h(t-u)\,du$. Now take the expectation value of both sides to write

$$E\,[\,Y(t)\,] = \int_{-\infty}^{\infty} h(t-u)\,E\,[\,F(u)\,]\,du$$

If $F(t)$ is stationary, then $E\,[\,F(u)\,] = \mu_F$ and

$$E\,[\,Y(t)\,] = \mu_F \int_{-\infty}^{\infty} h(z)\,dz = \mu_F \hat{H}(0)$$

21. We'll do Problem 20 first because we use the result of that problem. We wish to prove that $S_Y(\omega) = |\,\hat{H}(\omega)\,|^2 S_F(\omega)$. Take the Fourier transform of $R_Y(\tau)$ in Problem 19 to obtain

$$S_Y(\omega) = \int_{-\infty}^{\infty} du\, h(u) \int_{-\infty}^{\infty} dv\, h(v) \int_{-\infty}^{\infty} dt\, R_F(t+u-v)e^{i\omega t}$$

Now let $t+u-v = z$ to get

$$
\begin{aligned}
S_Y(\omega) &= \int_{-\infty}^{\infty} du\, h(u) \int_{-\infty}^{\infty} dv\, h(v) \int_{-\infty}^{\infty} dz\, R_F(z)e^{-i\omega(z-u+v)} \\
&= \int_{-\infty}^{\infty} du\, h(u)e^{i\omega u} \int_{-\infty}^{\infty} dv\, h(v)e^{-i\omega v} \int_{-\infty}^{\infty} dz\, R_F(z)e^{-i\omega z} \\
&= |\,\hat{H}(\omega)\,|^2 S_F(\omega)
\end{aligned}
$$

The system function given in Problem 21 is $\hat{H}(\omega) = \dfrac{1}{\alpha + i\omega}$ and the power spectrum corresponding to $R_F(\tau)$ is $S_F(\omega) = \dfrac{\beta}{\omega^2 + \beta^2}$. Using the result of Problem 20, we write

$$S_Y(\omega) = |\,\hat{H}(\omega)\,|^2 S_F(\omega) = \frac{\beta}{(\omega^2 + \alpha^2)(\omega^2 + \beta^2)}$$

Using partial fractions, we have that

$$S_Y(\omega) = \frac{\beta}{\beta^2 - \alpha^2}\left(\frac{1}{\omega^2 + \alpha^2} - \frac{1}{\omega^2 + \beta^2}\right)$$

Therefore,

$$
\begin{aligned}
R_Y(\tau) &= \frac{\beta}{\pi(\beta^2 - \alpha^2)}\left[\int_{-\infty}^{\infty} \frac{\cos\omega\tau}{\omega^2 + \alpha^2}\,d\tau - \int_{-\infty}^{\infty} \frac{\cos\omega\tau}{\omega^2 + \beta^2}\,d\tau\right] \\
&= \frac{\beta}{2(\beta^2 - \alpha^2)}\left(\frac{e^{-\alpha\tau}}{\alpha} - \frac{e^{-\beta\tau}}{\beta}\right)
\end{aligned}
$$

21.5   Stochastic Processes-Examples

1. Use $\lambda = 28$ min^{-1} and $t = 5$ min in Equation 3a and use any CAS to show

$$1 - e^{-140} \sum_{n=0}^{160} \frac{(140)^n}{n!} = 0.0439627$$

3. The derivative of Equation 8a yields

$$\frac{d}{dt}\left[1 - \sum_{k=0}^{m-1} \frac{(\lambda t)^k e^{-\lambda t}}{k!} \right] = -\sum_{k=1}^{m-1} \frac{\lambda^k t^{k-1}}{(k-1)!} e^{-\lambda t} + \sum_{k=0}^{m-1} \frac{\lambda^{k+1} t^k}{k!} e^{-\lambda t}$$

$$= \sum_{k=0}^{m-1} \frac{\lambda^{k+1} t^k}{k!} e^{-\lambda t} - \sum_{k=0}^{m-2} \frac{\lambda^{k+1} t^k}{k!} e^{-\lambda t} = \frac{\lambda^m t^{m-1}}{(m-1)!} e^{-\lambda t}$$

6. Start with Equation 11b and expand the factor $\displaystyle\int_{-\infty}^{\infty} (e^{isf(u)} - 1)\, du$ in powers of s to obtain

$$\lambda \int_{-\infty}^{\infty} (e^{isf(u)} - 1)\, du = i\lambda s \int_{-\infty}^{\infty} f(u)\, du - \frac{\lambda s^2}{2} \int_{-\infty}^{\infty} f^2(u)\, du + O(s^3)$$

$$= is\langle I \rangle - \frac{1}{2} s^2 \sigma_I^2 + O(s^3)$$

Equation 11b is now

$$p(I) = \frac{1}{2\pi} \int_{-\infty}^{\infty} \exp\left\{ -isI + is\langle I \rangle - \frac{s^2}{2}\sigma_I^2 + O(s^3) \right\} ds$$

$$\approx \frac{1}{2\pi} \int_{-\infty}^{\infty} \{\cos[\,s(I - \langle I \rangle)] + i\sin[\,s(I - \langle I \rangle)]\} e^{-s^2 \sigma_I^2/2}\, ds$$

$$= \frac{1}{\pi} \int_0^{\infty} \cos[\,s(I - \langle I \rangle)]\, e^{-s^2 \sigma_I^2/2}\, ds$$

$$= \frac{1}{\pi} \cdot \frac{(2\pi)^{1/2}}{2\sigma_I} e^{-(I - \langle I \rangle)^2/2\sigma_I^2} = \frac{1}{(2\pi\sigma_I^2)^{1/2}} e^{-(I - \langle I \rangle)^2/2\sigma_I^2}$$

9. Use Equation 14b:

$$R_0(\tau) = \frac{\lambda q^2}{2\pi \tau_c^2} \int_{-\infty}^{\infty} e^{-u^2/2\tau_c^2} e^{-(u+\tau)^2/2\tau_c^2}\, du$$

Complete the square in the exponential to write

$$R_0(\tau) = \frac{\lambda q^2}{2\pi \tau_c^2} e^{-\tau^2/4\tau_c^2} \int_{-\infty}^{\infty} e^{-(\sqrt{2}\,u + \tau/\sqrt{2})^2/2\tau_c^2}\, du$$

$$= \frac{\lambda q^2}{(4\pi\tau_c^2)^{1/2}} e^{-\tau^2/4\tau_c^2}$$

12. We'll use Equation 15b. First we calculate

$$
\begin{aligned}
\int_{-\infty}^{\infty} dz \, f(z) e^{iz\omega} &= \frac{2q}{(2\pi\tau_c^2)^{1/2}} \int_0^{\infty} \cos z\omega \, e^{-z^2/2\tau_c^2} \, dz \\
&= \left[\frac{2q}{(2\pi\tau_c^2)^{1/2}}\right] \left(\frac{\pi}{2}\right)^{1/2} \tau_c e^{-\omega^2\tau_c^2/2}
\end{aligned}
$$

Substitute this into Equation 15b to obtain $S_0(\omega) = \lambda q^2 e^{-\omega^2\tau_c^2}$.

15. We use Equation 20 of the previous section to write

$$
\begin{aligned}
R_0(\tau) &= \frac{1}{\pi} \int_0^{\infty} \cos\omega\tau \, S_0(\omega) \, d\omega \\
&= \frac{4\lambda}{\pi} \int_0^{\infty} \cos\omega\tau \left[\frac{\sin(\omega\tau_1/2)}{\omega}\right]^2 d\omega \\
&= \frac{\lambda}{2}\left[-2\tau \operatorname{Sgn}(\tau) + (\tau - \tau_1)\operatorname{Sgn}(\tau - \tau_1) + (\tau + \tau_1)\operatorname{Sgn}(\tau + \tau_1)\right] \\
&= \lambda \begin{cases} \tau_1 - |\tau| & |\tau| < \tau_1 \\ 0 & \text{otherwise} \end{cases}
\end{aligned}
$$

This result is shown in the following figure for $\tau_1 = 4$.

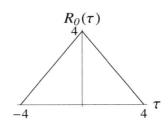

CHAPTER 22

Mathematical Statistics

22.1 Estimation of Parameters

1. First

$$\overline{x} = \frac{1 + 5 + 8 + 10}{4} = 6$$

and

$$\sigma_{\overline{x}}^2 = \frac{(1-6)^2 + (5-6)^2 + (8-6)^2 + (10-6)^2}{4} = \frac{23}{2}$$

Now, there are 16 possible samples:

1,1	5,1	8,1	10,1
1,5	5,5	8,5	10,5
1,8	5,8	8,8	10,8
1,10	5,10	8,10	10,10

with means

1	3	9/2	11/2
3	5	13/2	15/2
9/2	13/2	8	9
11/2	15/2	9	10

The mean of the means is equal to $96/16 = 6$. The variance of the sample means is given by subtracting 6 from each sample mean, squaring, and then dividing by 16 to give $92/16 = 5.75$, which is equal to $\sigma_{\overline{x}}^2/2 = 23/4$.

3. Start with Equation 15, $L(\lambda) = \prod_{j=1}^{n} \lambda^n e^{-\lambda x_j} = \lambda^n e^{-\lambda n \overline{x}}$. Then

$$\ln L(\lambda) = n \ln \lambda - \lambda n \overline{x} \qquad \text{and} \qquad \frac{\partial \ln L}{\partial \lambda} = \frac{n}{\hat{\lambda}} - n\overline{x} = 0$$

which gives $\hat{\lambda} = \dfrac{1}{\overline{x}}$. Note that $\dfrac{\partial^2 \ln L}{\partial \lambda^2} = -\dfrac{n}{\lambda^2} < 0$.

6. To prove that $\hat{\theta} = c_1 \hat{\theta}_1 + c_2 \hat{\theta}_2$ with $c_1 + c_2 = 1$ is an unbiased estimator of θ if $\hat{\theta}_1$ and $\hat{\theta}_2$ are unbiased estimators, note that

$$E[c_1 \hat{\theta}_1 + c_2 \hat{\theta}_2] = c_1 E[\hat{\theta}_1] + c_2 E[\hat{\theta}_2] = c_1 \theta + c_2 \theta = \theta$$

9. Start with $L(p) = \left(\dfrac{100!}{57! \, 43!} \right) p^{57} (1-p)^{43}$, or

$$\ln L = \ln \left(\frac{100!}{57! \, 43!} \right) + 57 \ln p + 43 \ln(1-p)$$

Now

$$\frac{\partial L}{\partial p} = \frac{57}{p} - \frac{43}{1-p}$$

and

$$\frac{\partial^2 L}{\partial p^2} = -\frac{57}{p^2} - \frac{43}{(1-p)^2} < 0$$

for $0 < p < 1$.

12. Start with

$$L(\mu, \sigma^2) = \prod_{j=1}^{n} (2\pi\sigma^2)^{-1/2} \exp \left\{ -(x_j - \mu)^2 / 2\sigma^2 \right\} = (2\pi\sigma^2)^{-n/2} \exp \left\{ -\sum_{j=1}^{n} (x_j - \mu)^2 / 2\sigma^2 \right\}$$

so that

$$\ln L = -\frac{n}{2} \ln 2\pi - \frac{1}{2\sigma^2} \sum_{j=1}^{n} (x_j - \mu)^2 - \frac{n}{2} \ln \sigma^2$$

Now,

$$\frac{\partial \ln L}{\partial \sigma^2} = \frac{1}{2\hat{\sigma}^4} \sum_{j=1}^{n} (x_j - \mu)^2 - \frac{n}{2\hat{\sigma}^2} = 0$$

which gives $\hat{\sigma}^2 = \dfrac{1}{n} \sum_{j=1}^{n} (x_j - \mu)^2$. The second derivative is

$$\frac{\partial^2 \ln L}{\partial \sigma^2 \partial \sigma^2} = -\frac{1}{(\hat{\sigma}^2)^3} \sum_{j=1}^{n} (x_j - \mu)^2 + \frac{n}{2(\hat{\sigma}^2)^2} = -\frac{n}{(\hat{\sigma}^2)^2} + \frac{n}{2(\hat{\sigma}^2)^2} = -\frac{n}{2(\hat{\sigma}^2)^2} < 0$$

15. We show on page 1081 that $\dfrac{\partial \ln L}{\partial \mu} = 0$ at $\hat{\mu} = \overline{x}$ and that $\dfrac{\partial \ln L}{\partial \sigma^2} = 0$ at $\hat{\sigma}^2 = \dfrac{1}{n} \sum_{j=1}^{n} (x_j - \overline{x})^2$.

Now

$$\frac{\partial^2 \ln L}{\partial \mu^2} = \frac{\partial}{\partial \mu} \left[\frac{1}{\sigma^2} \sum_{j=1}^{n} (x_j - \mu) \right] = -\frac{n}{\hat{\sigma}^2}$$

$$\frac{\partial^2 \ln L}{\partial \sigma^2 \partial \sigma^2} = \frac{\partial}{\partial \sigma^2} \left[-\frac{n}{2\sigma^2} + \frac{1}{2(\sigma^2)^2} \sum_{j=1}^{n} (x_j - \mu)^2 \right] = \frac{n}{2(\hat{\sigma}^2)^2} - \frac{n(\hat{\sigma}^2)}{(\hat{\sigma}^2)^3} = -\frac{n}{2(\hat{\sigma}^2)^2}$$

$$\frac{\partial^2 \ln L}{\partial \mu \partial \sigma^2} = \frac{\partial}{\partial \sigma^2} \left[\frac{1}{\sigma^2} \sum_{j=1}^{n} (x_j - \mu) \right] = 0$$

at the points $\hat{\mu}$ and $\hat{\sigma}^2$. Thus (in the notation of Section 6.8)

$$f_{\mu\mu}(\hat{\mu}, \hat{\sigma}^2) < 0 \qquad \text{and} \qquad D(\hat{\mu}, \hat{\sigma}^2) = f_{\mu\mu}(\hat{\mu}, \hat{\sigma}) f_{\sigma^2\sigma^2}(\hat{\mu}, \hat{\sigma}) > 0$$

and so $L(\mu, \sigma^2)$ is a maximum at its maximum liklihood estimators.

22.2   Three Key Distributions Used in Statistical Tests

1. Using the *CRC Standard Mathematical Tables*, for example, we have $1 - F(x) = 0.1587$ at $x = 1$. To find the area between $\mu \pm \sigma$ $(-1 \leq x \leq 1)$, we double 0.1587 and then subtract from one to get 0.6826. Similarly, $1 - F(x) = 0.0228$ at $x = 2$, so we obtain 0.9544. At $x = 3$, $1 - F(x) = 0.0013$, to yield 0.9974.

3. The characteristic function of $p(x_j) = (2\pi\sigma_j^2)^{-1/2} e^{-(x_j - \mu_j)^2 / 2\sigma_j^2}$ is $\phi_j(s) = e^{is\mu_j} e^{-\sigma_j^2 s^2 / 2}$. The characteristic function of $X = X_1 + X_2 + \cdots + X_n$ is

$$\prod_{j=1}^{n} \phi_j(s) = e^{-is(\mu_1 + \cdots + \mu_n)} e^{-(\sigma_1^2 + \cdots + \sigma_n^2)s^2 / 2}$$

This result is the characteristic function of a normal distribution with $\mu = \mu_1 + \mu_2 + \cdots + \mu_n$ and $\sigma^2 = \sigma_1^2 + \sigma_2^2 + \cdots + \sigma_n^2$.

6. Start with Equation 10:

$$F(x) = (2\pi)^{-1/2} \int_{-\infty}^{x} e^{-u^2/2} \, du$$

Then

$$\begin{aligned}
F(-x) &= (2\pi)^{-1/2} \int_{-\infty}^{-x} e^{-u^2/2} \, du = (2\pi)^{-1/2} \int_{x}^{\infty} e^{-u^2/2} \, du \\
&= (2\pi)^{-1/2} \left[\int_{-\infty}^{\infty} e^{-u^2/2} \, du - \int_{-\infty}^{x} e^{-u^2/2} \, du \right] = 1 - F(x)
\end{aligned}$$

9. We'll use the *CRC Standard Mathematical Tables*, although we could use any tables (or CAS, for that matter).
(a) The area between μ and $\mu + 2\sigma$ is given by $F(2) - F(0)$, where

$$F(x) = (2\pi)^{-1/2} \int_{-\infty}^{x} e^{-u^2/2} \, du$$

Thus, $F(2) - F(0) = 0.4772$.

(b) The area between $\mu - \sigma$ and $\mu + \sigma$ is $F(2) - F(-1)$, But $F(-1) = 1 - F(1)$ (see Problem 6), so the area is given by $F(1) + F(2) - 1 = 0.8185$.

(c) The area between $\mu - 2\sigma$ and $\mu + \sigma$ is the same as the area between $\mu - \sigma$ and $\mu + 2\sigma$ (see part b).

(d) The area between $\mu - \sigma$ and μ is given by $F(0) - F(-1) = F(0) + F(1) - 1 = 0.3413$.

12. We use Leibnitz's rule to differentiate

$$P(u) = \left(\frac{2}{\pi}\right)^{1/2} \int_0^{u^{1/2}} e^{-x^2/2}\, dx$$

in which case, we get

$$p(u) = \frac{dP}{du} = \left(\frac{2}{\pi}\right)^{1/2} e^{-u/2} \cdot \frac{1}{2u^{1/2}} = \frac{e^{-u/2}}{(2\pi u)^{1/2}} \qquad u > 0$$

To show that $p(u)$ is normalized

$$\int_0^\infty p(u)\, du \;=\; \frac{1}{(2\pi)^{1/2}} \int_0^\infty \frac{e^{-u/2}}{u^{1/2}}\, du = \frac{1}{\pi^{1/2}} \int_0^\infty \frac{e^{-x}}{x^{1/2}}\, dx$$

$$=\; \frac{1}{\pi^{1/2}} \Gamma(1/2) = 1$$

We could also show that $P(u) \to 1$ as $u \to \infty$:

$$\lim_{u\to\infty} P(u) = \lim_{u\to\infty} \left(\frac{2}{\pi}\right)^{1/2} \int_0^u e^{-x^2/2}\, dx = 1$$

15. Let $t = u/2$ in Equation 14 to write

$$P(x) \;=\; \frac{1}{2^{n/2}\Gamma(n/2)} \int_0^{x/2} 2^{(n-2)/2} t^{(n-2)/2} e^{-t} 2dt = \frac{1}{\Gamma(n/2)} \int_0^{x/2} t^{n/2-1} e^{-t}\, dt$$

$$=\; \frac{\gamma(n/2, x/2)}{\Gamma(n/2)}$$

As $x \to \infty$, $\gamma\left(\dfrac{n}{2}, \dfrac{x}{2}\right) \to \Gamma\left(\dfrac{n}{2}\right)$, and so $P(x) \to 1$ as $x \to \infty$.

18. We use the *CRC Standard Mathematical Tables*, for example.

(a) We choose $P(x) = 0.950$ to find that $x = 15.5$.

(b) We choose $P(x) = 0.050$ to find that $x = 2.73$.

(c) The answer is not unique.

21. From Example 4,

$$\sigma^2 \;=\; \frac{\Gamma\left[(n+1)/2\right]}{(n\pi)^{1/2}\Gamma(n/2)} \int_{-\infty}^\infty \frac{t^2\, dt}{(1 + t^2/n)^{(n+1)/2}}$$

$$=\; \frac{2n^{(n+1)/2}\Gamma\left[(n+1)/2\right]}{(n\pi)^{1/2}\Gamma(n/2)} \int_0^\infty \frac{t^2\, dt}{(n + t^2)^{(n+1)/2}}$$

To evaluate this integral, use the integral from the *CRC Standard Mathematical Tables*

$$\int_0^\infty \frac{x^a\, dx}{(m + x^b)^c} = \frac{m^{(a+1-bc)/b}}{b} \left[\frac{\Gamma\left(\dfrac{a+1}{b}\right)\Gamma\left(c - \dfrac{a+1}{b}\right)}{\Gamma(c)} \right]$$

with $a = 2$, $b = 2$, $m = n$, and $c = (n+1)/2$.

22.3   Confidence Intervals

1. Start with $-2.33 \le \sqrt{n}\,(\overline{x} - \mu)/\sigma \le 2.33$ with $n = 9$, $\overline{x} = 16$, and $\sigma^2 = 6$ to write $-2.33 \le \sqrt{9}\,(16 - \mu)/\sqrt{6} \le 2.33$, or

$$-5.707 \le \sqrt{9}\,(16 - \mu) \le 5.707$$
$$-1.902 \le 16 - \mu \le 1.902$$
$$14.10 \le \mu \le 17.90$$

3. Using the *CRC Standard Mathematical Tables*, for example, we find that $F(x) = 0.95$ when $x = 1.645$, or that $F(1.645) = 0.95 = \dfrac{1 + 0.90}{2}$, which is the first entry. Continuing for $\eta = 0.95$,

$$F(1.960) = \frac{1 + 0.95}{2} = 0.975, \text{ which gives the second entry}$$
$$F(2.326) = \frac{1 + 0.98}{2} = 0.999$$
$$F(2.576) = \frac{1 + 0.99}{2} = 0.995$$
$$F(2.810) = \frac{1 + 0.995}{2} = 0.9975$$

6. Consider the case, $\eta = 0.95$ first. Table 22.2 gives $a = 1.96$ and $\dfrac{L}{\sigma} = \dfrac{2(1.96)}{\sqrt{n}} = \dfrac{3.92}{\sqrt{n}}$. For $\eta = 0.99$, we have $a = 2.576$ and $\dfrac{L}{\sigma} = \dfrac{5.152}{\sqrt{n}}$.

9. Using the *CRC Standard Mathematical Tables*, for example, we find that $F(t) = 0.95$ when $t = 2.015$ for $n = 5$ or $F(2.015) = \dfrac{1 + 0.90}{2} = 0.95$ for $n = 5$.

Similarly, $F(2.571) = \dfrac{1 + 0.95}{2} = 0.975$ for $n = 5$.

Take $n = 20$. $F(1.725) = 0.95 = \dfrac{1 + 0.90}{2}$, and so on.

12. We want $2c = \dfrac{2a(n)(2.00)}{\sqrt{n}} = 1.50$ or $\dfrac{a(n)}{\sqrt{n}} = 0.375$. Using Table 22.4, we find that n should be a little larger than 25.

15. Using $\eta = 0.99$, we find that Table 22.6 gives us $a_1 = 7.43$ and $a_2 = 40.0$. Therefore,

$$c_1 = \frac{(20)(11.96)}{40.0} = 5.98 \quad \text{and} \quad c_2 = \frac{(20)(11.96)}{7.43} = 32.2$$

18. Square both sides of Equation 9 to get

$$\hat{p}^2 - 2\hat{p}p + p^2 = \frac{a^2 p(1 - p)}{n} = \frac{a^2 p}{n} - \frac{a^2 p^2}{n}$$

or

$$p^2 \left(1 + \frac{a^2}{n}\right) - p \left(2\hat{p} + \frac{a^2}{n}\right) + \hat{p}^2 = 0$$

The quadratic formula gives

$$p = \frac{\hat{p} + \dfrac{a^2}{2n} \pm \left[\dfrac{a^2}{n}\hat{p}(1 - \hat{p}) + \dfrac{a^4}{4n^2}\right]^{1/2}}{1 + \dfrac{a^2}{n}}$$

21. Using $n = \left(\dfrac{a}{\hat{p} - p}\right)^2 \hat{p}(1 - \hat{p})$, we have

$$n = \left(\frac{1.96}{0.010}\right)^2 \hat{p}(1 - \hat{p}) = \begin{cases} 9604 & \text{if} \quad \hat{p} = 1/2 \\ 9466 & \text{if} \quad \hat{p} = 0.44 \end{cases}$$

The result is fairly insensitive to the value of $\hat{p}$ that we choose for $\hat{p}$ close to $1/2$.

22.4   Goodness of Fit

1. The expected number of heads and tails is 25 if the coin is fair, and so

$$\chi^2 = \frac{(28 - 25)^2}{25} + \frac{(22 - 25)^2}{25} = 0.72$$

The value of χ^2 is less than c_α determined with 1 degree of freedom at $\alpha = 0.05$ and $\alpha = 0.01$. We do not reject the hypothesis that the coin is fair.

3. The expected value of each number is $50/6$ if the die is fair. The value of χ^2 is given by

$$\begin{aligned} \chi^2 &= \frac{6}{50}\left[\left(9 - \frac{50}{6}\right)^2 + \left(10 - \frac{50}{6}\right)^2 + \left(8 - \frac{50}{6}\right)^2 + \left(5 - \frac{50}{6}\right)^2 + \left(8 - \frac{50}{6}\right)^2 + \left(10 - \frac{50}{6}\right)^2\right] \\ &= 2.08 \end{aligned}$$

which is less than c_α determined with 5 degrees of freedom at $\alpha = 0.050$ ($c_\alpha = 11.1$) and $\alpha = 0.010$ ($c_\alpha = 15.1$.) We do not reject the hypothesis that the die is fair.

6. The expected frequencies based upon a Poisson distribution with $\lambda = 1.84$ are given by $50(1.84)^n e^{-1.84}/n!$, or 7.941, 14.61, 13.44, 8.245, 3.793, and 1.990. The value of χ^2 is 4.24, which is less than $c_\alpha = 9.49$ for $\alpha = 0.050$ with 4 degrees of freedom. Thus, we do not reject the hypothesis that the population distribution is Poisson with $\lambda = 1.84$.

9. The expected frequencies based upon a binomial distribution with $p = 1/2$ are given by $60 \cdot \dfrac{4!}{n!(4 - n)!} \left(\dfrac{1}{2}\right)^4 = \dfrac{90}{n!(4 - n)!}$. Thus, we have

x	0	1	2	3	4
Expected frequencies	30/8	15	45/2	15	30/8

The value of χ^2 is 4.53. The value of $c_{0.050}$ for 4 degrees of freedom is 9.49, so we do not reject the hypothesis that the population distribution is binomial with $p = 1/2$.

12. The expected frequencies based upon a normal distribution with $\mu = 62$ and $\sigma = 15$ are given by $(150)(450\pi)^{-1/2} \int_{x_1}^{x_2} e^{-(x-62)^2/450}\, dx$, where x_1 and x_2 are the end points of the given intervals. Thus, we have

Interval	< 50	50–60	60–70	70–90	80–90	90–100
Expected frequency	31.778	35.266	38.420	27.275	12.614	3.799

The corresponding value of χ^2 is 2.788. Using 5 degrees of freedom, we see that $\chi^2 < c_\alpha = 11.1$, and so we do not reject the hypothesis that the population is normally distributed with $\mu = 62$ and $\sigma = 15$.

15. The expected frequencies are given by $(300)(2\pi)^{-1/2} \int_{x_1}^{x_2} e^{-x^2/2}\, dx$, where x_1 and x_2 are the end points of the given intervals. Thus, we have

Interval	< -2	$(-2, -1)$	$(-1, 0)$	$(0, 1)$	$(1, 2)$	> 2
Expected frequency	6.83	40.77	102.4	102.4	40.77	6.83

The corresponding value of χ^2 is 41.48, which is larger than $c_\alpha = 11.1$ for 5 degrees of freedom. Consequently, we reject the hypothesis that the data are normally distributed with zero mean and unit variance.

22.5   Regression and Correlation

1. Differentiating $S(a, b)$ in Equation 2 with respect to a and b gives

$$\frac{\partial S}{\partial a} = -2\sum_{j=1}^{n}(y_j - a - bx_j) = -2n\overline{y} + 2n\hat{a} + 2n\hat{b}\overline{x} = 0$$

and

$$\frac{\partial S}{\partial b} = -2b\sum_{j=1}^{n}x_j(y_j - a - bx_j) = -2\hat{b}\sum_{j=1}^{n}x_j y_j + 2\hat{b}\hat{a}n\overline{x} + 2\hat{b}^2\sum_{j=1}^{n}x_j^2 = 0$$

Solving for $\hat{a}$ and $\hat{b}$ gives

$$\hat{b} = \frac{\sum x_i y_i - n\overline{x}\,\overline{y}}{\sum x_j^2 - n\overline{x}^2} \qquad \text{and} \qquad \hat{a} = \overline{y} - \hat{b}\overline{x}$$

3. Simply note that the first of Equations 3 is $\overline{y} = \hat{b}\overline{x} + \hat{a}$.

6. Multiply $y_j = a + bx_j + cx_j^2$ successively by 1, x_j, and x_j^2 to obtain

$$\sum_{j=1}^{n} y_j = na + b\sum_{j=1}^{n} x_j + c\sum_{j=1}^{n} x_j^2$$

$$\sum_{j=1}^{n} x_j y_j = a\sum_{j=1}^{n} x_j + b\sum_{j=1}^{n} x_j^2 + c\sum_{j=1}^{n} x_j^3$$

$$\sum_{j=1}^{n} x_j^2 y_j = a\sum_{j=1}^{n} x_j^2 + b\sum_{j=1}^{n} x_j^3 + c\sum_{j=1}^{n} x_j^4$$

Solve these linear equations for a, b, and c. It is easier first to substitute numerical values into the above equations and then solve for a, b, and c rather than to derive and use general equations.

9. We will first determine the least-squares regression line. It turns out that $\bar{x} = 2.000$ and $\bar{y} = 4.148$. In addition,

$$s_x^2 = \frac{1}{n-1}\sum_{j=1}^{n}(x_j - \bar{x})^2 = \frac{1}{16}\sum_{j=1}^{16}(x_j - 2.000)^2 = 1.5938$$

$$s_y^2 = \frac{1}{16}\sum_{j=1}^{16}(y_j - 4.148)^2 = 6.7862$$

$$s_{xy} = \frac{1}{16}\sum_{j=1}^{16}(x_j - 2.000)(y_j - 4.148) = 3.233$$

Equation 4 gives $b = \dfrac{s_{xy}}{s_x^2} = 2.029$ and the first of Equations 3 gives $a = \bar{y} - b\bar{x} = 0.09074$. The data and the least squares fit are shown in the following figure.

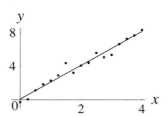

To calculate the confidence intervals, we need to calculate

$$\sigma_\alpha^2 = \frac{[(n-1)s_x^2 + n\bar{x}^2](s_y^2 - b^2 s_x^2)}{n(n-2)s_x^2}$$

$$= \frac{[(16)(1.5938) + (17)(4.000)][6.7862 - (4.1168)(1.5938)]}{(17)(15)(1.5938)} = 0.05200$$

and

$$\sigma_\beta^2 = \frac{s_y^2 - b^2 s_x^2}{(n-2)s_x^2} = 0.009455$$

We choose a 95% confidence interval, in which case $\eta = 1.96$ and

$$a = 0.09074 \pm (1.96)(0.2280) = 0.09074 \pm 0.4469$$
$$b = 2.029 \pm (1.96)(0.0972) = 2.029 \pm 0.1906$$

12. If X and Y are uncorrelated, then $r = 0$ in Example 10.6.5, and $f(x, y) = f(x)f(y)$.

15. We need to calculate $r = s_{ec}/s_c s_e$ from the data in Problem 13. It turns out that

$$s_e = 4.974 \qquad s_c = 0.3317 \qquad s_{ec} = 1.648$$

and so $r = 0.9988$. Following the procedure outlined in Table 22.12, we have

Step 1: $z = \dfrac{1}{2} \ln \dfrac{1+r}{1-r} = 3.709$

Step 2: $\eta = 0.99$

Step 3: $\xi = 2.576$

Step 4: $\rho_1 = \tanh\left(3.709 - \dfrac{2.576}{8^{1/2}}\right) = 0.9926$ and $\rho_2 = \tanh\left(3.709 + \dfrac{2.576}{8^{1/2}}\right) = 0.9998$

Thus, Conf $\{0.9926 \le \rho \le 1.000\}$.

18. We need to calculate $r = s_{\nu V}/s_\nu s_V$ from the data in Problem 16. It turns out that

$$s_\nu = 0.0490 \qquad s_V = 0.2452 \qquad s_{\nu V} = 0.007803$$

and so $r = 0.6496$. Following the procedure outlined in Table 22.12, we have

Step 1: $z = \dfrac{1}{2} \ln \dfrac{1+r}{1-r} = 0.775$

Step 2: $\eta = 0.90$

Step 3: $\xi = 1.6456$

Step 4: $\rho_1 = \tanh\left(0.775 - \dfrac{1.645}{5^{1/2}}\right) = 0.0393$ and $\rho_2 = \tanh\left(0.775 + \dfrac{1.645}{5^{1/2}}\right) = 0.907$

Thus, Conf $\{0.0393 \le \rho \le 0.907\}$.

21. Start with Equation 17, $s_f^2 = \dfrac{1}{n-1} \sum\limits_{j=1}^{n} (f_j - \overline{f})^2$. Now expand $f_j(x, y) = f(x_j, y_j)$ about $\overline{f} = f(\overline{x}, \overline{y})$ to write

$$
\begin{aligned}
s_f^2 &= \frac{1}{n-1} \sum_{j=1}^{n} \left[\left(\frac{\partial f}{\partial x}\right)_{\overline{x}, \overline{y}} (x_j - \overline{x}) + \left(\frac{\partial f}{\partial y}\right)_{\overline{x}, \overline{y}} (y_j - \overline{y}) + \cdots \right]^2 \\
&= \left(\frac{\partial f}{\partial x}\right)_{\overline{x}, \overline{y}}^2 \cdot \frac{1}{n-1} \sum_{j=1}^{n} (x_j - \overline{x})^2 + 2\left(\frac{\partial f}{\partial x}\right)_{\overline{x}, \overline{y}} \left(\frac{\partial f}{\partial y}\right)_{\overline{x}, \overline{y}} \cdot \frac{1}{n-1} \sum_{j=1}^{n} (x_j - \overline{x})(y_j - \overline{y}) \\
&\quad + \left(\frac{\partial f}{\partial y}\right)_{\overline{x}, \overline{y}}^2 \cdot \frac{1}{n-1} \sum_{j=1}^{n} (y_j - \overline{y})^2 + \cdots
\end{aligned}
$$

Now use Equations 5, 6, and 16 to write

$$s_f^2 = \left(\frac{\partial f}{\partial x}\right)_{\overline{x},\overline{y}}^2 s_x^2 + 2\left(\frac{\partial f}{\partial x}\right)_{\overline{x},\overline{y}}\left(\frac{\partial f}{\partial y}\right)_{\overline{x},\overline{y}} s_{xy}^2 + \left(\frac{\partial f}{\partial x}\right)_{\overline{x},\overline{y}} s_y^2 + \cdots$$

24. The volume is given by $V = \pi R^2 h$. The mean volume is

$$\overline{V} = \pi\overline{R}^2\overline{h} = \pi(3.751 \text{ cm})^2(16.06 \text{ cm}) = 709.9 \text{ cm}^3$$

The variance of the volume is given by Equation 22

$$
\begin{aligned}
s_V^2 &= \left(\frac{\partial V}{\partial R}\right)_{\overline{R},\overline{h}}^2 s_R^2 + \left(\frac{\partial V}{\partial h}\right)_{\overline{R},\overline{h}}^2 s_h^2 \\
&= (2\pi\overline{R}\overline{h})^2 s_R^2 + (\pi\overline{R}^2)^2 s_h^2 = (378.5 \text{ cm}^2)^2(0.018 \text{ cm})^2 + (44.2 \text{ cm}^2)^2(0.015 \text{ cm})^2 \\
&= 46.42 \text{ cm}^6 + 0.4396 \text{ cm}^6 = 46.86 \text{ cm}^6
\end{aligned}
$$

Therefore,

$$V = (709.9 \pm (46.86)^{1/2}) \text{ cm}^3 = (709.0 \pm 6.85) \text{ cm}^3$$

Cartesian Coordinates (x, y, z)

$$\nabla f = \text{grad } f = \frac{\partial f}{\partial x}\mathbf{i} + \frac{\partial f}{\partial y}\mathbf{j} + \frac{\partial f}{\partial z}\mathbf{k}$$

$$\nabla \cdot \mathbf{u} = \text{div } \mathbf{u} = \frac{\partial u_x}{\partial x} + \frac{\partial u_y}{\partial y} + \frac{\partial u_z}{\partial z}$$

$$\nabla \times \mathbf{u} = \text{curl } \mathbf{u} = \begin{vmatrix} \mathbf{i} & \mathbf{j} & \mathbf{k} \\ \dfrac{\partial}{\partial x} & \dfrac{\partial}{\partial y} & \dfrac{\partial}{\partial z} \\ u_x & u_y & u_z \end{vmatrix} = \left(\frac{\partial u_z}{\partial y} - \frac{\partial u_y}{\partial z} \right)\mathbf{i} + \left(\frac{\partial u_x}{\partial z} - \frac{\partial u_z}{\partial x} \right)\mathbf{j} + \left(\frac{\partial u_y}{\partial x} - \frac{\partial u_x}{\partial y} \right)\mathbf{k}$$

$$\nabla^2 f = \frac{\partial^2 f}{\partial x^2} = \frac{\partial^2 f}{\partial y^2} + \frac{\partial^2 f}{\partial z^2}$$

Cylindrical Coordinates (r, θ, z)

$$x = r\cos\theta \qquad y = r\sin\theta \qquad z = z$$

$$0 \le r < \infty \qquad 0 \le \theta \le 2\pi \qquad -\infty < z < \infty$$

$$dV = r\,dr\,d\theta\,dz$$

$$\nabla f = \frac{\partial f}{\partial r}\mathbf{e}_r + \frac{1}{r}\frac{\partial f}{\partial \theta}\mathbf{e}_\theta + \frac{\partial f}{\partial z}\mathbf{e}_z$$

$$\nabla \cdot \mathbf{u} = \frac{1}{r}\frac{\partial}{\partial r}(ru_r) + \frac{1}{r}\frac{\partial u_\theta}{\partial \theta} + \frac{\partial u_z}{\partial z}$$

$$\nabla^2 f = \frac{\partial^2 f}{\partial r^2} + \frac{1}{r}\frac{\partial f}{\partial r} + \frac{1}{r^2}\frac{\partial^2 f}{\partial \theta^2} + \frac{\partial^2 f}{\partial z^2}$$

Spherical Coordinates (r, θ, ϕ)

$$x = r\cos\phi\sin\theta \qquad y = r\sin\phi\sin\theta \qquad z = r\cos\theta$$

$$0 \le r < \infty \qquad 0 \le \theta \le \pi \qquad 0 \le \phi \le 2\pi$$

$$dV = r^2 \sin\theta\,dr\,d\theta\,d\phi$$

$$\nabla f = \frac{\partial f}{\partial r}\mathbf{e}_r + \frac{1}{r}\frac{\partial f}{\partial \theta}\mathbf{e}_\theta + \frac{1}{r\sin\theta}\frac{\partial f}{\partial \phi}\mathbf{e}_\phi$$

$$\nabla \cdot \mathbf{u} = \frac{1}{r^2}\frac{\partial}{\partial r}(r^2 u_r) + \frac{1}{r\sin\theta}\frac{\partial}{\partial \theta}(\sin\theta\, u_\theta) + \frac{1}{r\sin\theta}\frac{\partial u_\phi}{\partial \phi}$$

$$\nabla^2 f = \frac{1}{r^2}\frac{\partial}{\partial r}\left(r^2 \frac{\partial f}{\partial r} \right) + \frac{1}{r^2\sin\theta}\frac{\partial}{\partial \theta}\left(\sin\theta\frac{\partial f}{\partial r} \right) + \frac{1}{r^2\sin^2\theta}\frac{\partial^2 f}{\partial \phi^2}$$